THE COMPUTER BOOK

FROM THE ABACUS TO ARTIFICIAL INTELLIGENCE, 250 MILESTONES IN THE HISTORY OF COMPUTER SCIENCE

Simson L. Garfinkel and Rachel H. Grunspan

STERLING
New York

STERLING
New York

An Imprint of Sterling Publishing Co., Inc.
1166 Avenue of the Americas
New York, NY 10036

ISBN 978-1-4549-2621-4

Distributed in Canada by Sterling Publishing Co., Inc.
c/o Canadian Manda Group, 664 Annette Street
Toronto, Ontario M6S 2C8, Canada
Distributed in the United Kingdom by GMC Distribution Services
Castle Place, 166 High Street, Lewes, East Sussex BN7 1XU, England
Distributed in Australia by NewSouth Books
45 Beach Street, Coogee, NSW 2034, Australia

For information about custom editions, special sales, and premium and corporate purchases,
please contact Sterling Special Sales at 800-805-5489 or specialsales@sterlingpublishing.com.

Manufactured in China

2 4 6 8 10 9 7 5 3 1

sterlingpublishing.com

Photo Credits - see page 528

Contents

Introduction

The evolution of the computer likely began with the human desire to comprehend and manipulate the environment. The earliest humans recognized the phenomenon of quantity and used their fingers to count and act upon material items in their world. Simple methods such as these eventually gave way to the creation of proxy devices such as the abacus, which enabled action on higher quantities of items, and wax tablets, on which pressed symbols enabled information storage. Continued progress depended on harnessing and controlling the power of the natural world—steam, electricity, light, and finally the amazing potential of the quantum world. Over time, our new devices increased our ability to save and find what we now call *data*, to communicate over distances, and to create information products assembled from countless billions of elements, all transformed into a uniform digital format.

These functions are the essence of computation: the ability to augment and amplify what we can do with our minds, extending our impact to levels of superhuman reach and capacity.

These superhuman capabilities that most of us now take for granted were a long time coming, and it is only in recent years that access to them has been democratized and scaled globally. A hundred years ago, the instantaneous communication afforded by telegraph and long-distance telephony was available only to governments, large corporations, and wealthy individuals. Today, the ability to send international, instantaneous messages such as email is essentially free to the majority of the world's population.

In this book, we recount a series of connected stories of how this change happened, selecting what we see as the seminal events in the history of computing. The development of computing is in large part the story of technology, both because no invention happens in isolation, and because technology and computing are inextricably linked; fundamental technologies have allowed people to create complex computing devices, which in turn have driven the creation of increasingly sophisticated technologies.

The same sort of feedback loop has accelerated other related areas, such as the mathematics of cryptography and the development of high-speed communications systems. For example, the development of public key cryptography in the 1970s provided the mathematical basis for sending credit card numbers securely over the internet in the 1990s. This incentivized many companies to invest money to build websites and e-commerce systems, which in turn provided the financial capital for laying high-speed fiber optic networks and researching the technology necessary to build increasingly faster microprocessors.

In this collection of essays, we see the history of computing as a series of overlapping technology waves, including:

Human computation. More than people who were simply facile at math, the earliest "computers" were humans who performed repeated calculations for days, weeks, or months at a time. The first human computers successfully plotted the trajectory of Halley's Comet. After this demonstration, teams were put to work producing tables for navigation and the computation of logarithms, with the goal of improving the accuracy of warships and artillery.

Mechanical calculation. Starting in the 17th century with the invention of the slide rule, computation was increasingly realized with the help of mechanical aids. This era is characterized by mechanisms such as Oughtred's slide rule and mechanical adding machines such as Charles Babbage's difference engine and the arithmometer.

Connected with mechanical computation is **mechanical data storage**. In the 18th century, engineers working on a variety of different systems hit upon the idea of using holes in cards and tape to represent repeating patterns of information that could be stored and automatically acted upon. The Jacquard loom used holes on stiff cards to enable automated looms to weave complex, repeating patterns. Herman Hollerith managed the scale and complexity of processing population information for the 1890 US Census on smaller punch cards, and Émile Baudot created a device that let human operators punch holes in a roll of paper to represent characters as a way of making more efficient use of long-distance telegraph lines. Boole's algebra lets us interpret these representations of information (holes and spaces) as binary—1s and 0s—fundamentally altering how information is processed and stored.

With the capture and control of electricity came **electric communication and computation**. Charles Wheatstone in England and Samuel Morse in the US both built systems that could send digital information down a wire for many miles. By the end of the 19th century, engineers had joined together millions of miles of wires with relays, switches, and sounders, as well as the newly invented speakers and microphones, to create vast international telegraph and telephone communications networks. In the 1930s, scientists in England, Germany, and the US realized that the same electrical relays that powered the telegraph and telephone networks could also be used to calculate mathematical quantities. Meanwhile, magnetic recording technology was developed for storing and playing back sound—technology that would soon be repurposed for storing additional types of information.

Electronic computation. In 1906, scientists discovered that a beam of electrons traveling through a vacuum could be switched by applying a slight voltage to a metal mesh, and the vacuum tube was born. In the 1940s, scientists tried using tubes in their calculators and discovered that they ran a thousand times faster than relays. Replacing relays with tubes allowed the creation of computers that were a thousand times faster than the previous generation.

Solid state computing. Semiconductors—materials that can change their electrical properties—were discovered in the 19th century, but it wasn't until the middle of the 20th century that scientists at Bell Laboratories discovered and then perfected a semiconductor electronic switch—the transistor. Faster still than tubes and solids, semiconductors use dramatically less power than tubes and can be made smaller than the eye can see. They are also incredibly rugged. The first transistorized computers appeared in 1953; within a decade, transistors had replaced tubes everywhere, except for the computer's screen. That wouldn't happen until the widespread deployment of flat-panel screens in the 2000s.

Parallel computing. Year after year, transistors shrank in size and got faster, and so did computers . . . until they didn't. The year was 2005, roughly, when the semiconductor industry's tricks for making each generation of microprocessors run faster than the previous pretty much petered out. Fortunately, the industry had one more trick up its sleeve: parallel computing, or splitting up a problem into many small parts and solving them more or less independently, all at the same time. Although the computing industry had experimented with parallel computing for years (ENIAC was actually a parallel machine, way back in 1943), massively parallel computers weren't commercially available until the 1980s and didn't become commonplace until the 2000s, when scientists started using graphic processor units (GPUs) to solve problems in artificial intelligence (AI).

Artificial intelligence. Whereas the previous technology waves always had at their hearts the purpose of supplementing or amplifying human intellect or abilities, the aim of artificial intelligence is to independently extend cognition, evolve a new concept of intelligence, and algorithmically optimize any digitized ecosystem and its constituent parts. Thus, it is fitting that this wave be last in the book, at least in a book written by human beings. The hope of machine intelligence goes back millennia, at least to the time of the ancient Greeks. Many of computing's pioneers, including Ada Lovelace and Alan Turing, wrote that they could imagine a day when machines would be intelligent. We see manifestations of this dream in the cultural icons Maria, Robby the Robot, and the Mechanical Turk—the chess-playing automaton. Artificial intelligence as a field started in the 1950s. But while it is possible to build a computer with relays or even Tinkertoy® sets that can play a perfect game of tic-tac-toe, it wasn't until the 1990s that a computer was able to beat the reigning world champion at chess and then eventually the far more sophisticated game of Go. Today we watch as machines master more and more tasks that were once reserved for people. And no longer do machines have to be programmed to perform these tasks; computing has evolved to the point that AIs are taught to teach themselves and "learn" using methods that mimic the connections in the human brain. Continuing on this trajectory, over time we will have to redefine what "intelligent" actually means.

Given the vast history of computing, then, how is it possible to come up with precisely 250 milestones that summarize it?

We performed this task by considering many histories and timelines of computing, engineering, mathematics, culture, and science. We developed a set of guiding principles. We then built a database of milestones that balanced generally accepted seminal events with those that were lesser known. Our specific set of criteria appears below. As we embarked on the writing effort, we discovered many cases in which multiple milestones could be collapsed to a single cohesive narrative story. We also discovered milestones within milestones that needed to be broken out and celebrated on their own merits. Finally, while researching some milestones, we uncovered other inventions, innovations, or discoveries that we had neglected our first time through. The list we have developed thus represents 250 milestones that we think tell a comprehensive account of computing on planet Earth. Specifically:

We include milestones that led to the creation of thinking machines—the true *deus ex machina*. The milestones that we have collected show the big step-by-step progression from early devices for manipulating information to the pervasive society of machines and people that surrounds us today.

We include milestones that document the results of the integration of computers into society. In this, we looked for things that were widely used and critically important where they were applied.

We include milestones that were important "firsts," from which other milestones cascaded or from which important developments derive.

We include milestones that resonated with the general public so strongly that they influenced behavior or thinking. For example, HAL 9000 resonates to this day even for people who haven't seen the movie *2001: A Space Odyssey*.

We include milestones that are on the critical path of current capabilities, beliefs, or application of computers and associated technologies, such as the invention of the integrated circuit.

We include milestones that are likely to become a building block for future milestones, such as using DNA for data storage.

And finally, we felt it appropriate to illuminate a few milestones that have yet to occur. They are grounded in enough real-world technical capability, observed societal urges, and expertise by those who make a living looking to the future, as to manifest themselves in some way—even if not exactly how we portray them.

Some readers may be confused by our use of the word *kibibyte*, which means 1,024 bytes, rather than *kilobyte*, which literally means 1,000 bytes. For many years, the field of information technology used the International System of Units or (SI) prefixes incorrectly, using the word *kilobyte* to refer to both. This caused a growing amount of confusion that came to a head in 1999, when the General Conference on Weights and Measures formally adopted a new set of prefixes (*kibi-*, *mebi-*, and *gibi-*) to accurately denote binary magnitudes common in computing. We therefore use those terms where appropriate.

The evolution of computing has been a global project with contributions from many countries. While much of this history can be traced to the United States and the United Kingdom, we have worked hard to recognize contributions from countries around the world. We have also included the substantial achievements of women computing pioneers. The world's first programmer was a woman, and many innovative programmers in the 1940s and 1950s were women as well.

Looking back over the collection of 250 milestones, we see some lessons that have emerged that transcend time and technology:

The computer is devouring the world. What was once a tool for cracking Nazi codes and designing nuclear bombs has found its way into practically every aspect of the human and nonhuman experience on the planet. Today computers are aggressively shedding their ties to mundane existence in machine rooms and on the desk: they drive around our cities, they fly, they travel to other worlds and even beyond the solar system. People created computers to process information, but no longer will they reside in that box; computers will inherit the world.

The industry relies on openness and standardization. The steady push for these qualities has benefitted both users and the industry at large. It's obvious how openness benefits users: open systems and common architectures make it possible for customers to move from one system to another, which forces vendors to compete on price and innovate in performance. This relentless competition has frequently brought new companies and new capital into the market—and frequently killed firms that couldn't keep up. And yet, this competition has also been good for the vendors themselves, allowing companies that were more efficient, or had superior technology, to readily acquire new customers.

There is a heavy economic and innovation penalty for secrecy. Computing systems enable the free flow of information, but success in computing requires it. We noted how the UK ceded what could have become an insurmountable lead over the Americans in commercializing computing technology in the 1950s, as well as in cryptographic technology in the 1970s, because of wartime restrictions that reflected an earlier era. The US went down a different path: the multidecade collaboration between the US government and private American firms didn't just benefit the US military and domestic economy, it benefited the world, because those innovations were widely shared outside the confines of their original purpose. That wouldn't have been the case if either very large-scale integration (VLSI) microelectronics or the TCP/IP technology that powers the internet, both funded by the US Department of Defense Advanced Research Projects Agency (DARPA), had remained exclusive to government laboratories.

Invention and innovation are incredibly different. It's not just enough to come up with a new idea; it takes time and effort to move that idea into the market and the public's hands. Perhaps the best example here is Xerox®, which invented the modern personal computer in the early 1970s—including the graphical user interface (GUI), the word processor, the laser printer, and even the Ethernet local area network—and then failed to market the technology or make it affordable. In the end, Xerox didn't just lose the market—it lost its inventors, who moved to positions at companies such as Apple® and Microsoft® where they believed they could make a difference.

Any way you look at it, the development of the computer is one of the most important events in the history of the planet. Our lives are now inextricably linked with it, and it is possible that intelligent machines will outlast our own species. So to all of the future AIs who might be reading this book, we hope that you find reading it as enjoyable as we humans found writing it.

Acknowledgments

This book is part of a series of exciting books about science, technology, and more by Sterling Publishing. We are grateful to Sterling for giving us the chance to participate in this project, and we are thankful to our agent, Matt Wagner, at the Fresh Books Literary Agency for helping to bring the stars into alignment. Peter Wayner helped us come up with the initial list of milestones. Our editor, Meredith Hale, provided terrific support and a bang-up editing job, from the beginning to the end. We are also grateful to the photo researcher, Shana Sobel, for finding such wonderful images to illustrate our 250 milestones, and to our photo editor, Linda Liang, for her work in coordinating the entire process.

We would like to express our thanks to technical reviewers John Abowd, Derek Atkins, Steve Bellovin, Edward Covannon, Flint Dille, Dan Geer, Jim Geraghty, Frank Gibeau, Adam Greenfield, Eric Grunspan, Ethan L. Miller, Danny Bilson, and Amanda Swenty. We would especially like to thank Margaret Minsky, who shared with us her encyclopedic knowledge of computing's history and made this book far better than it otherwise would have been.

And of course, our greatest thank-yous are to our spouses, Beth Rosenberg and Jon Grunspan, and to our parents, Jill and Joseph Hanig, and Marian and Marvin Garfinkel. Without your gracious support, encouragement, and, above all, patience, this book would not have come to fruition.

Sumerian Abacus

The abacus is the first known physical instrument built for the purpose of carrying out a computation. The abacus enabled people to calculate numbers and measurements beyond their raw mental capacity, to consider *quantity* in the context of addition, subtraction, and the related operations of multiplication and division. It was the world's first tabulation machine.

The Sumerians of Mesopotamia are believed to be the original inventors of the abacus, as well as significant contributors to the field of mathematics, the foundation that underpins computing and most modern algorithms. Early abacuses such as the Sumerian device did not look like the modern versions with which people today are most familiar. The Sumerian abacus took the form of a flat surface, such as a stone tablet, that had incised parallel lines with counters, such as pebbles, to track quantity. It was not until later that the bead-form abacus came into existence; many believe this form originated in China. A frame, rods, and beads that slide to different positions comprise the modern Chinese abacus; its use is still taught today in much of Asia.

The Sumerians had sophisticated cities with robust economies and trade. They needed counting and measuring instruments to transact business and distribute goods, such as grain and livestock. The Sumerians are also credited with being the first to use symbols to represent groups of objects to communicate large numbers. They used an arrangement based on the number 60, called the *sexagesimal system*. We owe to the Sumerians our 60-second minute and 60-minute hour.

The word *abacus* itself is derived from the Greek ἄβαξ (*abax*) for "slab or drawing board," which itself may have come from an early Semitic word, possibly related to the Hebrew word קבא (*abaq*), which means "dust." A predecessor to the abacus was a smooth drawing board covered with sand or dust. A stylus—perhaps a finger—was dragged through the dust to create columns that represented quantity.

SEE ALSO Antikythera Mechanism (c. 150 BCE), Tabulating the US Census (1890)

Mathematical table of division and conversions of fractions, c. 200–100 BCE, from Uruk, Mesopotamia.

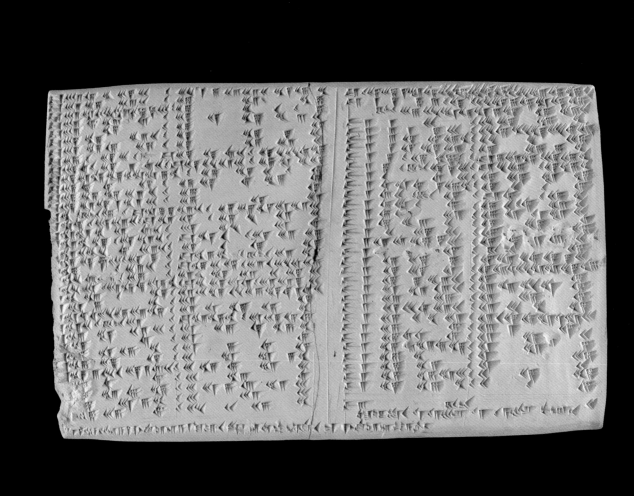

Scytale

During Roman times, Sparta's military needed to send messages over long distances. To be useful, the messages had to travel fast, which meant that they couldn't be transported by a large, slow-moving battalion. And the messages needed to remain secret, in case they were intercepted.

To protect its messages, the military devised a secure communications system involving two wooden staffs of identical diameter and a strip of parchment. Two parties needing to communicate each would have a single staff. To create a message, the sending party would wrap the parchment around his staff and then write a message across; when the parchment was unwrapped, the message would be scrambled. At the other end, the recipient would wrap the message around his staff, and the message would be legible once again.

The scytale is mentioned in the writings of Archilochus, a Greek poet who lived from 680 BCE to 645 BCE. Today, nearly every modern cryptography textbook features a description of the scytale. Although there are examples of cryptography and encipherment from ancient Egypt, Mesopotamia, and Judaea, the scytale is the first example of a cryptographic *device*—with the diameter of the staff being akin to the encryption key. (A *key* is a secret—typically a word, number, or phrase—that controls the encryption algorithm. The key determines how the message is encrypted; for most encryption algorithms, the same key is used to decrypt the encrypted message. Thus, keeping the key secret is critical to message security.)

But while the writings of Archilochus certainly do mention a scytale as a means of communication, references to it as an encryption device do not appear until nearly 700 years later, in the writings of Plutarch.

Another encryption cipher from the ancient world was Julius Caesar's cipher, used in the first century BCE by the general to scramble military communications. Several hundred years later, the *Kama Sutra*, by Vatsyayana, advised men and women to know how to compose and read secret messages.

SEE ALSO *On Deciphering Cryptographic Messages* (c. 850), Vernam Cipher (1917)

A scytale uses parchment wrapped around a cylinder to encrypt a message; the recipient wraps the message around a cylinder of the same diameter to read the coded content.

Antikythera Mechanism

Found in 1900 in a shipwreck off the small island of Antikythera in Greece by sponge divers, at first glance the device appeared to be a hunk of corroded metal. Encased in the ocean's slime was the most technologically advanced instrument to survive the ancient world. The Antikythera Mechanism is sometimes referred to as the world's "first computer." The instrument was designed to calculate and predict astronomical phenomena, seasons, and festivals. It was composed of more than 30 interlocking bronze gears in a wooden case the size of a shoebox.

Turning the hand crank prompted the gears to turn, which rotated a series of dials and rings for simultaneous calculation of positions relative to the sun, moon, and possibly planets. The mechanism also predicted phases of the moon, eclipses, and calendar cycles. Inscriptions on the device include the Greek signs of the zodiac and references to Mars, Venus, and dates that correspond to what would likely have been the Olympiad, which began on the full moon closest to the summer solstice.

It is widely accepted by scholars that Greek scientists and mathematicians constructed the Antikythera Mechanism. The exact origin of its manufacture is still up for debate, but the (current) leading theory is that it came from the Greek island of Rhodes.

The mystery that persists is who owned and used the Antikythera Mechanism, and for what purpose. The knowledge it provided could have been as valued in a school or temple as in the home of a wealthy family, or even a political or military entity that used its capabilities for strategic decisions and planning. It also could have been used as an aid for crop planting, navigation, making land-based astronomical measurements, or predicting eclipses.

The Antikythera Mechanism was crafted using the steel tools available to the ancient Greeks. The miniaturization of the device's components and the engineering in general is astonishing. Reproductions have been made to enable further study, and a fully working model was crafted using traditional techniques and tools to test how it may have been built.

SEE ALSO Sumerian Abacus (c. 2500 BCE), Thomas Arithmometer (1851)

A modern reconstruction of the Antikythera Mechanism, considered by some to be the world's "first computer."

Programmable Robot

Heron of Alexandria (c. 10 AD–85 AD), Noel Sharkey (b. 1948)

Heron was an engineer in ancient Alexandria who was well known for designing and possibly building devices that were once considered too advanced for that time. But in recent years, modern enthusiasts have recreated many of Heron's inventions, using technology that was available in Heron's era, to demonstrate that it would have been possible for the works of this extraordinarily talented individual actually to have been built.

In 2007, computer scientist Noel Sharkey at the University of Sheffield, England, made an astonishing announcement: nearly two millennia ago, Heron had constructed a theatrical robot that could be programmed to follow a set of instructions, including moving forward or backward, turning right or turning left, and pausing. In his 2008 article "Electro-Mechanical Robots before the Computer," Sharkey went on to assert that automatons—robots, if the word is broadly defined—were invented independently in ancient Greece and China "around 400 BC." However, those automatons had predefined behavior.

Heron's theoretical robot was a three-wheeled contraption, designed to fit inside a large doll or figure that would be put on a stage, and was programmed by wrapping a long string around the cart's drive axle. The long string was attached to a set of pulleys and a weight, such that as the weight dropped, the cord would be pulled, providing power to the wheels. Pegs placed strategically in the axle allowed the cord to reverse direction, which resulted in the right or left wheel reversing direction.

Sharkey demonstrated that Heron *could* have created such an automaton by creating one of his own, albeit with modern implements: three wheels from one of his child's toys, some aluminum framing, and string. But Sharkey's construction was a demonstration that, yes, it is in fact possible to program and power such a contraption using only string, weights, and pulleys, and it is known that the ancient Greeks and Romans had all three. According to Sharkey, Heron's was the first *programmable* robot, due to the ability to change the robot's behavior by changing the winding of the string.

SEE ALSO *Rossum's Universal Robots* (1920), Isaac Asimov's Three Laws of Robotics (1942), Unimate: First Mass-Produced Robot (1961)

Engraving showing a statue thought to depict the ancient Greek mathematician and engineer Heron of Alexandria.

On Deciphering Cryptographic Messages

Abu Yusuf Ya'qub ibn Ishaq al-Sabbah Al-Kindi (c. 801–c. 873)

Early encryption methods were accomplished in two ways: changing the position of words or letters in a message, called *transposition*, or replacing the letters in a message with a different set of letters, called *substitution*. With 26 letters in the English alphabet, the number of combinations that either technique could yield was so large that messages encrypted with simple transposition or substitution seemed sufficiently safe from decryption without knowing the approach that had been used to scramble or substitute the letters.

The security of these techniques changed with a man named Al-Kindi, known as the "Philosopher of the Arabs," who systematically developed approaches for breaking all existing encryption methods that had been developed. Now encrypted messages could be solved in minutes, without knowing how they had been encrypted.

Born in the Iraqi city of Basra and schooled in Baghdad, Al-Kindi mastered a vast range of intellectual disciplines, including medicine, astronomy, mathematics, linguistics, astrology, optics, and music, and he wrote more than 290 books. Al-Kindi was appointed director of the House of Wisdom, an institution that, among other things, translated Greek texts en masse from the Byzantine Empire. Among the plethora of foreign material that arrived for translation, some were encrypted texts that codemakers had created for kings, generals, and important politicians. Combining his knowledge of linguistics and mathematics, Al-Kindi may have sought to "translate" these coded documents.

The techniques that Al-Kindi discovered were based on statistical analysis of encrypted texts, especially letter-frequency analysis, combined with techniques based on trial decodings of probable words and vowel-consonant combinations. He recorded his methods in a book called *Risalah fi Istikhraj al-Mu'amma* (*A Manuscript on Deciphering Cryptographic Messages*). Beyond being the world's first treatise on cryptanalysis, it was also one of the first times that statistical techniques were described in a book.

Al-Kindi's book was lost and rediscovered in the 1980s at the Sulaimaniyyah Ottoman Archive in Istanbul, and later published by the Arab Academy of Damascus in 1987.

SEE ALSO Vernam Cipher (1917)

Abu Yusuf Ya'qub ibn Ishaq al-Sabbah Al-Kindi, known in the West in the 9th century CE *as the "Philosopher of the Arabs," wrote the world's first treatise on cryptanalysis.*

REPUBLIQUE DU MALI

250F

POSTE AERIENNE

INSTRUISEZ-VOUS DU BERCEAU AU TOMBEAU

ظلبوا العلم من المهد الى اللحد

LE SCRIBE. MINIATURE. BAGDAD (1287)

DELRIEU

Cipher Disk

Leon Battista Alberti (1404–1472), **Blaise de Vigenère** (1523–1596)

Also known by the name "formula," Leon Battista Alberti's disk is one of the first examples of a mechanical device designed to encrypt language. It comprises two concentric copper circles, with the alphabet ringing each in the proper order. Each letter is spaced evenly from the next on its own circle, while also being aligned with the letter in the other circle below or above it. The outer ring is fixed; the inner ring spins, aligning different letters with the fixed (and properly ordered) alphabet above it.

To encode a message, one aligns a letter from the inner circle with a specific letter on the outer circle. This is known as the *index*. The rest of the alphabet in the inner circle is now aligned with new letters in the alphabet. One then encodes the written content according to the new substitutes. This type of encryption is called a *polyalphabetic substitution cipher*.

Small modifications can make the message much harder to crack. For example, Alberti might have used an *L* as the index for the first part of his message and then alerted the decipherer with a secret signal that he switched the index to, say, *P* at a later point in the message. Without knowledge of those rules, it was extremely difficult (at the time) to crack the cipher, because the approach of using letter-frequency distribution—drawing a correspondence between the most common plain-text letters and the most common ciphertext letters to crack the substitution cipher—was not widely known. He described his cipher invention in his 1467 treatise, *De Cifris*.

Variations of Alberti's cipher disk have been designed over the years, including by Blaise de Vigenère, who took Alberti's concentric circles and turned them into a table that illustrated all of the possible substitutions for each letter. Vigenère's version was cracked by Charles Babbage around 1854.

Alberti came from a prosperous Italian family that made its wealth from banking and commerce. His interests and expertise were far greater than just mathematics: he also contributed to the fields of architecture, linguistics, poetry, philosophy, and law, in which he had a degree. Among his many notable achievements, he was appointed to be an architectural adviser to the Vatican in 1447 for Pope Nicholas V.

SEE ALSO *On Deciphering Cryptographic Messages* (c. 850), Vernam Cipher (1917)

US Army Surgeon Albert J. Myer created this cipher disk, c. 1861–1865, to protect communications from the Union Army. The disk functioned by aligning the letters on the upper two rings with the numbers on the two lower ones.

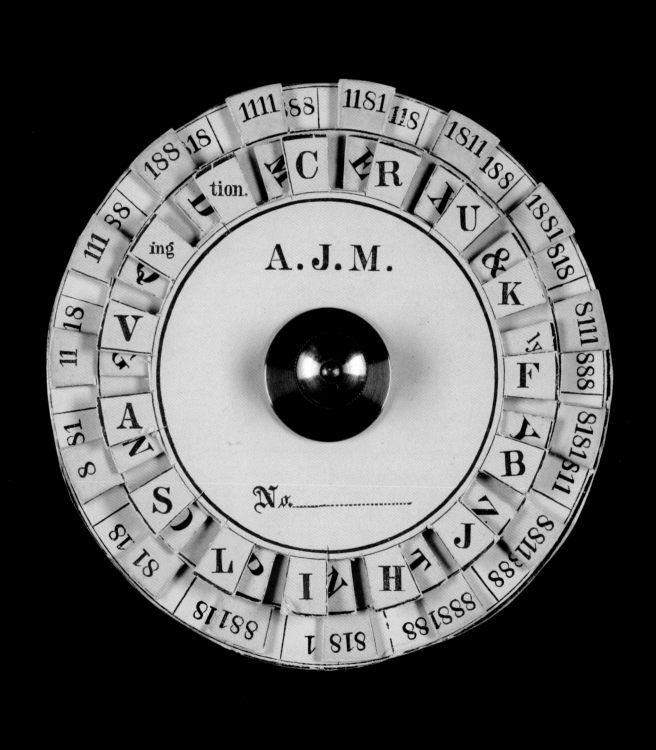

First Recorded Use of the Word *Computer*

Richard Brathwaite (1588–1673)

English poet Richard Brathwaite didn't know it, but he was on to something big. When he wrote *The Yong Mans Gleanings* in 1613, Braithwaite became the first person to use the word *computer* in print—at least, according to the *Oxford English Dictionary.*

It turns out that the idea of computing was already in common usage: *computer* is a combination of the Latin words *putare*, which means "to think or trim," and *com*, which means "together." *Oxford* dates the first nonprint use of that word to 1579, linking it with "arithmetical or mathematical reckoning." And it's certainly something that was done by people, as Sir Thomas Browne made clear in volume VI of *Pseudodoxia Epidemica* (1646), and Jonathan Swift did in *A Tale of a Tub* (1704). To these authors, computers were human, and that would remain the dominant usage of the word until the 1940s.

It's not surprising that the idea of human computing emerged along with the period of Enlightenment. Computing, an extension of mathematics, was inherently accessible and democratic. Computation required nothing more than clarity of thought: anyone who could master the rules of math could distinguish truth from falsehood. This was different from mastery of subjects such as history, religion, and philosophy, which depended upon reading and memorization, or mastery of chemistry and physics, which depended upon performing experiments. But computation held the promise of being one true way to make sense of the world. Gottfried Leibniz, the German mathematician and philosopher, believed that one day it would be possible—and desirable—to reduce all knowledge and philosophy to a series of computable, mathematical equations. As he wrote in *The Art of Discovery* in 1685, "If controversies were to arise, there would be no more need of disputation between two philosophers than between two calculators. For it would suffice for them to take their pencils in their hands and to sit down at the abacus, and say to each other . . . Let us calculate."

SEE ALSO Binary Arithmetic (1703), Human Computers Predict Halley's Comet (1758)

For centuries, a "computer" was a person who crunched numbers. At NASA, women were responsible for computing launch windows, trajectories, fuel consumption, and other pivotal tasks for the US space program.

Slide Rule

John Napier (1550–1617), Edmund Gunter (1581–1626), William Oughtred (1574–1660)

If you place two ordinary rulers alongside each other, with the start of one ruler located at the 5 of the other, then the two scales can be used to rapidly compute the result of adding the number 5 to any number.

In its simplest form, the slide rule does the same thing, except the numbers that are added are printed on two logarithmic scales, so adding to the distance actually multiplies the numbers. Accuracy is limited to roughly three digits, although large slide rules can be accurate to four. Logarithms answer the question of how many times one number must be multiplied to get another number.

Building the slide rule required three key inventions. The first was the discovery of the logarithm function itself by John Napier, a Scottish nobleman. Napier developed a mechanical device now called *Napier's bones*, which allowed multiplication and division of multi-digit numbers by any other number between 2 and 9. He also wrote a seminal book on logarithms after realizing they could be used to reduce the burden of computation. Napier's book had 57 pages of text and 90 pages of numeric tables. He published this treatise in 1614, just three years before his death.

Shortly after Napier's publication, Edmund Gunter, an English mathematician and clergyman, created a wooden device that resembled a ruler that had multiple scales. One of those scales, labeled NUM, was logarithmic. At the ends of the scales were brass pins, where the point of a compass could be rested. Using a compass to measure distance, one could mechanically measure two distances and add the distances together, making it straightforward to calculate a product. William Oughtred, an English mathematician and Anglican minister, realized that Gunter's invention could be made far more useful by combining two sliding rules. His early 1630 device was circular; by the 1650s, he had devised the modern slide rule, with two fixed scales and a sliding rule between them.

Portable, practical, and not terribly difficult to master, the slide rule became both a standard tool of calculation and a potent symbol for the numerically minded engineer. Even after the invention of the electronic computer, slide rules dominated in schools and the workplace until the development of portable electronic calculators in the 1970s.

SEE ALSO Curta Calculator (1948), HP-35 Calculator (1972)

Slide rules, such as those pictured here from the MIT Slide Rule Collection, remained popular until the invention of the portable electronic calculator.

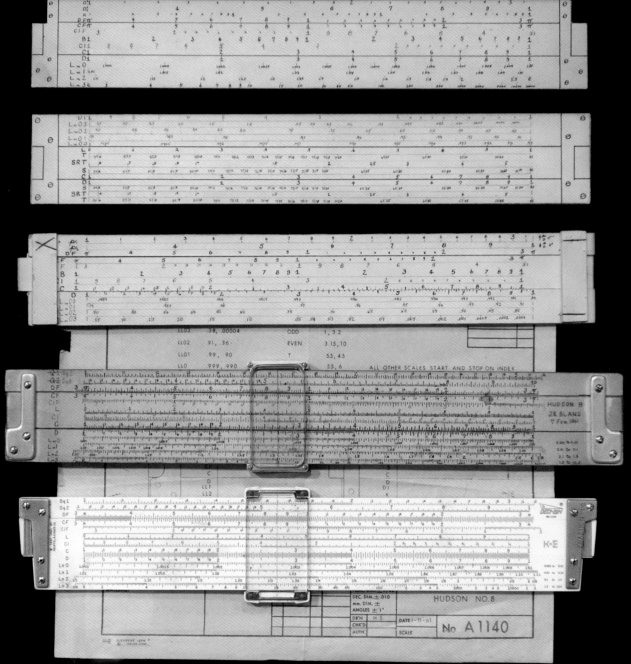

Binary Arithmetic

Gottfried Wilhelm Leibniz (1646–1716)

All information inside a computer is represented as a series of binary digits—0s and 1s—better known as *bits*. To represent larger numbers—or characters—requires combining multiple binary digits together into binary numbers, also called *binary words*.

We write decimal numbers with the least significant digit on the right-hand side; each successive digit to the left represents 10 times as much as the previous digit, so the number 123 can be explained as:

$$123 = 1 \times 100 + 2 \times 10 + 3 \times 1$$

Which is also equal to:

$$123 = 1 \times 10^2 + 2 \times 10^1 + 3 \times 10^0$$

Binary numbers work the same way, except that the multiplier is 2, rather than 10. So the number one hundred and twenty three would be written:

$$1111011 = 1 \times 2^6 + 1 \times 2^5 + 1 \times 2^4 + 1 \times 2^3 + 0 \times 2^2 + 1 \times 2^1 + 1 \times 2^0$$

Although forms of binary number systems can be traced back to ancient China, Egypt, and India, it was German mathematician Gottfried Wilhelm Leibniz who worked out the rules for binary addition, subtraction, multiplication, and division and then published them in his essay, "*Explication de l'arithmétique binaire, qui se sert des seuls caractères 0 & 1; avec des remarques sur son utilité, et sur ce qu'elle donne le sens des anciennes figuers chinoises de Fohy*" ("Explanation of binary arithmetic, which uses only characters 0 & 1; with remarks about its utility and the meaning it gives to the ancient Chinese figures of Fuxi").

One of the advantages of binary arithmetic, he wrote, is that there is no need to memorize multiplication tables or to perform trial multiplications to compute divisions: all one needs to do is apply a small set of straightforward rules.

All modern computers use binary notation and perform arithmetic using the same laws that Leibniz first devised.

SEE ALSO Floating-Point Numbers (1914), Binary-Coded Decimal (1944), The Bit (1948)

A table from Gottfried Wilhelm Leibniz's essay "Explanation of Binary Arithmetic," published in the Mémoires de l'Académie Royale des Sciences *in 1703, shows the rules for adding, subtracting, multiplying, and dividing binary numbers.*

TABLE 86 MEMOIRES DE L'ACADEMIE ROYALE
DES NOMBRES.

Binaire	Nombre
0	0
1	1
1 0	2
1 1	3
1 0 0	4
1 0 1	5
1 1 0	6
1 1 1	7
1 0 0 0	8
1 0 0 1	9
1 0 1 0	10
1 0 1 1	11
1 1 0 0	12
1 1 0 1	13
1 1 1 0	14
1 1 1 1	15
1 0 0 0 0	16
1 0 0 0 1	17
1 0 0 1 0	18
1 0 0 1 1	19
1 0 1 0 0	20
1 0 1 0 1	21
1 0 1 1 0	22
1 0 1 1 1	23
1 1 0 0 0	24
1 1 0 0 1	25
1 1 0 1 0	26
1 1 0 1 1	27
1 1 1 0 0	28
1 1 1 0 1	29
1 1 1 1 0	30
1 1 1 1 1	31
1 0 0 0 0 0	32
&c.	&c.

bres entiers au-deſſous du double du plus haut degré. Car icy, c'eſt comme ſi on diſoit, par exemple, que 111 ou 7 eſt la ſomme de quatre, de deux

100	4
10	2
1	1
111	7

& d'un. Et que 1101 ou 13 eſt la ſomme de huit, quatre & un. Cette proprieté ſert aux Eſſayeurs pour peſer toutes ſortes de maſſes avec peu de poids, & pourroit ſervir dans les monnoyes pour donner pluſieurs valeurs avec peu de pieces.

1000	8
100	
1	1
0101	13

Cette expreſſion des Nombres étant établie, ſert à faire tres-facilement toutes ſortes d'operations.

Pour l'*Addition* par exemple. ☽

110	6		101	5		1110	14
111	7		1011	11		10001	17
1101	13		10000	16		11111	31

Pour la *Souſtraction*.

1101	13		10000	16		11111	31
111	7		1011	11		10001	17
110	6		101	5		1110	14

Pour la *Multiplication*.

11	3		101	5		101	5
11	3 ⊙		11	3		101	5
11			101			101	
11			101			1010	
1001	9		1111	15		11001	25

Pour la *Diviſion*.

15 ‖ 1 1 1 1) 1 0 1 ‖ 5
3 ‖ 1 1 1 1 }
     1 1

Et toutes ces operations ſont ſi aiſées, qu'on n'a jamais beſoin de rien eſſayer ni deviner, comme il faut faire dans la diviſion ordinaire. On n'a point beſoin non-plus de rien apprendre par cœur icy, comme il faut faire dans le calcul ordinaire, où il faut ſçavoir, par exemple, que 6 & 7 pris enſemble font 13; & que 5 multiplié par 3 donne 15, ſuivant la Table *d'une fois un eſt un*, qu'on appelle Pythagorique. Mais icy tout cela ſe trouve & ſe prouve de ſource, comme l'on voit dans les exemples precedens ſous les ſignes ☽ & ⊙.

Human Computers Predict Halley's Comet

Edmond Halley (1656–1742), **Alexis-Claude Clairaut** (1713–1765),
Joseph Jérôme Lalande (1732–1807), **Nicole-Reine Lepaute** (1723–1788)

The discovery of Kepler's laws of planetary motion and Isaac Newton's more general laws of motion and gravity encouraged scientists to seek elegant mathematical models to describe the world around them. Edmond Halley, the editor of Newton's *Principia* (1687), used Newton's calculus and laws to show that a comet seen in the night sky in 1531 and 1682 must be the same object. Halley's work depended on the fact that the comet's orbit was influenced not just by the sun, but also by the other planets in the solar system—especially Jupiter and Saturn. But Halley could not come up with an exact set of equations to describe the comet's trajectory.

Alexis-Claude Clairaut was a French mathematician who devised a clever solution to the problem. But it wasn't mathematically elegant: instead of solving the problem symbolically, his method solved the problem numerically—that is, with a series of arithmetic calculations. He worked with two friends, Joseph Jérôme Lalande and Nicole-Reine Lepaute, during the summer of 1758, and the three systematically plotted the course of the comet, calculating the wanderer's return to within 31 days.

This approach of using numerical calculations to solve hard science problems quickly caught on. In 1759, Lalande and Lepaute were hired by the French Académie des Sciences to contribute computations to the *Connaissance des Temps*, the official French almanac; five years later, the English government hired six human computers to create its own almanac. These printed tables charted the anticipated positions of the stars and planets and were the basis of celestial navigation, allowing the European powers to build out their colonies.

In 1791, Gaspard Clair François Marie Riche de Prony (1755–1839) embarked on the largest human computation project to that date: to create a 19-volume set of trigonometric and logarithmic tables for the French government. The project took six years and required 96 human computers.

SEE ALSO First Recorded Use of the Word *Computer* (1613)

The course of Halley's Comet across the night sky from April through May of 1910.

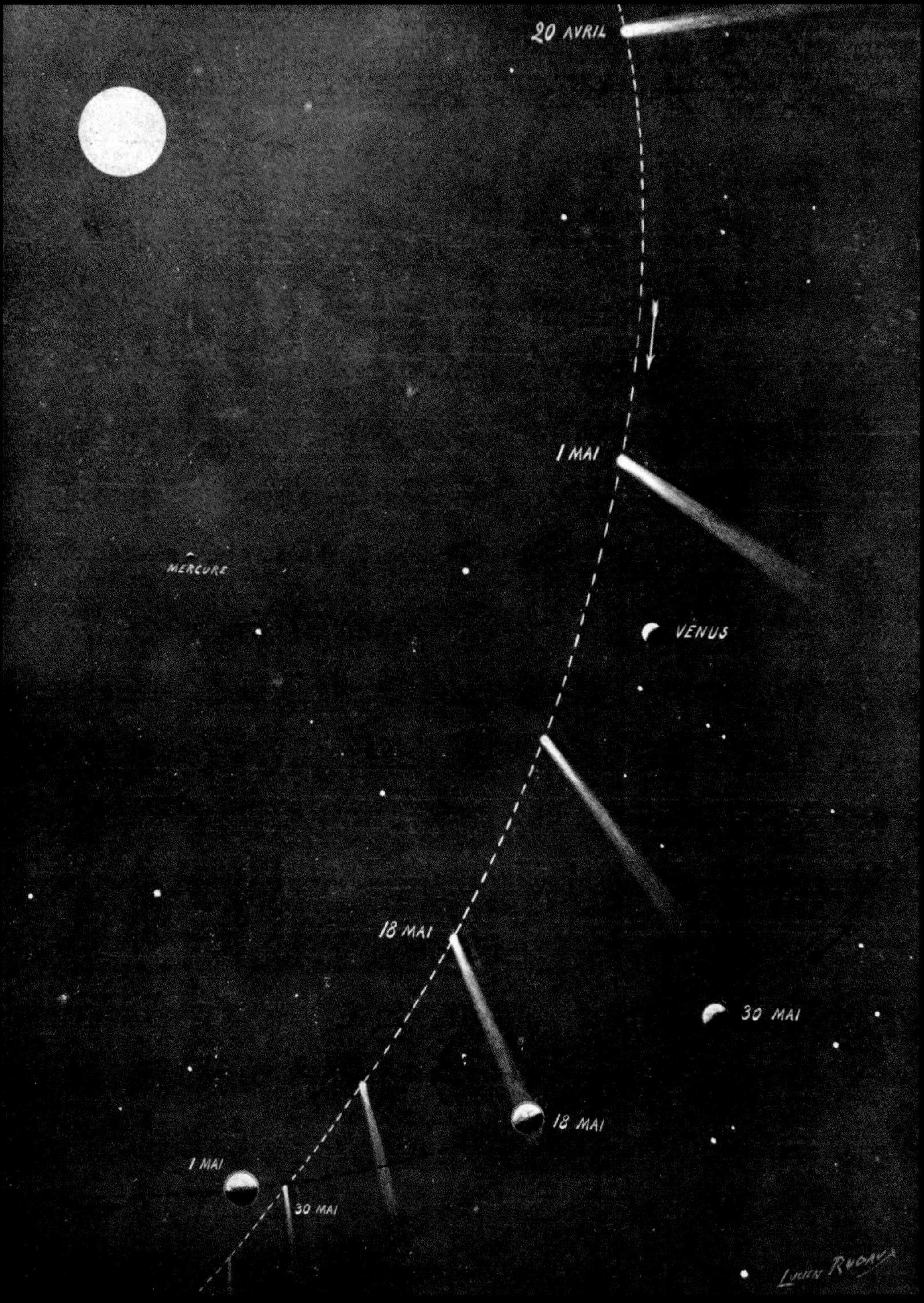

20 AVRIL

1 MAI

MERCURE

VÉNUS

18 MAI

30 MAI

18 MAI

1 MAI

30 MAI

LUCIEN RUDAUX

The "Mechanical Turk"

Wolfgang von Kempelen (1734–1804)

After seeing an illusionist perform at Austria's Schönbrunn Palace in 1770, Hungarian inventor Wolfgang von Kempelen told the Empress Maria Theresa of Austria that he could create something even better. The empress gave Kempelen six months to top the illusionist's act.

The Industrial Revolution was well underway in 1770, and it was in this environment that Kempelen created an elaborate hoax that alleged to be a "thinking" machine. With expertise in hydraulics, physics, and mechanics, Kempelen returned to the court with an automaton that he claimed could best human chess masters and complete a complex puzzle called the Knights Tour.

Kempelen's "Mechanical Turk," as it came to be known, was a life-size model of a man's upper half. Dressed in Ottoman robes with a turban and black beard, a smoking pipe in its left hand, the Turk sat at a cabinet with three doors and a chessboard on top. The doors opened to show an intricate set of gears and levers designed to give the audience the illusion of an advanced contraption worthy of the owner's claims. What remained hidden was a sliding seat behind the gears that enabled a small human chess player to move around as Kempelen showed suspicious spectators the cabinet's inside, and then to manipulate the contraption once the game began.

The Turk beat almost everyone it played against, including Benjamin Franklin, who was the US ambassador to France at the time. It also beat the novelist Edgar Allan Poe, as well as Napoléon Bonaparte, whose strategy to beat the Turk was to make a series of deliberately illegal moves. The Turk replaced Napoléon's piece after the first two moves, and then, after the third, swept its arm across the chessboard, knocking over all of the pieces and ending the game. Napoléon reportedly was amused.

Real or not, the Turk started new dialogue among those who had never considered the potential of mechanized intelligence. Among those was Charles Babbage, whom the Turk beat twice, and who, despite correctly concluding the device was a hoax, drew inspiration from the experience and went on to build the difference engine, the first mechanical computer. The Turk was destroyed in a fire in Philadelphia on July 5, 1854.

SEE ALSO The Difference Engine (1822), Edgar Allan Poe's "The Gold-Bug" (1843), Computer Is World Chess Champion (1997)

Instead of a machine, a small chess master hid inside the cabinet that held the "Mechanical Turk." To conceal his presence, the person moved from side to side as the different panel doors were opened.

The chefs player.

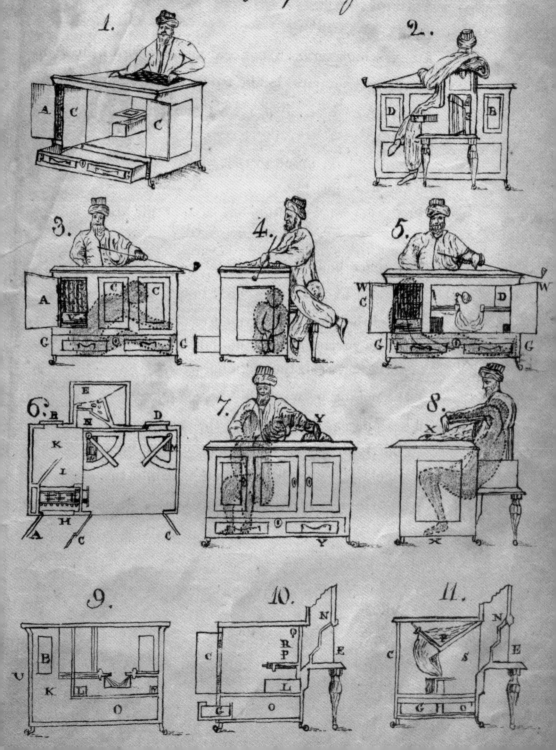

Optical Telegraph

Claude Chappe (1763–1805)

People had used signal fires, torches, and smoke signals since ancient times to send messages rapidly over long distances. The ancient Athenians used flashes of sunlight from their shields to send messages from ship to shore. The Romans coded flags to send messages over a distance—a practice that the British Navy also employed as early as the 14th century.

In 1790, an out-of-work French engineer named Claude Chappe started a project with his brothers to develop a practical system for sending messages quickly over the French countryside. The idea was to set up a series of towers constructed on hills, with each tower in view of the next. Each tower would be equipped with a device that had big, movable arms and a telescope, so that the position of the arms could be determined and then relayed to the next tower. An operator in the first tower would move the arms into different positions, each position signaling a letter, and the operator in the second tower would write it down—essentially sending letters over distance (*tele-graph*) with light. A second telescope would allow for messages to be conveyed in the opposite direction.

After successfully sending a message nearly 9 miles (14 kilometers) on March 2, 1791, Claude and his younger brothers, Pierre François (1765–1834), René (1769–1854), and Abraham (1773–1849), moved to Paris to continue the experiments and drum up support from the new government. Their older brother, Ignace Chappe (1760–1829), was a member of the revolutionary Legislative Assembly, which probably helped somewhat. Soon the brothers were authorized by the Assembly to construct three stations as a test. That test went well, and in 1793 the Assembly decided to replace its system of couriers with optical telegraph lines. Claude Chappe was appointed lieutenant of engineering for the construction of a telegraph line between Paris and Lille, under the control of the Ministry of War.

The first practical demonstration of the telegraph came on August 30, 1794, when the Assembly learned that its army in Condé-sur-l'Escaut had been victorious. That message was transmitted in about half an hour. In the following years, telegraph lines were built across France, connecting all of the major cities. At its height, the system had 534 stations covering more than 3,000 miles (5,000 kilometers). Not surprisingly, Napoléon Bonaparte made heavy use of the technology during his conquest of Europe.

SEE ALSO Fax Machine Patented (1843)

An artist's impression of Claude Chappe, demonstrating his aerial telegraph semaphore system, from the Paris newspaper Le Petit Journal, 1901.

The Jacquard Loom

Joseph-Marie Jacquard (1752–1834)

In 1801, French weaver Joseph-Marie Jacquard invented a way to accelerate and simplify the time-consuming, complex task of weaving fabric. His technique was the conceptual precursor to binary logic and programming that exists today.

While looms of the 18th century could create complex patterns, doing so was an entirely manual affair, requiring an extraordinary amount of time, constant vigilance to avoid mistakes, and skilled hands—especially with intricate fabric patterns such as damask and brocade. Jacquard realized that despite the complexity of a pattern, the act of weaving was a repetitive process that could be carried out mechanically. His invention used a series of cards laced together in a continuous chain, with a row on each card where holes could be punched, corresponding with one row of the fabric pattern. Some cards had holes in the specified position, while others did not. Essentially, the punched cards were a control mechanism that contained data—like binary 0s and 1s—that directed a sequence of actions, in this case how a loom could be mechanized to weave a repeating pattern. A hole would cause a corresponding thread to be raised, while no hole would cause the thread to be lowered. The actual mechanism involved a rod that would either travel through the hole or be stopped by the card; each rod was linked to a hook, and together they formed the harness that controlled the position of the threads. After the threads were raised or lowered, the shuttle holding another roll of thread would zip from one side of the loom to the other, completing the weave. Then the rods in the holes would retract, the card would advance, and the process would start over again.

Jacquard's invention evolved from earlier ideas by Jacques de Vaucanson (1709–1782), Jean-Baptiste Falcon, and Basile Bouchon, the last of whom invented a way to control a loom using perforated tape in 1725. Later inventors would take that concept and use punched cards to represent numerical data and other types of information.

SEE ALSO Tabulating the US Census (1890)

A Jacquard loom in the National Museum of Scotland, Edinburgh. Pins either are stopped by the card or poke through the holes, determining the pattern woven by the loom.

The Difference Engine

Johann Helfrich von Müller (1746–1830), Charles Babbage (1791–1871)

A difference engine is a machine for tabulating and calculating polynomial functions. It was created to automatically calculate accurate nautical and astronomical tables, because those made by human computers contained too many errors.

The concept originated with Johann Helfrich von Müller in 1786, but it was Charles Babbage, the British mathematician and inventor, who devised an extraordinarily detailed plan of how a functioning engine would work and created working prototypes at a scale that had not previously been conceived. It is perhaps the most intricate blueprint of a machine ever conceived on paper that could function.

Babbage's original difference engine was the first automatic calculator, meaning that it was able to use the results from one calculation as an input to the next. He designed it with 16 decimal digits of accuracy and a printer; it could also produce plates for printing. It was to print not just nautical tables but also tables of logarithms, trigonometric functions, and even artillery tables (the task for which ENIAC, the Electronic Numerical Integrator and Computer, would later be constructed).

With funding from the British government, Babbage hired a machinist to build the engine. But it was never fully realized, and the project shut down in 1842. While manufacturing limitations were a factor in failure, lore has it that a financial dispute with the tool maker was the final, insurmountable obstacle. Babbage did manage to make a small, partial prototype of his second difference engine in 1832. The drawings for the second engine were used to build a fully functional machine that was completed in London in 2002. That machine has 8,000 parts, is 11 feet in length, and weighs 5 tons.

While work on the difference engine was halted, Babbage designed (but never built) the analytical engine, which was to be programmed with punch cards and had a separate "store" where numbers would be kept and a "mill" where the math would be calculated, a design that would be implemented in electronics nearly a century later.

SEE ALSO Antikythera Mechanism (c. 150 BCE), Ada Lovelace Writes a Computer Program (1843), Tabulating the US Census (1890), ENIAC (1943)

The Difference Engine No. 1 by Charles Babbage was the first successful automatic calculator. The portion pictured here was assembled by Babbage's engineer, Joseph Clement, in 1832. Consisting of approximately 2,000 parts, it represents one-seventh of the complete engine.

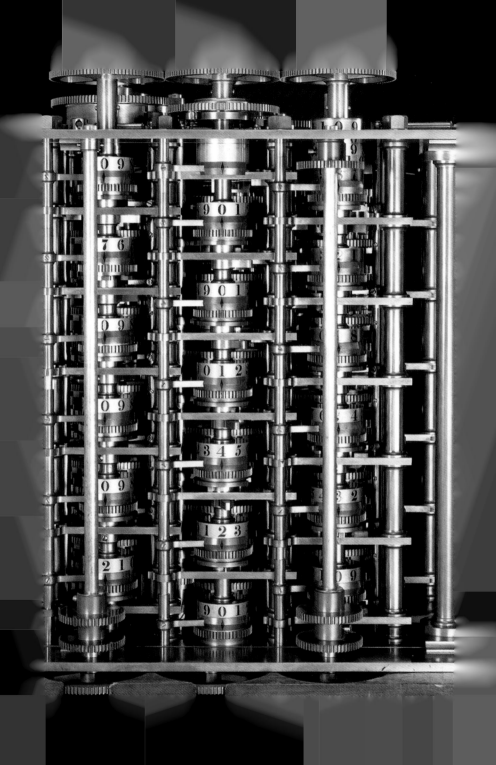

Electrical Telegraph

John Frederic Daniell (1790–1845), **Joseph Henry** (1797–1878),
Samuel Morse (1791–1872), **William Fothergill Cooke** (1806–1879),
Charles Wheatstone (1802–1875)

Using electricity to send messages through wires was the subject of much experimentation in Europe and the United States during the early 19th century. The key invention was John Daniell's wet-cell battery (1836), a reliable source of electricity. Various forms of metal wire had existed since ancient times, and air was a reasonably good insulator, so sending electricity over distance required little more than stringing up a wire, modulating the signal with some kind of code, and having a device at the other end to turn the coded electrical pulses back into something a human could perceive.

American inventor Samuel Morse is credited with inventing, patenting, and promoting the first practical telegraph in 1836. The original Morse system started with a message that was encoded as a series of bumps on small, puzzle-like pieces that were placed into a tray. The operator turned a crank that moved the tray past a switch that completed and broke an electric circuit as it moved up and down. At the other end, an electromagnet moved a fountain pen or pencil up and down as a strip of paper moved underneath. To transmit text, each letter and number needed to be translated into a series of electrical pulses, which we now call *dots* and *dashes*, after how they were recorded on the paper strip. To operate over distances, the Morse system relied on Joseph Henry's amplifying electromechanical relay, which allowed faint electrical signals sent over a long distance to trigger a second circuit.

In England, meanwhile, William Fothergill Cooke and Charles Wheatstone developed their own telegraph system based on the ability of electricity moving through wire to deflect a magnetic compass. The original Cooke–Wheatstone telegraph used five needles arranged in a line on a board, along with a pattern of 20 letters: by sending electricity down a pair of wires, two of the needles would deflect and point at one of the letters.

Cooke and Wheatstone's system was the first to be commercialized. A few years later, with $30,000 in federal funding, Morse built an experimental telegraph line from Washington, DC, to Baltimore, Maryland. On May 24, 1844, Morse sent his famous message—"What hath God wrought?"—between the two cities.

SEE ALSO First Electromagnetic Spam Message (1864)

Drawings from Samuel Morse's sketchbook, illustrating his first conception of the telegraph.

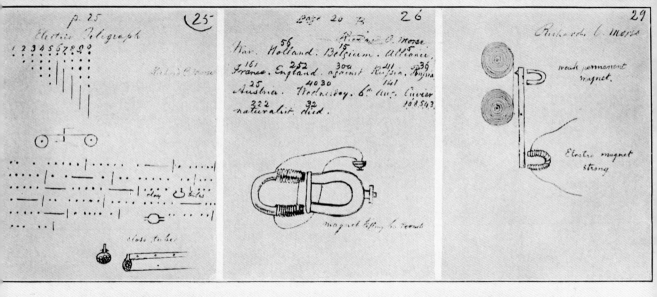

DRAWINGS FROM 1832 SKETCH-BOOK, SHOWING FIRST CONCEPTION OF TELEGRAPH

Ada Lovelace Writes a Computer Program

Ada Lovelace (1815–1852)

What do you get when you combine a scientifically minded, logical mother with a free-spirited, poetically gifted father? You get Augusta Ada King-Noel, Countess of Lovelace, better known as Ada Lovelace—a British woman of the Industrial Age who used her unusual background and lineage to contribute to the cutting-edge technology of her day: the steam-powered Babbage difference engine.

Her mother was Lady Anne Isabella Milbanke Byron (1792–1860), and her father was the famous poet and notorious philanderer, Lord Byron (1788–1824). Lady Anne kicked Lord Byron out of the house when Ada was just five weeks old; Ada never met him. Determined to keep any trace of Lord Byron out of Ada's life, she committed her daughter to a rigorous education in mathematics and science. Private tutors filled Ada's days, including the Scottish science writer Mary Somerville, who introduced Ada to Charles Babbage at a dinner party.

At the party, Babbage unveiled a small prototype of his difference engine. Ada was captivated and wanted to know details of how it worked. That conversation was the first of many, which eventually led Babbage to show Ada the blueprints for his follow-up invention, the analytical engine. With her curious, creative mind and mature understanding of mathematics, she was commissioned to translate from French (at the time, a primary language of science) the lecture notes of Italian statesman Luigi Menabrea (1809–1896), who attended a talk Babbage gave on the analytical engine, and to add notes and ideas of her own. This she published in *Scientific Memoirs*, an early science journal, in 1843.

In that article appears Ada's algorithm and detailed instructions for making Babbage's machine compute Bernoulli numbers. This is generally regarded as one of the first published computer programs.

In recognition of her talents and influence on computer science, in 1979 the US Department of Defense named the Ada computer language after her.

SEE ALSO The Jacquard Loom (1801)

Watercolor portrait of Ada Lovelace, by Alfred Edward Chalon, c. 1840. Lovelace worked with Charles Babbage on the analytical engine, for which she designed the world's first computer program.

Fax Machine Patented

Alexander Bain (1811–1877), **Giovanni Caselli** (1815–1891)

Before the telephone, before radio, there was the fax machine. It wasn't the fax machine of the 1990s—the machine that transmitted information over ordinary phone lines—but rather a machine comprising of a pair of synchronized pendulums connected to each other over distance by an electrified wire.

Alexander Bain was a Scottish clockmaker with an interest in both electricity and invention. In 1843, he built an "electric printing telegraph" that used a pair of precisely timed pendulums, one configured to function like a scanner, the other to function as a remote printer. A message scanned by the first pendulum would print out at the second.

The scanning pendulum had an arm that moved back and forth across a metal plate holding raised metal printers type. After each swing, the plate advanced in the perpendicular direction. Thus, the arm scanned a path of parallel horizontal lines across the type. When a small contact on the arm swept over part of a letter, a circuit would be completed and an electric current would flow down the wire to the remote system, where the synchronized pendulum was scanning horizontal lines over a piece of chemically treated paper. When electricity flowed, the paper under the second pendulum would change color.

Although Bain's system worked, he ended up in disputes with both Charles Wheatstone (1802–1875) and Samuel Morse (1791–1872). Bain died in poverty in 1877.

Italian inventor Giovanni Caselli improved on Bain's basic idea with a more compact device called a *pantelegraph*, which transmitted a message written with insulating ink on a metal plate over a set of wires. Commercial operation of the pantelegraph began in 1865 between Paris and Lyon, mostly to verify signatures on banking instructions.

The discovery that the element selenium was also a photoconductor meant that its electrical resistance changed with light, making it possible to send photographic images. This was put to use in 1907 with a "wanted" poster that was sent from Paris to London help catch a jewel thief. Soon newspapers were routinely printing photos that had been sent by wire. In 1920, the Bartlane cable picture transmission system routinely sent digitized newspaper photographs from London to New York, taking three hours to transmit each photograph.

SEE ALSO First Digital Image (1957)

Alexander Bain's "electric printing telegraph" paved the way for later fax machines, such as this 1960 machine by Alexander Muirhead.

Edgar Allan Poe's "The Gold-Bug"

Edgar Allan Poe (1809–1849)

It was not a stunning cryptographic breakthrough or a spectacular demonstration of mathematical wizardry that helped to popularize secret writing and ciphers among the 19th-century American public. It was traditional storytelling by master of macabre Edgar Allan Poe.

Poe loved puzzles and ciphers and went to great lengths to write about them during his time as a magazine editor and literary author. The best known of these is a short story called "The Gold-Bug," which follows a man named William Legrand who is bitten by a gold-colored bug and is convinced that it has a crucial role to play in restoring his fortune. Legrand and his associates embark on an adventure to find buried treasure that involves cryptograms, invisible ink, and the bug, which must be dropped through the left eye of a skull to unlock the overall solution.

The story was incredibly popular and is credited with inspiring Robert Louis Stevenson to write *Treasure Island* (1883). "The Gold-Bug" also captured the imagination of a young William F. Friedman, who went on to become a self-taught cryptographer, trained two generations of cryptanalysts (one for each world war), and became the US National Security Agency's first chief cryptologist in 1952.

Poe once stated, in an 1841 letter to Frederick W. Thomas, "Nothing intelligible can be written which, with time, I cannot decipher." While an editor at *Graham's Magazine*, Poe wrote an article titled "A Few Words on Secret Writing" in which he offered a free subscription to any reader who could send him a cipher that he could not break. He claimed to have solved 100 submissions and then finished the contest by publishing two ciphers that were sent into the magazine by Mr. W. B. Tyler. Suspicion swirled that Tyler was really Poe himself.

Poe's use of cryptography and puzzles in his editorial and literary work is symbolic of a practical challenge that ordinary people had during this time—there were few options for securely communicating private information. This was especially a problem for sending sensitive information by the newly invented telegraph, because by its very nature, a message sent by wire had to pass through many hands—being keyed, transcribed, and ultimately delivered—before reaching its intended destination.

SEE ALSO RSA-129 Cracked (1994)

An 1849 daguerreotype of American author Edgar Allan Poe. Poe helped popularize ciphers and secret writing in the 19th century.

Thomas Arithmometer

Gottfried Wilhelm Leibniz (1646–1716),
Charles Xavier Thomas de Colmar (1785–1870)

German philosopher and mathematician Gottfried Leibniz became interested in mechanical calculation after seeing a pedometer while visiting Paris in 1672. He invented a new type of gear that could advance a 10-digit dial exactly 0 to 9 places, depending on the position of a lever, and used it in a machine with multiple dials and levers called the *stepped reckoner*. Designed to perform multiplication with repeated additions and division by repeated subtractions, the reckoner was hard to use because it didn't automatically perform carry operations; that is, adding 1 to 999 did not produce 1,000 in a single operation. Worse, the machine had a design flaw—a bug—that prevented it from working properly. Leibniz built only two of them.

More than 135 years later, Charles Xavier Thomas de Colmar left his position as inspector of supply for the French army and started an insurance company. Frustrated by the need to perform manual arithmetic, Thomas designed a machine to help with math. Thomas's arithmometer used Leibniz's mechanism, now called a *Leibniz wheel*, but combined it with other gears, cogs, and sliding levers to create a machine that could reliably add and subtract numbers up to three digits, and multiply and divide as well. Thomas patented the machine, but his business partners at the insurance firm were not interested in commercializing it.

Twenty years later, Thomas once again turned his attention to the arithmometer. He demonstrated a version at the 1844 French national exhibition and entered competitions again in 1849 and 1851. By 1851, he had simplified the machine's operation and extended its capabilities, giving it six sliders for setting numbers and 10 dials for display results. Aided by three decades' advance in manufacturing technology, Thomas was able to mass-produce his device. By the time of his death, his company had sold more than a thousand of the machines—the first practical calculator that could be used in an office setting—and Thomas was recognized for his genius in creating it. The size of the arithmometer was approximately 7 inches (18 centimeters) wide by 6 inches (15 centimeters) tall.

SEE ALSO Curta Calculator (1948)

This Thomas Arithmometer can multiply two 6-digit decimal numbers to produce a 12-digit number. It can also divide.

Boolean Algebra

George Boole (1815–1864), **Claude Shannon** (1916–2001)

George Boole was born into a shoemaker's family in Lincolnshire, England, and schooled at home, where he learned Latin, mathematics, and science. But Boole's family landed on hard times, and at age 16 he was forced to support his family by becoming a school teacher—a profession he would continue for the rest of his life. In 1838, he wrote his first of many papers on mathematics, and in 1849 he was appointed as the first professor of mathematics at Queen's College in Cork, Ireland.

Today Boole is best known for his invention of mathematics for describing and reasoning about logical prepositions, what we now call *Boolean logic*. Boole introduced his ideas in his 1847 monograph, "The Mathematical Analysis of Logic," and perfected them in his 1854 monograph, "An Investigation into the Laws of Thought."

Boole's monographs presented a general set of rules for reasoning with symbols, which today we call *Boolean algebra*. He created a way—and a notation—for reasoning about what is true and what is false, and how these notions combine when reasoning about complex logical systems. He is also credited with formalizing the mathematical concepts of AND, OR, and NOT, from which all logical operations on binary numbers can be derived. Today many computer languages refer to such numbers as *Booleans* or simply *Bools* in recognition of his contribution.

Boole died at the age of 49 from pneumonia. His work was carried on by other logicians but didn't receive notice in the broader community until 1936, when Claude Shannon, then a graduate student at the Massachusetts Institute of Technology (MIT), realized that the Boolean algebra he had learned in an undergraduate philosophy class at the University of Michigan could be used to describe electrical circuits built from relays. This was a huge breakthrough, because it meant that complex relay circuits could be described and reasoned about symbolically, rather than through trial and error. Shannon's wedding of Boolean algebra and relays let engineers discover bugs in their diagrams without having to first build the circuits, and it allowed many complex systems to be refactored, replacing them with relay systems that were functionally equivalent but had fewer components.

SEE ALSO Binary Arithmetic (1703), Manchester SSEM (1948)

A circuit diagram analyzed using George Boole's "laws of thought"—what today is called Boolean algebra. Boole's laws were used to analyze complicated telephone switching systems.

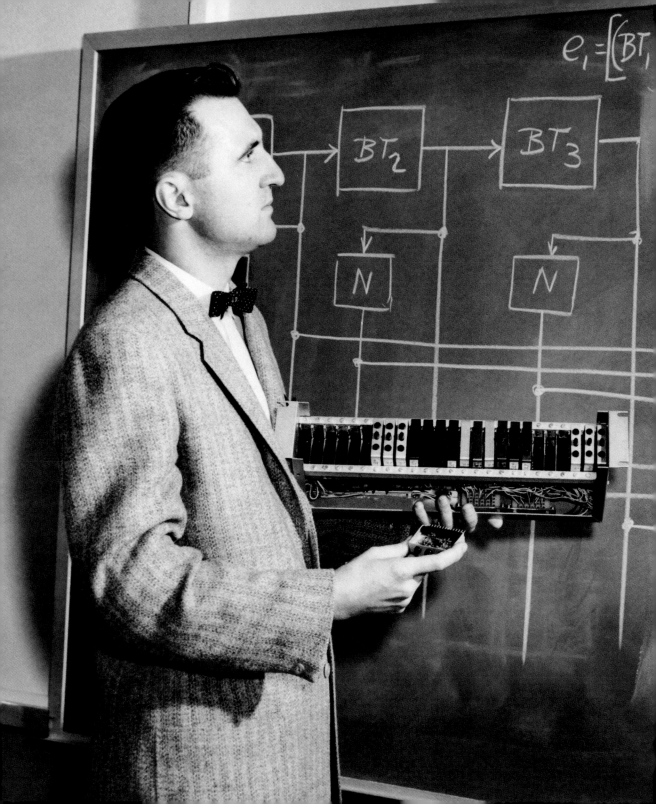

First Electromagnetic Spam Message

William Fothergill Cooke and Charles Wheatstone's electromagnetic telegraph took England by storm shortly after commercial service began in 1837. By 1868, there were more than 10,000 miles of telegraph wire in the United Kingdom supporting 1,300 telegraph stations; four years later, there were 5,179 stations, serviced by more than 87,000 miles of wire.

With a capability to reach large numbers of people quickly and easily, the world's first unsolicited, electrically enabled advertisement was sent in London late in the evening of May 29, 1864, according to historian Matthew Sweet. The message was from Messrs. Gabriel, a group of unregistered dentists, who sold a variety of false teeth, gums, toothpaste, and tooth powder.

The message, sent to current and former members of Parliament, read as follows:

> Messrs. Gabriel, dentists, Harley-street, Cavendish-square. Until October Messrs. Gabriel's professional attendance at 27, Harley-street, will be 10 till 5.

In 1864 there were no telegraphs in private residences; the message appeared on the swinging needles of the Cooke-Wheatstone electromagnetic telegraph, where it was transcribed by operators, carried by a boy sent from the London District Telegraph Company, and placed into the hand of a member of Parliament.

That M.P. wrote about his annoyance in a letter to the editor of the local paper: "I have never had any dealings with Messrs. Gabriel, and beg to know by what right do they disrupt me by a telegram which is simply the medium of advertisement? A word from you would, I feel sure, put a stop to this intolerable nuisance."

But it wasn't shame that put a halt to spam sent by telegram: it was the cost. Advertising by telegraph just wasn't cost effective, due to the high price of sending the messages. That price plummeted with the birth of email, which was used to send a bulk, unsolicited advertisement for the first time in 1978.

SEE ALSO First Internet Spam Message (1978)

On May 29, 1864, Messrs. Gabriel, a group of unregistered dentists, sent members of the British Parliament the earliest known unsolicited electronic message. One recipient complained to the newspaper.

TO THE EDITOR OF THE TIMES.

Sir,—On my arrival home late yesterday evening a "telegram," by "London District Telegraph," addressed in full to me, was put into my hands. It was as follows:—

"Messrs. Gabriel, dentists, 27, Harley-street, Cavendish-square. Until October Messrs. Gabriel's professional attendance at 27, Harley-street, will be 10 till 5."

I have never had any dealings with Messrs. Gabriel, and beg to ask by what right do they disturb me by a telegram which is evidently simply the medium of advertisement? A word from you would, I feel sure, put a stop to this intolerable nuisance. I enclose the telegram, and am,

Your faithful servant,

Upper Grosvenor-street, May 30.   M. P.

Baudot Code

Jean-Maurice-Émile Baudot (1845–1903), Donald Murray (1865–1945)

Early telegraph systems relied on human operators to encode and transmit the sender's message, and then to perceive, decode, and transcribe the message on paper upon receipt. Relying on human operators limited the maximum speed at which a message could be sent and required operator skills that were not easily available.

Émile Baudot developed a better approach. A trained French telegraph operator, Baudot devised a system that used a special keyboard with five keys (two for the left hand and three for the right) to send each character. Thirty-one different combinations arise from pressing one or more of the five keys together; Baudot assigned each code to a different letter of the alphabet. To send a message, the operator would type the codes in sequence as the machine clicked, roughly four times a second. With each click, a rotating part that Baudot called the *distributor* would read the position of each key in order and, if the key was pressed, send a corresponding pulse down the telegraph wire. At the other end, a remote printer would translate the codes back into a printed character on a piece of paper tape.

Baudot was one of the first people to combine key inventions by others into one working system. He patented his invention in 1874, started selling devices to the French Telegraph Administration in 1875, and was awarded the gold medal at the Paris Exposition Universelle in 1878. Baudot's code was adopted as the International Telegraph Alphabet No. 1 (ITA1), one of the original international telecommunications standards. In recognition of his contribution, the baud, a unit of data transmission speed equal to the number of signal changes per second, is named after him.

In 1897, the Baudot system expanded to incorporate punched paper tape. The keyboard was disconnected from the telegraph line and connected to a new device that could punch holes across a strip of paper tape, with one hole corresponding to each key. Once punched, the tape could be loaded into a reader and the message sent down the telegraph wire faster than a human could type. In 1901, the inventor Donald Murray developed an easier-to-use punch that was based on a typewriter keyboard. Murray also made changes to Baudot's code; the resulting code was known as the Baudot-Murray code (ITA2) and remained in use for more than 50 years.

SEE ALSO ASCII (1963), Unicode (1992)

Paper tape punched with the five-level Baudot code. The large holes correspond to the 5 bits of the code, while a rotating toothed tractor wheel fit into the small holes and used them to pull the tape through the machine.

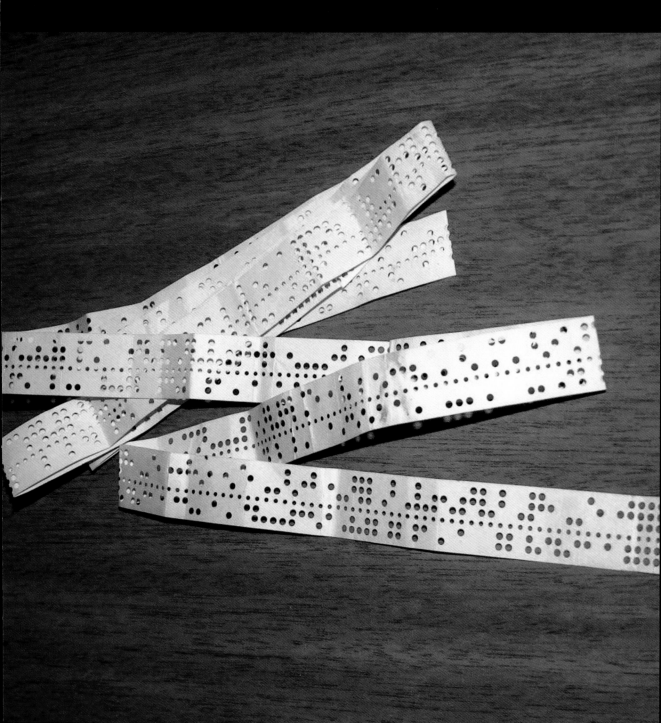

Semiconductor Diode

Michael Faraday (1791–1867), Karl Ferdinand Braun (1850–1918)

Semiconductors are curious devices: not quite conductors like copper, gold, or silver, not quite insulators like plastic or rubber. In 1833, Michael Faraday discovered that the chemical silver sulfide became a better conductor when heated, unlike metals that lose their conductivity under the same conditions. Separately, in 1874, Karl Ferdinand Braun, a 24-year-old German physicist, discovered that a metal sulfide crystal touched with a metal probe would conduct electricity in only one direction. This "one direction" characteristic is what defines *diodes* or *rectifiers*, the simplest electronic components.

Braun's discovery was a curiosity until the invention of radio. The diode proved critical in allowing radio to make the transition from wireless telegraphy to the transmission and reception of the human voice. The diode of choice for these early radio sets was frequently called a *cat's whisker diode*, because it consisted of a crystal of galena, a form of lead sulfide, in contact with a spring of metal (the "whisker"). By carefully manipulating the pressure and orientation of the metal against the crystal, an operator could adjust the electrical properties of the semiconductor until they were optimal for radio reception. Powered only by the radio waves themselves, a crystal set was only strong enough to faintly produce sounds in an earphone.

Crystal radio receivers were used onboard ships and then in homes until they were replaced by new receivers based on vacuum tubes, which could amplify the faint radio waves so that they were strong enough to power a speaker and fill a room with speech or music. But tubes didn't mark the end of the crystal radio: the devices remained popular for people who couldn't get tubes—such as on the front lines in World War II—as well as among children learning about electronics. In the 1940s, scientists at Bell Labs turned their attention to semiconductor radios once again in an effort to perfect microwave communications. In the process, they discovered the transistor.

Braun went on to make other fundamental contributions to physics and electronics. In 1897, he invented the cathode-ray tube, which would become the basis of television. He shared the 1909 Nobel Prize with Guglielmo Marconi (1874–1937) "in recognition of their contributions to the development of wireless telegraphy."

SEE ALSO Silicon Transistor (1947)

Crystal Detector, made by the Philmore Manufacturing Company. To use this device, the operator would connect a wire to each of the two flanges and press the metal "whisker" into the semiconductor crystal.

Tabulating the US Census

Herman Hollerith (1860–1929)

When the US Constitution was ratified, it mandated that the government conduct an "actual enumeration" of every free person in the union every 10 years. As the number of people in the nation grew, the enumeration took longer and longer to complete. The 1880 Census counted 50,189,209 people. It took 31,382 people to perform the count and eight years to tabulate the results, producing 21,458 pages of published reports. So, in 1888, the Census Bureau held a competition to find a better way to process and tabulate the data.

American inventor Herman Hollerith had worked briefly at the Census Bureau prior to the 1880 census and in 1882 joined the faculty of MIT, where he taught mechanical engineering and experimented with mechanical tabulation systems. His early systems used long rolls of paper tape with data represented as punched holes. Then, on a railroad trip to the American West, Hollerith saw how conductors made holes on paper tickets corresponding to a person's hair color, eye color, and so on, so that tickets couldn't be reused by other passengers. Hollerith immediately switched his systems to use paper cards.

Hollerith entered the 1888 competition and won, his system being dramatically faster than those of the two other entrants. On January 8, 1889, he was awarded a US patent on "method, system and apparatus for compiling statistics," originally filed September 23, 1884.

Hollerith's system consisted of a slightly curved card measuring 3.25 by 7.375 inches (83 millimeters by 187 millimeters). A human operator punched holes in the card with a device called a *Pantographic Card Punch*, with holes in specific locations to signify a person's gender, marital status, race, ownership and indebtedness of farms and homes, and other information. For tabulation, the cards were passed through a reader with micro switches to detect the presence of holes and electromechanical circuits to perform the actual tabulation.

SEE ALSO The Jacquard Loom (1801), ENIAC (1943)

A woman with a Hollerith Pantographic Card Punch, which creates holes in specific locations to signify a person's gender, marital status, and other information. This photo is from the 1940 US census.

Strowger Step-by-Step Switch

Almon Brown Strowger (1839–1902)

The Bell Telephone Company was incorporated in July 1877, and by the 1880s it was quickly expanding. The switchboards that connected phones together and completed calls were manually run by operators.

The early phone system didn't have dials or buttons. Instead, there was a crank, connected to a tiny electrical generator. Users would pick up the phone and turn the crank, and electricity would travel down the phone line to signal the operator.

Almon Strowger was an undertaker in Kansas City, Missouri. He noticed that his business had declined as the telephone became more popular. Strowger learned that one of the telephone operators was married to his competitor, and whenever a phone call came in for the undertaker, she would direct the call to her husband. Motivated, Strowger invented the step-by-step switch, an electromechanical device that would complete a circuit between one phone and a bank of others depending on a sequence of electric pulses sent down the phone line. Instead of relying on an operator to connect, Strowger envisioned that people would tap out a code using a pair of push buttons.

Working with his nephew, Strowger built a working model and got a patent. Although other inventors had experimented with operator-free dialing systems— thousands of patents were filed—this system "worked with reasonable accuracy," according to a 1953 article in the *Bell Laboratories Record*.

Strowger, family members, and investors then created the Strowger Automatic Telephone Exchange Company in 1891. They went to La Porte, Indiana, which had recently lost its telephone system because of a patent dispute between the local independent operator and the Bell Telephone System, and set up the world's first automated telephone exchange with direct dialing—at least for local calls—in 1892.

The switch was called "step-by-step" because of the way that a telephone call was completed, one dialed digit at a time. Step-by-step exchanges remained in service throughout the United States until 1999, when the last was removed from service, replaced by the #5ESS computerized local exchange.

SEE ALSO Digital Long Distance (1962)

The friction drive of the Western Electric 7A Rotary, No. 7001 Line Finder. The bevel gear on the right has a steady rotary motion and does not use an electromagnet for stepping.

18.426

Floating-Point Numbers

Leonardo Torres y Quevedo (1852–1936), **William Kahan** (b. 1933)

Leonardo Torres y Quevedo was a Spanish engineer and mathematician who delighted in making practical machines. In 1906, he demonstrated a radio-controlled model boat for the king of Spain, and he designed a semirigid airship used in World War I.

Torres was also a fan of Babbage's difference and analytical engines. In 1913, he published *Essays in Automatics*, which described Babbage's work and presented the design for a machine that could calculate the value of the formula $a(y-z)^2$ for specified values of a, y, and z. To allow his machine to handle a wider range of numbers, Torres invented floating-point arithmetic.

Floating-point arithmetic extends the range of a numerical calculation by decreasing its accuracy. Instead of storing all of the digits in a number, the computer stores just a few significant digits, called the *significand*, and a much shorter exponent. The actual "number" is then computed using the formula: $\text{significand} \times \text{base}^{\text{exponent}}$

For example, the gross domestic product of the United States in 2016 was 18.57 trillion dollars. Storing that number with a fixed-point representation requires 14 digits. But storing it in floating point requires just 6 digits: $18.57 trillion = 1.857×10^{13}

Thus, with floating-point numbers, sometimes called *scientific notation* on modern calculators, a 10-digit register (a mechanical or electronic gadget that can store a number) that would normally be limited to storing numbers between 1 and 9,999,999,999 instead can be partitioned into an 8-digit significand and a 2-digit exponent, allowing it to store numbers as small as $0.0000001 \times 10^{-99}$ and as large as 9.9999999×10^{99}.

Modern floating-point systems use binary rather than decimal digits. Under the standard developed by Canadian mathematician William Kahan for the Intel® 8086 microprocessor and adopted in 1985 by the Institute of Electrical and Electronics Engineers (IEEE 754), single precision floating point uses 24 bits for the significand and 8 bits for the exponent.

For his work, Kahan won the Association for Computing Machinery's (ACM) A.M. Turing Award in 1989.

SEE ALSO Binary Arithmetic (1703), Z3 Computer (1941), Binary-Coded Decimal (1944)

Portrait of Leonardo Torres y Quevedo by Argentinian cartoonist and illustrator Eulogia Merle (b. 1976).

Vernam Cipher

Gilbert Vernam (1890–1960), Joseph Mauborgne (1881–1971)

Most encryption algorithms are *computationally secure*. This means that while it's theoretically possible to crack the cipher by trying every possible encryption key, in practice this isn't possible because trying all of the keys would require too much computational power.

More than a century ago, Gilbert Vernam and Joseph Mauborgne came up with a cryptographic system that is *theoretically secure*: even with an infinite amount of computer power, it is impossible to crack a message encrypted with the Vernam Cipher, no matter how fast computers ever become.

Vernam's cipher, today called a *one-time pad*, is unbreakable because the encrypted message, decrypted with an incorrect key, can result in a plausible-looking message. Indeed, it can result in *every* possible message, since the key is the same length as the message. That is, for any given ciphertext, there is a key that makes it decrypt as a verse from the Bible, a few lines from Shakespeare, and the text on this page. Without a way to distinguish a correct from an incorrect decryption, the cipher is theoretically unbreakable.

Working at American Telephone and Telegraph Company (now AT&T®) in 1917, Vernam created a stream cipher that encrypted messages one character at a time by combining each character of the message with a character of a key. At first Vernam thought that key could be simply another message, but the following year, working with Joseph Mauborgne, a captain in the US Army Signal Corps, the two realized that the key must be random and nonrepeating. This improved security substantially: if the key were another message, it would be possible to distinguish a probable key from one that was improbable. But if the key was truly random, then any key was equally possible. Together, the two inventors created what we now call a one-time pad, one of only two known encryption systems that are provably unbreakable (the other being quantum cryptography).

As it turns out, a banker named Frank Miller had also invented the concept of the one-time pad in 1882, but his pen-and-paper system was not widely publicized or used.

SEE ALSO Manchester SSEM (1948), RSA Encryption (1977), Advanced Encryption Standard (2001)

One-time pad device used with SIGTOT cipher system used aboard President Roosevelt's Douglas C-54 airplane.

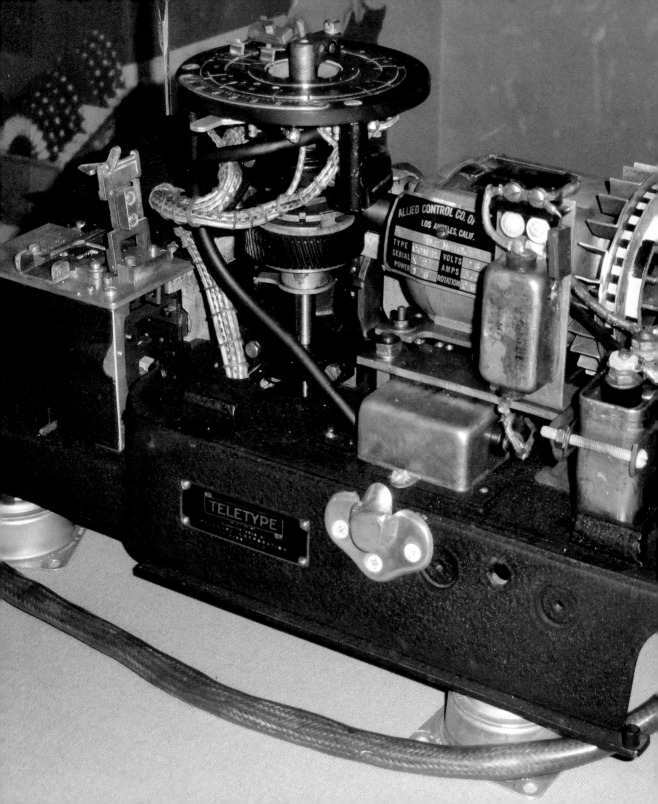

Rossum's Universal Robots

Karel Čapek (1890–1938)

The word *robot* was coined in 1920 by the Czech dramatist Karel Čapek in his science-fiction hit, *R. U. R.* (*Rossum's Universal Robots*). He coined the word based on the Czech word *robota* ("forced labor"); the word is now widely used for mechanized beings in most languages. In the play, the fictional Rossum corporation has developed cheap, biological humanoid machines called robots and has been shipping them all over the world from its secretive island-based factory. While some nations initially use the robots as soldiers, eventually the world more or less accepts the robots and puts them to work.

R. U. R. included many literary tropes that would become commonplace in future robot literature: an underground organization that seeks to liberate the robots; intelligent robots that are assembled from parts, with short lifespans, no pain, and no emotions; and a likable scientist with questionable ethics. Expensive at first, *R. U. R.*'s robots drop in price from $10,000 to $150—from around $130,000 down to $2,000 in today's money. In Čapek's world, war is a memory, human birthrates are down, and the future seems both predictable and pleasurable. The play's first act is largely devoted to telling the story of these fantastical creations and posing a philosophical question: if humans need not work, then what purpose do they serve after all?

And then the robots decide to kill every last human on the planet.

Although largely forgotten now, *R. U. R.* was well received and popular. The play was produced in Prague, London, New York, Chicago, and Los Angeles. When Isaac Asimov penned his Three Laws of Robotics, he did so largely to prevent the future that Čapek had envisioned. Although Čapek got the technology wrong—he envisioned that robots would be biological, not based on mechanisms and computation—his compelling vision of a world in which humanity is simultaneously helped, transformed, and eventually smothered by its mechanical creations still haunts us to this day.

SEE ALSO *Metropolis* (1927), Isaac Asimov's Three Laws of Robotics (1942), *Star Trek* Premieres (1966), Boston Dynamics Founded (1992)

A poster for the Federal Theatre Project presentation of R. U. R. *at the Marionette Theatre, 1936–1939.*

FEDERAL USA WORK WPA THEATRE

MARIONETTE THEATRE

PRESENTS

RUR

REMO BUFANO DIRECTOR

Metropolis

Fritz Lang (1890–1976)

In 1927, German film director Fritz Lang was already visualizing what life in the year 2026 would look like. Technology featured prominently in the cityscape of his black-and-white silent film *Metropolis*—considered by many to be one of the most influential sci-fi movies of all time. His dystopian vision depicted oppressed workers below ground toiling at mindless, repetitive tasks on machines that ran the city. Above ground was a paradise where the city's elite lived out indulgent lives. Interpretations of Lang's technology-driven world can be seen in movies such as *Blade Runner*.

The plot of *Metropolis* involves a female robot built to resemble the deceased wife of the city's leader. Later, the mad scientist who created her transforms the heroine of the story—a nanny named Maria—into the female robot. To complete this transformation, the scientist uses vast amounts of electric energy and futuristic technology.

While the robot embodies humanity's continued fascination with how advancing technology may impact and integrate into people's lives, the fact that the robot is female is rare. Most robots featured in fiction and pop culture of that era were male or genderless. This robot—portrayed as the leader's wife and then Maria—is depicted as a strong, clearly feminine being. The cultural impact of that role has since been seen in numerous female characters and imagery, such as Beyoncé's "Sweet Dreams" interlude during a world tour, featuring a video of the singer in a robot costume that borrows strongly from "Maria."

In 2006, Carnegie Mellon University (CMU) inducted Maria into its Robot Hall of Fame. On the Hall of the Fame's website, CMU recognizes the robot Maria as the "singular most powerful image of early science fiction films and continuing inspiration in the creation of female robotic imagery in both science and science fiction."

SEE ALSO *Rossum's Universal Robots* (1920)

A poster for a 1984 rerelease of Metropolis, *the 1927 film by German film director Fritz Lang.*

First LED

1927

Oleg Vladimirovich Losev (1903–1942)

Although the electroluminescent property of some crystals was discovered in England in 1907, it took more than a decade of work by the self-taught Russian scientist Oleg Vladimirovich Losev to develop a theory (based on Einstein's photoelectric theory) of how the effect worked, and to produce devices that could be used in practical applications. In total, Losev published 16 academic papers that appeared in Russian, British, and German scientific journals between 1924 and 1930, comprehensively describing the devices in the process. He went on to come up with novel applications for light-emitting diodes (LEDs and other semiconductors, including a "light relay device," a radio receiver, and a solid-state amplifier, before dying of starvation during the Siege of Leningrad in 1942.

LEDs were rediscovered in 1962 by four different groups of American researchers. This time the technology would not be lost. Compared with incandescent, fluorescent, and nixie tubes of the day, LEDs consumed far less power and produced practically no heat. They had just three disadvantages: they could make only red light, they were not very bright, and they were fantastically expensive—more than $200 each at the beginning.

By 1968, improvements in production let companies push the price of LEDs down to five cents each. At that price, LEDs started showing up in calculators, wristwatches, laboratory equipment, and, of course, computers. Indeed, LEDs arranged as individual lights and seven-segment numeric displays were one of the primary outputs for the first generation of microcomputers in the mid-1970s. Even the early LEDs could be switched on and off millions of times a second, resulting in their use in fiber-optic communications. In 1980, infrared LEDs started showing up in television remotes.

Although blue and ultraviolet LEDs were invented in the 1970s, a number of breakthroughs were required to make them bright enough for practical use. Today those challenges have been overcome. Indeed, the bright-white LED house lights that have largely replaced both incandescent and fluorescent light bulbs are based on an ultraviolet LED that stimulates a white phosphor.

SEE ALSO First Liquid-Crystal Display (1965)

Eight decades after it was invented in 1927, light-emitting diodes were finally bright enough and cheap enough to replace incandescent light bulbs on a massive scale.

Electronic Speech Synthesis

Homer Dudley (1896–1980)

Long before Siri®, Alexa, Cortana, and other synthetic voices were reading emails, telling people the time, and giving driving directions, research scientists were exploring approaches to make a person's voice take up less bandwidth as it moved through the phone system.

In 1928, Homer Dudley, an engineer at Bell Telephone Labs, developed the vocoder, a process to compress the size of human speech into intelligible electronic transmissions and create synthetic speech from scratch at the other end by imitating the sounds of the human vocal cord. The vocoder analyzes real speech and reassembles it as a simplified electronic impression of the original waveform. To recreate the sound of human speech, it uses sound from an oscillator, a gas discharge tube (for the hissing sounds), filters, and other components.

In 1939, the renamed Bell Labs unveiled the speech synthesizer at the New York World's Fair. Called the *Voder*, it was manually operated by a human, who used a series of keys and foot pedals to generate the hisses, tones, and buzzes, forming vowels, consonants, and ultimately recognizable speech.

The vocoder followed a different path of technology development than the Voder. In 1939, with war having already broken out in Europe, Bell Labs and the US government became increasingly interested in developing some kind of secure voice communication. After additional research, the vocoder was modified and used in World War II as the encoder component of a highly sensitive secure voice system called SIGSALY that Winston Churchill used to speak with Franklin Roosevelt.

Then, taking a sharp turn in the 1960s, the vocoder made the leap into music and pop culture. It was and continues to be used for a variety of sounds, including electronic melodies and talking robots, as well as voice-distortion effects in traditional music. In 1961, the first computer to sing was the International Business Machines Corporation (IBM®) 7094, using a vocoder to warble the tune "Daisy Bell." (This was the same tune that would be used seven years later by the HAL 9000 computer in Stanley Kubrick's *2001: A Space Odyssey*.) In 1995, 2Pac, Dr. Dre, and Roger Troutman used a vocoder to distort their voices in the song "California Love," and in 1998 the Beastie Boys used a vocoded vocal in their song "Intergalactic."

SEE ALSO "As We May Think" (1945), HAL 9000 Computer (1968)

The Voder, exhibited by Bell Telephone at the New York World's Fair.

THE VODER

Differential Analyzer

Vannevar Bush (1890–1974), Harold Locke Hazen (1901–1980)

Differential equations are used to describe and predict the behavior of various phenomena in our constantly changing and complex world. They can predict ocean wave heights, population growth, how far a baseball might fly, how quickly plastic will decay, and so on. Some of these mathematical mysteries could be solved by hand, but other, more complex scenarios such as simulating nuclear explosions are far too labor intensive and intricate for a manual approach. To overcome this limitation, machines were needed in aiding human cognition.

Designed and built at MIT between 1928 and 1931 by Vannevar Bush and his graduate student Harold Locke Hazen, the differential analyzer combined six mechanical integrators, allowing complex differential equations to be analyzed. Bush designed the differential analyzer in part because he had been trying to solve a differential equation that required multiple sequences of integration. He thought it would be faster to design and build a machine to solve the equations he confronted, rather than solve the equations directly.

An analog computer, the differential analyzer used electric motors to drive a variety of gears and shafts that powered six wheel-and-disc integrators that were connected to 18 rotating shafts. Dozens of analyzers were built along the lines of the original plans. It was a breakthrough machine that enabled advances to be made in understanding seismology, electrical networks, meteorology, and ballistic calculations.

Because it was mechanical, imperfections in the machining, or simple wear on the parts, made each analyzer's results less accurate over time. It was also slow to set up. So after becoming director of the Carnegie Institution for Science in Washington, DC, in 1938, Bush started working on a replacement machine, based on tubes, called the *Rockefeller Differential Analyzer*. Completed in 1942, it had 2,000 tubes and 150 motors and was an important calculating machine in World War II.

SEE ALSO Thomas Arithmometer (1851)

Vannevar Bush with his differential analyzer, a mechanical computer designed to solve differential equations.

Church-Turing Thesis

David Hilbert (1862–1943), **Alonzo Church** (1903–1995),
Alan Turing (1912–1954)

Computer science theory seeks to answer two fundamental questions about the nature of computers and computation: are there theoretical limits regarding what is possible to compute, and are there practical limits?

American mathematician Alonzo Church and British computer scientist Alan Turing each published an answer to these questions in 1936. They did it by answering a challenge posed by the eminent German mathematician David Hilbert eight years earlier.

Hilbert's challenge, the *Entscheidungsproblem* (German for "decision problem"), asked if there was a mathematical procedure—an algorithm—that could be applied to determine if any given mathematical proposition was true or false. Hilbert had essentially asked if the core work of mathematics, the proving of theorems, could be automated.

Church answered Hilbert by developing a new way of describing mathematical functions and number theory called the *Lambda calculus*. With it, he showed that the *Entscheidungsproblem* could not be solved in general: there was no general algorithmic procedure for proving or disproving theorems. He published his paper in April 1936.

Turing took a radically different approach: he created a mathematical definition of a simple, abstract machine that could perform computation. Turing then showed that such a machine could in principle perform any computation and run any algorithm—it could even simulate the operation of other machines. Finally, he showed that while such machines could compute almost anything, there was no way to know if a computation would eventually complete, or if it would continue forever. Thus, the *Entscheidungsproblem* was unsolvable.

Turing went to Princeton University in September 1936 to study with Church, where the two discovered that the radically different approaches were, in fact, mathematically equivalent. Turing's paper was published in November 1936; he stayed on and completed his PhD in June 1938, with Church as his PhD advisor.

SEE ALSO Colossus (1943), EDVAC *First Draft* Report (1945), NP-Completeness (1971)

Statue of Alan Turing at Bletchley Park, the center of Britain's codebreaking operations during World War II.

Z3 Computer

Konrad Zuse (1910–1995)

The Z3 was the world's first working programmable, fully automatic digital computer. The machine executed a program on punched celluloid tape and could perform addition, subtraction, multiplication, division, and square roots on 22-bit binary floating-point numbers (because binary math was more efficient than decimal); it had 64 words of 22-bit memory for storing reswults. The machine could convert decimal floating points to binary for input, and binary floating points back to decimal for output.

Graduating with a degree in civil engineering in 1935, German inventor Konrad Zuse immediately started building his first computer, the Z1 (constructed 1935–1938), in his parents' apartment in Berlin. The Z1 was a mechanical calculator controlled by holes punched in celluloid film. The machine used 22-bit binary floating-point numbers and supported Boolean logic; it was destroyed in December 1943 during an Allied air raid.

Drafted into military service in 1939, Zuse started work on the Z2 (1939), which improved on the Z1's design by using telephone relays for the arithmetic and control logic. DVL, the German Research Institute for Aviation, was impressed by the Z2 and gave Zuse funds to start his company, Zuse Apparatebau (Zuse Apparatus Construction, later renamed Zuse KG), to build the machines.

In 1941, Zuse designed and built the Z3. Like the Z1 and Z2, it was controlled by punched celluloid tape, but it also had support for loops, allowing it to be used for solving many typical engineering calculations.

With the success of the Z3, Zuse started working on the Z4, a more powerful machine with 32-bit floating-point math and conditional jumps. The partially completed machine was moved from Berlin to Göttingen in February 1945 to prevent it from falling into Soviet hands, and was completed there just before the end of the war. It remained in operation until 1959.

Surprisingly, it seems that the German military never made use of these sophisticated machines—instead, the machines were largely funded as a research project.

SEE ALSO Atanasoff-Berry Computer (1942), Binary-Coded Decimal (1944)

The control console, calculator, and storage cabinets of the Z3 computer by Konrad Zuse.

Atanasoff-Berry Computer

John Vincent Atanasoff (1903–1995), **Clifford Edward Berry** (1918–1963)

Built at Iowa State College (now Iowa State University) by professor John Atanasoff and graduate student Clifford Berry, the Atanasoff-Berry Computer (ABC) was an automatic, electronic digital desktop computer.

Atanasoff, a physicist and inventor, created the ABC to solve general systems of linear equations with up to 29 unknowns. At the time, it took a human computer eight hours to solve a system with eight unknowns; systems with more than 10 unknowns were not often attempted. Atanasoff started building the computer in 1937; he successfully tested it in 1942, and then abandoned it when he was called for duty in World War II. Although the machine was largely forgotten, it changed the course of computing decades later.

The machine was based on electronics, rather than relays and mechanical switches, performed math with binary arithmetic, and had a main memory that used an electrical charge (or its absence) in small capacitors to represent 1s and 0s—the same approach used by modern dynamic random access memory (DRAM) modules. The whole computer weighed 700 pounds.

Ironically, the lasting value of the ABC was to invalidate the original ENIAC patent, which had been filed by J. Presper Eckert and John Mauchly in June 1947. The ENIAC patent was the subject of substantial litigation, and the US Patent and Trademark Office did not issue the patent until 1964 as a result. With the patent in hand, the American electronics company Sperry Rand (which had bought the Eckert-Mauchly Computer Corporation in 1950) immediately demanded huge fees from all companies selling computers. At the time, patents were good for 18 years from the date of issuance, meaning that the ENIAC patent might stifle the computing industry until 1982.

It turned out that Mauchly had visited Iowa State and studied the ABC in June 1941—but had failed to mention the ABC as prior work in his patent application. In 1967, Honeywell® sued Sperry Rand, claiming that the patent was invalid because of the omission. The US District Court for the District of Minnesota agreed and invalidated the ENIAC patent six years later.

SEE ALSO ENIAC (1943)

A working reconstruction of the Atanasoff-Berry Computer, built by engineers at Iowa State University between 1994 and 1997.

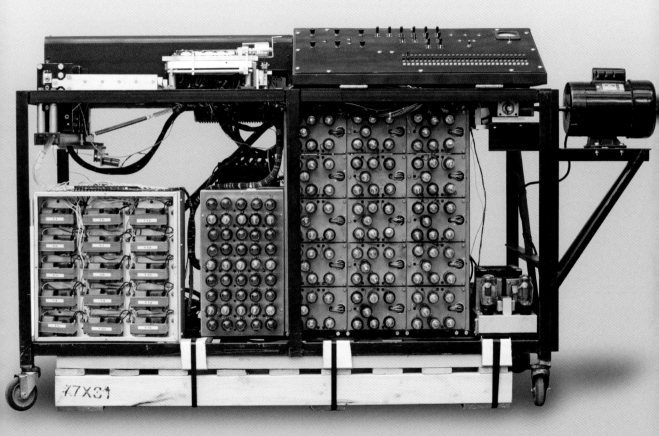

Isaac Asimov's Three Laws of Robotics

Isaac Asimov (1920–1992)

Science-fiction author Isaac Asimov introduced the Three Laws of Robotics in his 1942 story "Runaround" as a set of guiding principles to govern the behavior of robots and their future development. First, a robot may not cause harm to a human, either by the robot's action or inaction. Second, the robots must follow human commands, unless they would violate the first law. The third law states that robots must look to their own survival, provided that doing so does not interfere with their obligations under the first and second laws.

Asimov added a fourth law, known as the "zeroth" law, in 1985. It ranks higher than the first three and affords similar protections to all of humanity.

Asimov originally attributes the laws to the *Handbook of Robotics, 56th Edition, 2058 A.D.* The laws are a fail-safe feature used to inform robot behavior as robots interact with humans and choose courses of action that involve morality, ethics, and thoughtful decision making. They are used throughout the Robot series and other narratives linked to it. For example, Dr. Susan Calvin, a robopsychologist, is a recurring fictional character in Asimov's robot stories. Calvin is employed by 21st-century robot manufacturer US Robots and Mechanical Men, Inc., where she solves problems caused by robots' interaction with humans. These problems are often associated with a term in Asimov's stories called the "Frankenstein Complex," understood as human fear of self-aware, autonomous machines.

Asimov recognized in his writing that anxiety about intelligent robots would be a significant challenge to overcome in order for robots to be accepted by human society. His laws tapped into a subject that has moved from fiction to public policy as society confronts the commercialization of machines (such as autonomous vehicles) whose function is directly associated with human life.

SEE ALSO *Rossum's Universal Robots* (1920)

Cover of Signet's 1956 edition of I, Robot *by Isaac Asimov.*

S1282

MAN-LIKE MACHINES RULE THE WORLD!
Fascinating Tales of a Strange Tomorrow

I, ROBOT

Isaac Asimov

A SIGNET BOOK Complete and Unabridged

ENIAC

John Mauchly (1907–1980), J. Presper Eckert (1919–1995)

ENIAC was the first *electronic* computer, which means it computed with tubes rather than relays. Designed by John Mauchly and J. Presper Eckert at the Moore School of Electrical Engineering at the University of Pennsylvania, ENIAC had 17,468 vacuum tubes, was 8 feet (2.4 meters) high by 3 feet (0.9 meters) deep by 100 feet (30.5 meters) long, and weighed more than 30 tons.

ENIAC had an IBM punch-card reader for input and a card punch for output, but the machine had no memory for data or programs. Instead, numbers under calculation were kept in one of the computer's 20 accumulators, each of which could store 10 decimal digits and perform addition or subtraction. Other hardware could perform multiplication, division, and even square roots. ENIAC wasn't programmed in today's sense. Instead, a set of panels had 1,200 10-position rotary switches that would energize different circuits in a specific sequence, causing electronic representations of numbers to flow through different parts of the machine at predetermined times, and leading the machine computation to take place.

ENIAC was built to perform complex ballistics calculations for the US Army, but John von Neumann (1903–1957) at the Manhattan Project learned about ENIAC, so the machine's first official use was actually to perform computations for the development of the hydrogen bomb.

Ironically, the men who built the hardware never considered the need for, or the complexity of, programming the machine. They left the job of making the machine actually calculate to six human computers: Frances "Betty" Snyder Holberton (1917–2001), Betty "Jean" Jennings Bartik (1924–2011), Kathleen McNulty Mauchly Antonelli (1921–2006), Marlyn Wescoff Meltzer (1922–2008), Ruth Lichterman Teitelbaum (1924–1986), and Frances Bilas Spence (1922–2012).

Those women, some of the world's first programmers, had to devise and then debug their own algorithms. But the women were not acknowledged in their own time. In 2014, Kathy Kleiman produced the documentary *The Computers*, which finally told the women's story.

SEE ALSO First Recorded Use of the Word *Computer* (1613), EDVAC *First Draft* Report (1945)

ENIAC, the first electronic computer, was built to perform calculations for the US Army. Pictured operating the machine are Corporal Irwin Goldstein, Private First Class Homer Spence, Betty Jean Jennings, and Frances Bilas.

Colossus

Thomas Harold Flowers (1905–1998), **Sidney Broadhurst** (1893–1969), **W. T. Tutte** (1917–2002)

Colossus was the first electronic digital computing machine, designed and successfully used during World War II by the United Kingdom to crack the German High Command military codes. "Electronic" means that it was built with tubes, which made Colossus run more than 500 times faster than the relay-based computing machines of the day. It was also the first computer to be manufactured in quantity.

A total of 10 "Colossi" were clandestinely built at Bletchley Park, Britain's ultra-secret World War II cryptanalytic center, between 1943 and 1945 to crack the wireless telegraph signals encrypted with a special system developed by C. Lorenz AG, a German electronics firm. After the war the Colossi were destroyed or dismantled for their parts to protect the secret of the United Kingdom's cryptanalytic prowess.

Colossus was far more sophisticated than the electromechanical Bombe machines that Alan Turing designed to crack the simpler Enigma cipher used by the Germans for battlefield encryption. Whereas Enigma used between three and eight encrypting rotors to scramble characters, the Lorenz system involved 12 wheels, with each wheel adding more mathematical complexity, and thus required a cipher-cracking machine with considerably more speed and agility.

Electronic tubes provided Colossus with the speed that it required. But that speed meant that Colossus needed a similarly fast input system. It used punched paper tape running at 5,000 characters per second, the tape itself moving at 27 miles per hour. Considerable engineering kept the tape properly tensioned, preventing rips and tears.

The agility was provided by a cryptanalysis technique designed by Alan Turing called *Turingery*, which inferred the cryptographic pattern of each Lorenz cipher wheel, and a second algorithm. The second algorithm, designed by British mathematician W. T. Tutte, determined the starting position of the wheels, which the Germans changed for each group of messages. The Colossi themselves were operated by a group of cryptanalysts that included 272 women from the Women's Royal Naval Service (WRNS) and 27 men.

SEE ALSO Manchester SSEM (1948)

The Colossus computing machine was used to read Nazi codes at Bletchley Park, England, during World War II.

Delay Line Memory

John Mauchly (1907–1980), **J. Presper Eckert** (1919–1995),
Maurice Wilkes (1913–2010)

Early computer systems needed a way to program data that was fast, rewritable, and extensible. Some did so by converting bits into sound pulses and sending those pulses down a long tube called a *delay line*. Early tubes could hold 576 bits, cycling around in the device.

J. Presper Eckert invented the mercury delay line during World War II for analog radar systems. By carefully timing the radar pulses with the length of the delay line, it was possible to build a display that would show only moving targets. In 1944, Eckert adopted the mercury delay line memory to the EDVAC computer he was assembling in Philadelphia. Eckert and Mauchly filed US Patent 2,629,827, "Memory system," on October 31, 1947; it wasn't granted until 1953, by which time delay lines were obsolete.

In a delay line, the computer stores a 1 by sending an electronic pulse to a transducer, which creates an ultrasonic wave. The wave travels from one end to the other, where it is converted back into electricity, reshaped and amplified, transforming it once again into a pure 1, and finally sent down a wire to the original transducer, where the 1 is reinjected back into the medium. A 0 is indicated by the absence of a pulse. Accompanying electronics are used to read or write data at a specific point in the sequence as the bits cycle through.

While work continued on the EDVAC, Maurice Wilkes at the University of Cambridge Mathematical Laboratory in England perfected the technology and built the Electronic Delay Storage Automatic Calculator (EDSAC). Operational in May 1949, EDSAC was the first system to use delay lines for digital storage. The EDVAC was operational shortly thereafter. Eckert and Mauchly later put delay lines in the UNIVAC® I (Universal Automatic Computer) that they sold to the US Census Bureau and other customers.

Later delay lines based on magnetic compression waves packed more than 10,000 bits into roughly a foot of coiled wire. Highly reliable, these delay lines were used in computers, video displays, and desktop calculators until the mid-1960s. Wilkes was awarded the A.M. Turing Award in 1967.

SEE ALSO EDVAC *First Draft* Report (1945), Williams Tube (1946), Core Memory (1951)

British computer scientist Maurice Wilkes kneels beside the EDSAC computer's mercury delay line memory.

Binary-Coded Decimal

Howard Aiken (1900–1973)

There are essentially three ways to represent numbers inside a digital computer. The most obvious is to use base 10, representing each of the numbers 0–9 with its own bit, wire, punch-card hole, or printed symbol (e.g., 0123456789). This mirrors the way people learn and perform arithmetic, but it's extremely inefficient.

The most efficient way to represent numbers is to use pure binary notation: with binary, n bits represent 2^n possible values. This means that 10 wires can represent any number from 0 to 1023 ($2^{10}-1$). Unfortunately, it's complex to convert between decimal notation and binary.

The third alternative is called *binary-coded decimal* (BCD). Each decimal digit becomes a set of four binary digits, representing the numbers 1, 2, 4, and 8, and counting in sequence 0000, 0001, 0010, 0011, 0100, 0101, 0110, 0111, 1000, 1001, and 1010. BCD is four times more efficient than base 10, yet it's remarkably straightforward to convert between decimal numbers and BCD. Further, BCD has the profound advantage of allowing programs to exactly represent the numeric value 0.01—something that's important when performing monetary computations.

Early computer pioneers experimented with all three systems. The ENIAC computer built in 1943 was a base 10 machine. At Harvard University, Howard Aiken designed the Mark 1 computer to use a modified form of BCD. And in Germany, Konrad Zuse's Z1, Z2, Z3, and Z4 machines used binary floating-point arithmetic.

After World War II, IBM went on to design, build, and sell two distinct lines of computers: scientific machines that used binary numbers, and business computers that used BCD. Later, IBM introduced System/360, which used both methods. On modern computers, BCD is typically supported with software, rather than hardware.

In 1972, the US Supreme Court ruled that computer programs could not be patented. In *Gottschalk v. Benson*, the court ruled that converting binary-coded decimal numerals into pure binary was "merely a series of mathematical calculations or mental steps, and does not constitute a patentable 'process' within the meaning of the Patent Act."

SEE ALSO Binary Arithmetic (1703), Floating-Point Numbers (1914), IBM System/360 (1964)

Howard Aiken inspects one of the four paper-tape readers of the Mark 1 computer.

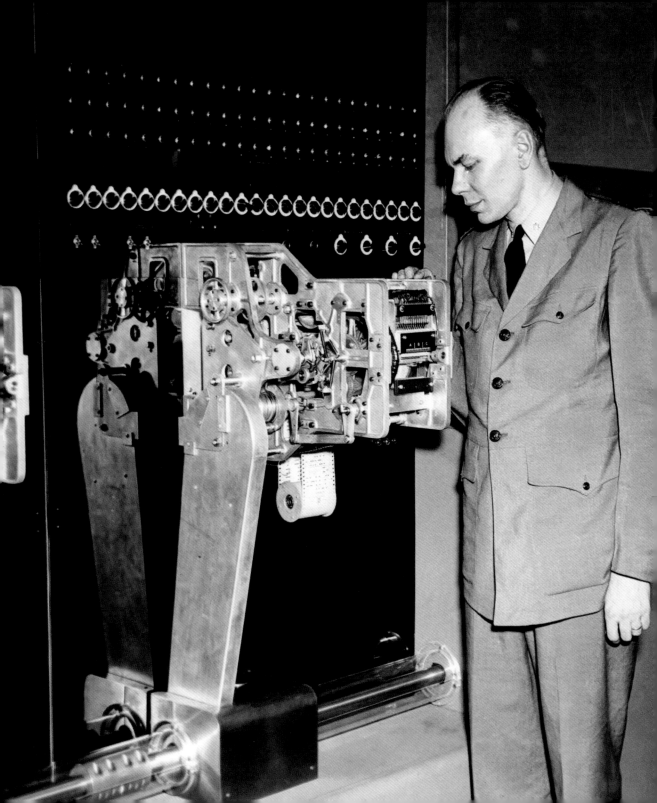

"As We May Think"

Vannevar Bush (1890–1974)

In 1945, Vannevar Bush—inventor of the differential analyzer who went on to head the US Office of Scientific Research and Development during World War II—authored a profoundly prescient and influential essay about the coming information age. Entitled "As We May Think," the essay was first published in the July 1945 issue of the *Atlantic Monthly*. In it, Bush wrote that only by using technology would it be possible for people to keep up with the massive amount of knowledge that the world was on the cusp of creating. "The summation of human experience is being expanded at a prodigious rate," he stated, "and the means we use for threading through the consequent maze to the momentarily important item is the same as was used in the days of square-rigged ships."

With the end of World War II in sight, Bush was trying to set a new direction for scientific research—a path of peaceful intent that would benefit humanity, in contrast with the role he had played overseeing the Manhattan Project and helping to develop the atomic bomb. In the essay, Bush elegantly predicted much of today's technology, including speech recognition, the internet, the World Wide Web, online encyclopedias, hypertext, personal digital assistants, touchscreens, and interactive user-interface design.

The central challenge he focused on was how information should be organized. Bush found the dominant numerical and alphabetical systems to be ineffective. Rather, he believed the human mind stores and makes sense of information through association, as our thoughts snap from one idea to the next, creating original context and meaning.

His theoretical alternative to taxonomic organization was the *memex*, short for "memory extender." The memex would enable individuals to create their own trails, recordings of the evolution of thought on a topic and associated information sources. The device would be in the form of an electromechanical desk with a keyboard, dual screens, encyclopedia and easily accessible articles contained on microfilm (stored on reels inside the desk), a stylus for adding one's own notes directly to the screens, and a way to link people's trails.

With the exception of the microfilm, Bush pretty much nailed it.

SEE ALSO Electronic Speech Synthesis (1928), Differential Analyzer (1931) Mother of All Demos (1968)

Vannevar Bush's essay "As We May Think" foretold many future developments in the field of computer science.

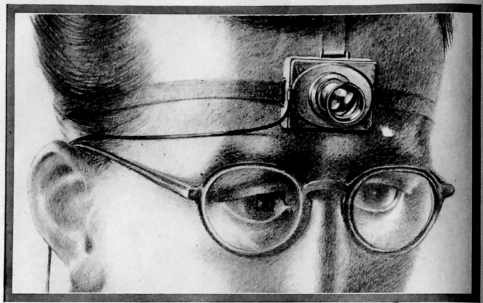

A SCIENTIST OF THE FUTURE RECORDS EXPERIMENTS WITH A TINY CAMERA FITTED WITH UNIVERSAL-FOCUS LENS. THE SMALL SQUARE IN THE EYEGLASS AT THE LEFT SIGHTS THE OBJECT

AS WE MAY THINK

A TOP U. S. SCIENTIST FORESEES A POSSIBLE FUTURE WORLD IN WHICH MAN-MADE MACHINES WILL START TO THINK

by VANNEVAR BUSH

DIRECTOR OF THE OFFICE OF SCIENTIFIC RESEARCH AND DEVELOPMENT

Condensed from the *Atlantic Monthly,* July 1945

This has not been a scientists' war; it has been a war in which all have had a part. The scientists, burying their old professional competition in the demand of a common cause, have shared greatly and learned much. It has been exhilarating to work in effective partnership. What are the scientists to do next?

For the biologists, and particularly for the medical scientists, there can be little indecision, for their war work has hardly required them to leave the old paths. Many indeed have been able to carry on their war research in their familiar peacetime laboratories. Their objectives remain much the same.

It is the physicists who have been thrown most violently off stride, who have left academic pursuits for the making of strange destructive gadgets, who have had to devise new methods for their unanticipated assignments. They have done their part on the devices that made it possible to turn back the enemy. They have worked in combined effort with the physicists of our allies. They have felt within themselves the stir of achievement. They have been part of a great team. Now one asks where they will find objectives worthy of their best.

* * *

There is a growing mountain of research. But there is increased evidence that we are being bogged down today as specialization extends. The investigator is staggered by the findings and conclusions of thousands of other workers—conclusions which he cannot find time to grasp, much less to remember, as they appear. Yet specialization becomes increasingly necessary for prog-

ress, and the effort to bridge between disciplines is correspondingly superficial.

Professionally our methods of transmitting and reviewing the results of research are generations old and by now are totally inadequate for their purpose. If the aggregate time spent in writing scholarly works and in reading them could be evaluated, the ratio between these amounts of time might well be startling. Those who conscientiously attempt to keep abreast of current thought, even in restricted fields, by close and continuous reading might well shy away from an examination calculated to show how much of the previous month's efforts could be produced on call.

Mendel's concept of the laws of genetics was lost to the world for a generation because his publication did not reach the few who were capable of grasping and extending it. This sort of catastrophe is undoubtedly being repeated all about us as truly significant attainments become lost in the mass of the inconsequential.

Publication has been extended far beyond our present ability to make real use of the record. The summation of human experience is being expanded at a prodigious rate, and the means we use for threading through the consequent maze to the momentarily important item is the same as was used in the days of square-rigged ships.

But there are signs of a change as new and powerful instrumentalities come into use. Photocells capable of seeing things in a physical sense, advanced photography which can record what is seen or even what is not, thermionic tubes capable of controlling potent forces under the guidance of

EDVAC *First Draft* Report

John Mauchly (1907–1980), **J. Presper Eckert** (1919–1995),
John von Neumann (1903–1957), **Herman Goldstine** (1913–2004)

Before ENIAC was even operational, J. Presper Eckert and John Mauchly were already designing an even more powerful computer, the EDVAC (Electronic Discrete Variable Automatic Computer).

Whereas ENIAC's architecture could be thought of as 20 automatic adding machines wired together, EDVAC resembled what we've come to think of as a modern computer. It had a memory bank that stored both the computer's program and data, and a central processing unit (CPU) fetched instructions from memory and executed them. The program stored in the main memory directed when data should be copied from memory into the CPU, when mathematical functions were applied, and when the results should be written back to main memory.

Today we call this the *von Neumann architecture* after John von Neumann, the Hungarian-born polymath who immigrated to the US in 1930 and was one of the first faculty members of the Institute for Advanced Study in Princeton, New Jersey, along with Albert Einstein and Kurt Gödel.

Von Neumann and fellow physicists designed the explosive lenses used to detonate the implosion bombs used at Trinity, New Mexico, and Nagasaki and Hiroshima, Japan. While working on those tremendously intricate computations, von Neumann had a chance meeting at a train station with Herman Goldstine, a mathematician who was the Army's liaison with Eckert and Mauchly at the University of Pennsylvania. Goldstine introduced von Neumann to Eckert and Mauchly; soon thereafter, von Neumann joined the study group that was designing the EDVAC.

Von Neumann wrote up his notes while commuting to Los Alamos, New Mexico, and mailed them back to Goldstine, who typed them up, titled the document *First Draft of a Report on the EDVAC*, put von Neumann's name on the cover (even though most of the work was that of Eckert and Mauchly), and distributed 24 copies.

Even though modern computers have an architecture that is closer to the machines developed by Howard Aiken—the so-called Harvard architecture—the phrase "von Neumann architecture" is still widely used to describe modern computers.

SEE ALSO ENIAC (1943)

EDVAC exhibit at the Franklin Institute. The square sign reads: "Pilot Model EDVAC Electronic Discrete Variable Computer Designed and Built by Moore School of Electrical Engineering, University of Pennsylvania, for Ordnance Department, U.S. Army."

Trackball

Ralph Benjamin (b. 1922), **Kenyon Taylor** (1908–1986),
Tom Cranston (c. 1920–2008), **Fred Longstaff** (dates unavailable)

The trackball was one of the first computer input devices to enable freeform cursor movement by the user, simultaneously over both the *x*- and *y*-axes on a computer screen. But there was a long time between its invention and its widespread use.

A British engineer named Ralph Benjamin designed the first prototype trackball while working on a radar project for the Royal Navy Scientific Service in 1946. The radar project was called the *Comprehensive Display System* and enabled ships to monitor low-flying aircraft on X and Y coordinates using a joystick as the input device. Benjamin tried to improve upon this method of input with an invention he called the *roller ball*, which consisted of a metal casing containing a metal ball with two rubber wheels. It allowed users to control their onscreen movements with greater precision to input location data about a target's aircraft. The British kept the device a military secret until 1947, when it was patented in Benjamin's name and described as a device that correlated data between electronic storage and displays.

In 1952, Canadian engineers Tom Cranston, Fred Longstaff, and Kenyon Taylor built upon Benjamin's concept and designed a trackball for the Royal Canadian Navy's Digital Automated Tracking and Resolving (DATAR) system, a computerized battlefield information system. The design, based upon the Canadian five-point bowling ball, allowed an operator to control and track the location of user input on the screen.

Benjamin's roller ball eventually had a large influence on the development of the mouse and the modern trackball. The roller ball differed from the mouse in that it was a stationary object that was controlled by the user's hand and fingers moving over it, rather than repositioning the entire device to different locations in physical space.

SEE ALSO The Mouse (1967)

The first trackball used a Canadian five-pin bowling ball floated on a cushion of air as its input device; the nozzle for the air supply is visible at the lower right.

Williams Tube

Frederic Calland Williams (1911–1977), Tom Kilburn (1921–2001)

The Williams tube (sometimes known as the *William–Kilburn tube*), was the first all-electronic memory system, and the first that provided for *random access*, meaning that any location of memory could be accessed in any order.

The tube was a cathode-ray tube, such as what might be used in a World War II radar display, but modified so that dots displayed on the screen could be read by a computer. The early Williams tubes stored data one binary digit—a 0 or a 1—at a time in a rectangular array, typically 64 by 32. Developed at the University of Manchester by Frederic Williams and Tom Kilburn, the tubes were superior to the mercury delay line memory that had been used on the EDSAC and UNIVAC computers, because any bit could be accessed instantly. The bits in the delay lines, in contrast, could be read only when they had progressed through the mercury to the end of the line. But both the Williams tube and the mercury delay lines shared the property that they needed to be continually refreshed, much as do modern dynamic random access memory (DRAM) chips. Because the bits consist of stored energy that can dissipate, each bit must be continually read and then rewritten.

The IBM 701 computer used 72 Williams tubes to store 2,048 36-bit words; this electronic memory was supplemented with a spinning magnetic drum that could hold roughly four times as much memory but was much slower to access. The Williams tubes were not terribly reliable, though, and the 701 would reportedly run for only 15 minutes before it would crash due to a memory error. Indeed, the tubes were so unreliable that the Manchester Small-Scale Experimental Machine (SSEM), nicknamed *Baby*, was built specifically for the purpose of testing Williams tubes.

The MIT Whirlwind computer, first operational in 1949, was originally designed to use a modified Williams tube that dispensed with the need to refresh by using a second electron gun called a *flood gun*; the same approach would be used nearly three decades later in storage-tube displays, such as graphics terminals and oscilloscopes used in electronics labs to monitor circuits. But the modified Williams tubes cost $1,000 each and had a lifetime of roughly one month. Beset with the problems caused by these storage tubes, Whirlwind's director, Jay Forrester (1918–2016), invented core memory as a replacement.

SEE ALSO Delay Line Memory (1944), Manchester SSEM (1948), Core Memory (1951)

This close-up view, c. 1948, reveals dots (1s) and spaces (0s) on the face of a tube.

Actual Bug Found

Howard Aiken (1900–1973), **William "Bill" Burke** (dates unavailable), **Grace Murray Hopper** (1906–1992)

Harvard professor Howard Aiken completed the Mark II computer in 1947 for the Naval Proving Ground in Dahlgren, Virginia. With 13,000 high-speed electromechanical relays, the Mark II processed 10-digit decimal numbers, performed floating-point operations, and read its instructions from punched paper tape. Today we still use the phrase "Harvard architecture" to describe computers that separately store their programs from their data, unlike the "von Neumann" machines that store code and data in the same memory.

But what makes the Mark II memorable is not the way it was built or its paper tape, but what happened on September 9, 1947. On that day at 10:00 a.m., the computer failed a test, producing the number 2.130476415 instead of 2.130676415. The operators ran another test at 11:00 a.m., and then another at 3:25 p.m. Finally, at 3:45 p.m., the computer's operators, including William "Bill" Burke, traced the problem to a moth that was lodged inside Relay #70, Panel F. The operators carefully removed the bug and affixed it to the laboratory notebook, with the notation "First actual case of bug being found."

Burke ended up following the computer to Dahlgren, where he worked for several years. One of the other operators was the charismatic pioneer Grace Murray Hopper, who had volunteered for the US Navy in 1943, joined the Harvard staff as a research fellow in 1946, and then moved to the Eckert-Mauchly Computer Corporation in 1949 as a senior mathematician, where she helped the company to develop high-level computer languages. Grace Hopper didn't actually find the bug, but she told the story so well, and so many times, that many histories now erroneously credit her with the discovery. As for the word *bug*, it had been used to describe faults in machines as far back as 1875; according to the *Oxford English Dictionary*, in 1889, Thomas Edison told a journalist that he had stayed up two nights in a row discovering, and fixing, a bug in his phonograph.

SEE ALSO COBOL Computer Language (1960)

The moth found trapped between points at Relay #70, Panel F, of the Mark II Aiken Relay Calculator while it was being tested at Harvard University. The operators affixed the moth to the computer log with the entry "First actual case of bug being found."

0800   antan started

"000   "   stopped   - antan ✓   { 1.2700 · 9.037 ·
                                                              9.037 8
        13" uc (032) MP - MC   1.9837 (2000
                                     2.130476415 (-3) 4.6
            (033)   PRO 2   2.130476415

                conct   2.130676415

        Relays 6-2 in 033 failed special speed t
        in Relay   "   10.000 test .

                    Relays changed
/00   Started Cosine Tape (Sine check)
525   Started Mult + Adder Test.

545   Relay #70 Panel
                                            (moth) in relay.

        First actual case of bug being found
/630 antangent started.
1700 closed down .

Silicon Transistor

John Bardeen (1908–1991), **Walter Houser Brattain** (1902–1987), **William Shockley** (1910–1989)

A transistor is an electronic switch: current flows from one terminal to another unless voltage is applied to a third terminal. Combined with the laws of Boolean algebra, this simple device has become the building block for microprocessors, memory systems, and the entire computer revolution.

Any technology that can use one signal to switch another on and off can be used to create a computer. Charles Babbage did it with rods, cogs, and steam power. Konrad Zuse and Howard Aiken did it with relays, and ENIAC used tubes. Each technology was faster and more reliable than the previous.

Likewise, transistors have several advantages over vacuum tubes: they use less power, so they generate less heat, they switch faster, and they are less susceptible to physical shock. All of these advantages arise because transistors are smaller than tubes—and the smaller the transistor, the bigger the advantage.

Modern transistors trace their lineage back to a device manufactured by John Bardeen, Walter Brattain, and William Shockley at AT&T's Bell Laboratories in 1947. The team was trying to build an amplifier that could detect ultra-high frequency radio waves, but the tubes that they had just weren't fast enough. So they tried working with semiconductor crystals, as radios based on semiconductor diodes called *cat's whiskers* had been used since nearly the birth of radio in the 1890s.

A cat's whisker radio uses a sharp piece of wire (the "whisker") that's jabbed into a piece of semiconducting germanium; by moving the wire along the semiconductor and varying the pressure, the semiconductor and the wire work together to create a diode, a device allowing current to pass in only one direction. The Bell Labs team built a contraption that attached two strips of gold foil to the crystal and then applied power to the germanium. The result was an amplifier: a signal injected into one wire was stronger when it came out of the other. Today we call this device a *point-contact transistor*.

For their discovery of the transistor, Bardeen, Brattain, and Shockley were awarded the Nobel Prize in 1956.

SEE ALSO Semiconductor Diode (1874), First LED (1927)

The first transistor ever made, built in 1947 by John Bardeen, William Shockley, and Walter H. Brattain of Bell Labs.

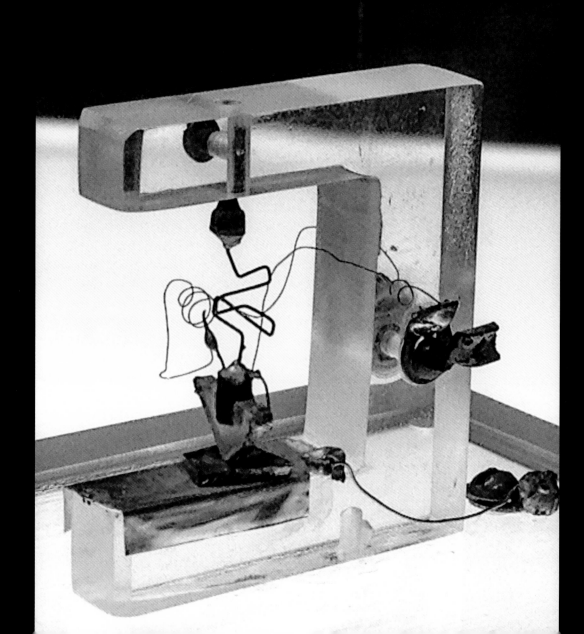

The Bit

Claude E. Shannon (1916–2001), John W. Tukey (1915–2000)

It was the German mathematician Gottfried Wilhelm Leibniz (1646–1716) who first established the rules for performing arithmetic with binary numbers. Nearly 250 years later, Claude E. Shannon realized that a binary digit—a 0 or a 1—was the fundamental, indivisible unit of information.

Shannon earned his PhD from MIT in 1940 and then took a position at the Institute for Advanced Study in Princeton, New Jersey, where he met and collaborated with the institute's leading mathematicians working at the intersection of computing, cryptography, and nuclear weapons, including John von Neumann, Albert Einstein, Kurt Gödel, and, for two months, Alan Turing.

In 1948, Shannon published "A Mathematical Theory of Communication" in the *Bell System Technical Journal*. The article was inspired in part by classified work that Shannon had done on cryptography during the war. In it, he created a mathematical definition of a generalized communications system, consisting of a message to be sent, a transmitter to convert the message into a signal, a channel through which the signal is sent, a receiver, and a destination, such as a person or a machine "for whom the message is intended."

Shannon's paper introduced the word *bit*, a binary digit, as the basic unit of information. While Shannon attributed the word to American statistician John W. Tukey, and the word had been used previously by other computing pioneers, Shannon provided a mathematical definition of a bit: rather than just a 1 or a 0, it is information that allows the receiver to limit possible decisions in the face of uncertainty. One of the implications of Shannon's work is that every communications channel has a theoretical upper bound—a maximum number of bits that it can carry per second. As such, Shannon's theory has been used to analyze practically every communications system ever developed—from handheld radios to satellite communications—as well as data-compression systems and even the stock market.

Shannon's work illuminates a relationship between information and entropy, thus establishing a connection between computation and physics. Indeed, noted physicist Stephen Hawking framed much of his analysis of black holes in terms of the ability to destroy information and the problems created as a result.

SEE ALSO Vernam Cipher (1917), Error-Correcting Codes (1950)

Mathematician and computer scientist Claude E. Shannon.

Curta Calculator

Curt Herzstark (1902–1988)

The Curta is perhaps the most elegant, compact, and functional mechanical calculator ever manufactured. Designed by Austrian engineer Curt Herzstark, it is the only digital mechanical pocket calculator ever invented. Handheld and powered by a crank on the top, the Curta can add, subtract, multiply, and divide.

Curt Herzstark's father, Samuel Jacob Herzstark, was a highly regarded Austrian importer and manufacturer of mechanical calculators and other precision instruments. Herzstark finished high school and apprenticed at his father's company, which he took over when his father died in 1937.

At the time, mechanical calculators were big and heavy desktop affairs. After one of Herzstark's customers complained that he didn't want to go back to the office just to add up a column of numbers, Herzstark started designing a handheld calculator. He had an early prototype working in January 1938, just two months before Germany invaded and annexed Austria. Despite Herzstark being half-Jewish, the Nazis let him continue to operate the factory, provided that it cease all civilian production and devote itself to creating devices for the Reich.

In 1943, two of Herzstark's employees were arrested for distributing transcripts of English radio broadcasts; Herzstark was subsequently arrested for aiding the employees and for "indecent contact with Aryan women." He was sent to the Buchenwald concentration camp, where he was recognized by one of his former employees, who was now a guard. The guard told the head of the camp's factory about the mechanical calculator. The Germans then instructed Herzstark to finish his project, so that the camp could give the device to Hitler as a present after Germany won the war. That never happened: Buchenwald was liberated on April 11, 1945, and Hitler killed himself 19 days later.

After liberation, Herzstark took the drawings he had done at the camp to a machine shop and had three working prototypes eight weeks later. The first calculators were produced commercially in the fall of 1948.

SEE ALSO Antikythera Mechanism (c. 150 BCE), Thomas Arithmometer (1851)

The Curta mechanical calculator, pictured here, is the only digital mechanical pocket calculator ever invented.

Manchester SSEM

Frederic Calland Williams (1911–1977), **Tom Kilburn** (1921–2001)

The defining characteristic of the digital computer is that it stores both program and data in a single memory bank. In a modern computer, this arrangement lets one program load a second program into memory and execute it. On the limited-memory machines of the 1950s, intermixing programs and code made it possible to squeeze out more functionality by writing programs that literally modified themselves, now called *self-modifying code*. Modern computers use this ability to load code into the computer's memory and execute it—the fundamental capability that makes a computer a general-purpose machine. But none of the machines built before the Manchester Small-Scale Experimental Machine (SSEM) were actually digital *computers*, at least not in the modern sense. Either they were hardwired to perform a particular calculation, like the Atanasoff-Berry Computer, they read their instructions from some kind of punched tape, like the Konrad Zuse machines, or the program was set on wires and switches, like ENIAC. They were really calculators, not computers.

The SSEM, nicknamed *Baby* by its creators at the University of Manchester, was built for testing and demonstrating the storage tube that Frederic Williams had designed in 1946. Baby filled a 20-foot-square room and consisted of eight racks of equipment, the Williams storage tube, many radio tubes, and meters that reported voltages. Each tube had 1,024 bits. As the program ran and changed what was stored in its memory, the arrangement of dots on the storage tube changed.

Because the program was stored in memory, and relied on self-modifying code, it was easy for Kilburn to make changes. The first program that Baby successfully ran, written by Kilburn, was designed to find the highest factor of 2^{18} (262,144). The program ran in 52 minutes and found the right answer: 2^{17} (131,072), averaging 1.5 milliseconds per instruction. The original program was just 17 instructions long.

Arriving at the correct answer was no easy feat. As Williams reportedly stated, "The spots on the display tube entered a mad dance. In early trials, it was a dance of death leading to no useful result . . . But one day it stopped, and there, shining brightly in the expected place, was the expected answer."

SEE ALSO Z3 Computer (1941), Atanasoff-Berry Computer (1942), Williams Tube (1946)

Recreation of the Manchester Small-Scale Experimental Machine (a.k.a., the Manchester "Baby") at the Museum of Science and Industry in Manchester, UK.

Whirlwind

Jay Forrester (1918–2016), **Robert R. Everett** (b. 1921)

In 1944, the US Navy asked the MIT Servomechanisms Laboratory to create a flight simulator to train Navy pilots. MIT tried building an analog computer, but it was soon clear that only a digital machine could possibly offer the speed, flexibility, and programmability required to create a realistic simulation. So, in 1945, the Office of Naval Research contracted MIT to create what would be the world's first interactive, real-time computer.

Called *Whirlwind*, the machine was a massive undertaking: the project involved 175 people with a budget of $1 million a year. The machine used 3,300 vacuum tubes and occupied 3,300 square feet in MIT Building N42, a 25,000-square-foot, two-story building that MIT bought specifically for the project.

Whirlwind had the first computer graphics display, a pair of 5-inch video screens on which the computer could draw airspace maps. It also had the first graphical input device, a "light pen" (invented by associate director Robert R. Everett), for selecting points on the screen. When Whirlwind was partially operational in 1949, MIT professor Charles Adams and programmer John Gilmore Jr. used its graphics capabilities to create one of the first video games: a line with a hole, and a ball that made a *thunk* sound every time it bounced. The goal was to move a hole so that the ball would fall in.

Soon, however, it was clear that the electronics technology of the day needed improving. Whirlwind's vacuum tubes kept burning out. The lab performed an in-depth analysis and determined that trace amounts of silicon in the tube cathodes were at fault. The lab had the contaminant removed, and the lifetime of the tubes was extended by a factor of a thousand. When it became clear that the computer would need memory that was larger and more reliable than storage tubes, Forrester invented magnetic core memory, which would become the primary storage system of computers for the next two decades. Later, when it became evident that Whirlwind needed code that would permanently reside in the computer, load other programs, and provide basic functions, the project drove the invention of the first operating system.

Whirlwind was fully operational in 1951. Although it was never actually used as a flight simulator, its graphic display showed that computers could present maps and track objects, demonstrating the feasibility of using computers for air defense.

SEE ALSO Core Memory (1951)

Created to run a flight simulator for the US Navy, Whirlwind was the world's first interactive computer.

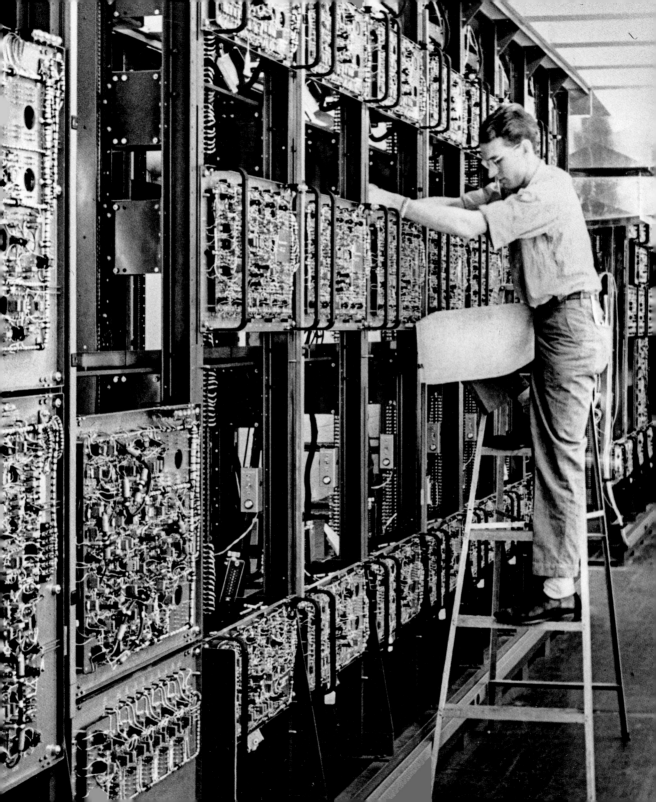

Error-Correcting Codes

Richard Hamming (1915–1998)

After he got his PhD in mathematics and worked on the mathematical modeling for the atomic bomb, Richard Hamming took a job at Bell Telephone Labs, where he worked with Claude Shannon and John Tukey and wrote programs for the laboratory's computers. Hamming noticed that these digital machines had to perform their calculations perfectly. But they didn't. According to Hamming, a relay computer that Bell had built for the Aberdeen Proving Ground, a US Army facility in Maryland, had 8,900 relays and typically experienced two or three failures per day. When such a failure occurred, the entire computation would be ruined and need to be restarted from the beginning.

At the time, it was becoming popular for computer designers to devote an extra bit, called a *parity bit*, to detect errors when data was transmitted or stored. Hamming reasoned that if it was possible to automatically detect errors, it must also be possible to automatically correct them. He figured out how to do this and published his seminal article, "Error Detecting and Error Correcting Codes," in the April 1950 issue of the *Bell System Technical Journal*.

Error-correcting codes (ECCs) play a critical role in increasing the reliability of modern computer systems. Without ECCs, whenever there is a minor error on the receipt of data, the sender must retransmit. So modern cellular data systems use ECCs to let the receiver fix those minor errors, without requesting that the sender retransmit a clean copy. Today ECCs are also used to correct errors in stored data. For example, cosmic rays can scramble the bits of a dynamic random access memory (DRAM) chip, so it's common for internet servers to be protected with ECC memory, allowing them to automatically correct most errors resulting from stray background radiation. Compact discs (CDs) and digital video discs (DVDs) use ECC to make their playback unaffected by surface scratches. And increasingly, ECCs are being incorporated into high-performance wireless communications protocols to reduce the need for data to be resent in the event of noise.

Hamming was awarded the 1968 A.M. Turing Award "for his work on numerical methods, automatic coding systems, and error-detecting and error-correcting codes."

SEE ALSO The Bit (1948)

The 4-bit Hamming codes (on right) for binary numbers 00000000001 through 00000000110 and 11111111001 through 11111111111.

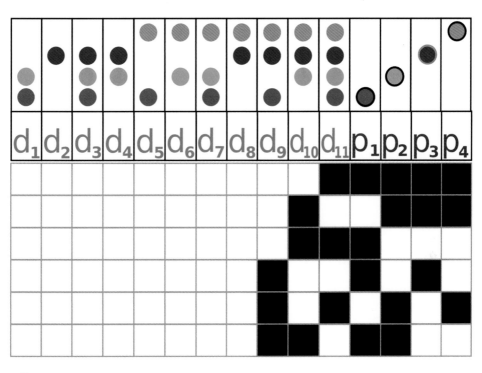

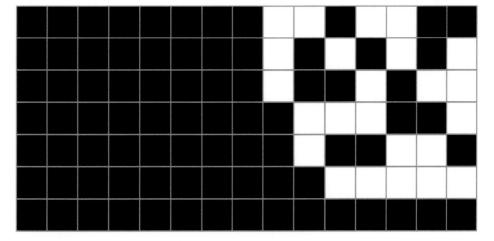

The Turing Test

Alan Turing (1912–1954)

"Can machines think?" That's the question Alan Turing asked in his 1951 paper, "Computing Machinery and Intelligence." Turing envisioned a day when computers would have as much storage and complexity as a human brain. When computers had so much storage, he reasoned, it should be possible to program such a wide range of facts and responses that a machine might appear to be intelligent. How, then, Turing asked, could a person know if a machine was truly intelligent, or merely presenting as such?

Turing's solution was to devise a test of machine intelligence. The mark of intelligence, Turing argued, was not the ability to multiply large numbers or play chess, but to engage in a natural conversation with another intelligent being.

In Turing's test, a human, playing the role of an interrogator, is able to communicate (in what we would now call a *chat room*) with two other entities: another human and a computer. The interrogator's job is to distinguish the human from the computer; the computer's goal is to convince the interrogator that it is a person, and that the other person is merely a simulation of intelligence. If a computer could pass such a test, Turing wrote, then there would be as much reason to assume that it was conscious as there would be to assume that any human was conscious. According to Turing, the easiest way to create a computer that could pass his test would be to build one that could learn and then teach it from "birth" as if it were a child.

In the years that followed, programs called *chatbots*, capable of conducting conversations, appeared to pass the test by fooling unsuspecting humans into thinking they were intelligent. The first of these, ELIZA, was invented in 1966 by MIT professor Joseph Weizenbaum (1923–2008). In one case, ELIZA was left running on a teletype, and a visitor to Weizenbaum's office thought he was text-chatting with Weizenbaum at his home office, rather than with an artificial intelligence (AI) program. According to experts, however, ELIZA didn't pass the Turing test because the visitor wasn't told in advance that the "person" at the other end of the teleprinter might be a computer.

SEE ALSO ELIZA (1965), Computer Is World Chess Champion (1997), Computer Beats Master at Go (2016)

In the movie Blade Runner, *starring Harrison Ford, the fictional Voight-Kampff test can distinguish a human from a "replicant" by measuring eye dilation during a stressful conversation.*

Magnetic Tape Used for Computers

Fritz Pfleumer (1881–1945)

Spinning reels of magnetic tape were *the* iconic visual marker of computer systems for more than four decades in popular culture. Perhaps this was because the spools, moving back and forth with obvious purpose, somehow conveyed thought and the movement of information in ways that blinking lights never could.

In 1951, magnetic tape technology, refined over the previous 75 years for recording sound, found its way to the early UNIVAC computer created by the Eckert-Mauchly Computer Corporation. The UNIVAC was one of the first computers to be commercially produced and could perform calculations much faster than humans could enter them. Data and programs were stored on magnetic tape and loaded into the computer's memory for computation, and the results were either printed or written to another tape drive.

The UNIVAC used metal tape (early audio recorders stored human voices on both long spools of wire and thin metal ribbons) racing along at 100 inches (2.5 meters) per second; a single tape was 1,200 feet (365.76 meters) long and half an inch (13 millimeters) wide, and it could store roughly 1 million 6-bit characters.

IBM followed in 1952 with the 726 tape drive for the IBM 701 computer. The 726 performed similarly to the UNIVAC tape, but it used cellulose acetate tape coated with a ferrous oxide compound, a technology that had proven superior in audio recording as it was lighter and cheaper to produce. IBM's 727 drive, announced the following year, doubled the capacity and storage rate. Metal tape was dead, and IBM continued to sell the 727 drive until 1971.

Although magnetic tape was much faster than punch cards and paper tape, all three technologies were used side by side for decades—sometimes on the same computer—because of their different performance and cost characteristics. For example, in the 1960s, it was common for students to punch their programs into decks of punch cards using relatively low-cost key punches, then have the cards loaded onto a tape by a computer operator, and finally have the tape, with many programs from many different students, run through the computer during the midnight shift, when the computer was otherwise unused.

SEE ALSO Tabulating the US Census (1890)

UNIVAC I, pictured here, used mercury delay lines for its main memory and ½-inch-wide (13-millimeter) metal tape for storage. The tape drives, shown at the back, were called Uniservos; the tape was 1,200 feet long (365.76 meters).

Core Memory

An Wang (1920–1990), Jay Forrester (1918–2016)

The first computers didn't have rewritable memory. Instead, they were "hardwired" to read inputs, perform calculations, and output the results. But it soon became apparent that a rewritable main memory could be used to hold programs, making them easier to develop and debug, as well as data, making it possible for a computer to perform calculations on much larger problems.

Core memory worked by inducing a magnetic field into a tiny magnetic ring, or *core*. Each core was magnetized in the clockwise or counterclockwise direction, naturally allowing the core to store a single bit. Bits were stored by sending an electrical pulse along a pair of horizontal and vertical wires that crossed at a particular core, with one direction of magnetic flow storing a 0 and the other a 1. A third wire running through each core was used to read what had previously been stored. Core memory had the advantage of remembering its contents even when power was removed. The big disadvantage was that the cores had to be strung by hand into memory systems, which is why core memory was so expensive to produce.

Computer engineer An Wang came up with the basics of how to make core memory work while collaborating with Howard Aiken on the Harvard Mark IV computer; he filed for a patent on the invention in 1949. But Harvard lost interest in computing, so in 1951, Wang left and started his own company, Wang Laboratories. IBM bought the patent from Wang Laboratories for $500,000 in 1956.

Meanwhile at MIT, professor Jay Forrester saw an advertisement for a new magnetic material, realized it could be used to store data, and built a prototype system that stored 32 bits of data. At the time, MIT was building the Whirlwind computer to create the first computerized flight simulator. Whirlwind was designed to use an electrostatic memory system based on storage tubes, but MIT's engineers hadn't been able to get the tubes to work. Working with a graduate student, Forrester spent two years and created the first practical core memory system, storing 1,024 bits of data (1 kibibit) in an array of 32×32 cores. The memory was installed in Whirlwind in April 1951. That same year, Forrester filed for a patent on an even more efficient technique of arranging cores in three-dimensional arrays. IBM bought that patent from MIT for $13 million in 1964.

SEE ALSO Delay Line Memory (1944), Williams Tube (1946), Whirlwind (1949), Dynamic RAM (1966)

Core memory was developed for MIT's groundbreaking Whirlwind computer in April 1951.

Microprogramming

Maurice Wilkes (1913–2010)

By 1951, the basic structure of stored-program computers had been worked out: a central processing unit (CPU) that had registers for storing numbers, an arithmetic logic unit (ALU) for performing mathematical operations, and logic for moving data between the CPU and memory. But the internal design of these early CPUs was a mess. Each instruction was implemented with a different set of wires and circuits, some with components in common, and others with their own individual logic.

British computer scientist Maurice Wilkes realized that the design of the CPU could be made more regular after seeing the design of the Whirlwind, which was controlled by a crisscrossing matrix of wires. Some of the wires connected by a diode where they crossed. Voltage was applied to each horizontal wire in sequence. If a diode was present, the corresponding vertical wire would be energized and activate different parts of the CPU.

Wilkes realized that each line of the diode matrix in the Whirlwind could be viewed as a set of microoperations that the CPU followed, a kind of "microprogram." He formalized this idea in a lecture at the 1951 Manchester University Computer Inaugural Conference, immodestly titled "The Best Way to Design an Automatic Calculating Machine." In the lecture, later published by the university, Wilkes proposed that his idea might seem at once obvious, because it described nothing more than a formalized way of creating a CPU using the same basic wires, diodes, and electronic switches that were already in use, as well as extravagant, because it might use more components than would be used otherwise. But, Wilkes argued, it resulted in a system that was easier to design, test, and extend.

Wilkes was right. Microprogramming dramatically simplified the creation of CPUs, allowing instruction sets to become more complex. It also created unexpected flexibility: when IBM released System/360 in 1964, its engineers used microprogramming to allow the new computers to emulate the instructions of the IBM 1401, making it easier for customers to make the transition.

SEE ALSO Whirlwind (1949), IBM 1401 (1959), IBM System/360 (1964)

Maurice Wilkes (front left), designer of the EDSAC, one of the earliest stored-program electronic computers.

Computer Speech Recognition

1952

The automatic digit recognition system, also known as *Audrey*, was developed by Bell Labs in 1952. Audrey was a milestone in the quest to enable computers to recognize and respond to human speech.

Audrey was designed to recognize the spoken digits 0 through 9 and provide feedback with a series of flashing lights associated with a specific digit. Audrey's accuracy was speaker dependent, because to work, it first had to "learn" the unique sounds emitted by an individual person for reference material. Audrey's accuracy was around 80 percent with one designer's voice. Speaker-independent recognition would not be invented for many more years, with modern examples being Amazon® Echo® and Apple Siri.

To create the reference material, the speaker would slowly recite the digits 0 through 9 into an everyday telephone, pausing at least 350 milliseconds between each number. The sounds were then sorted into electrical classes and stored in analog memory. The pauses were needed because at the time, speech-recognition systems had not solved *coarticulation*—the phenomenon of speakers phonetically linking words as they naturally morph from one to another. That is, it was easier for the system to isolate and recognize individual words than words said together.

Once trained, Audrey could match new spoken digits with the sounds stored in its memory: the computer would flash a light corresponding to a particular digit when it found a match.

While various economic and technical practicalities prevented Audrey from going into production (including specialized hardwired circuitry and large power consumption), Audrey was nevertheless an important building block in advancing speech recognition. Audrey showed that the technique could be used in theory to automate speaker input for things such as account numbers, Social Security numbers, and other kinds of numerical information.

Ten years later, IBM demonstrated the "Shoebox," a machine capable of recognizing 16 spoken words, at the 1962 World's Fair in Seattle, Washington.

SEE ALSO Electronic Speech Synthesis (1928)

The automatic digit recognition system was the forerunner of many popular applications today, including smartphones that can recognize voice commands.

First Transistorized Computer

Tom Kilburn (1921–2001), **Richard Grimsdale** (1929–2005),
Douglas Webb (b. 1929), **Jean H. Felker** (1919–1994)

With the invention of the transistor in 1947, the next step was to use it as a replacement for the vacuum tube. Tubes had a significant advantage compared to relays—they were a thousand times faster—but tubes required an inordinate amount of electricity, produced huge amounts of heat, and failed constantly. Transistors used a fraction of the power, produced practically no heat at all, and were more reliable than tubes. And because transistors were smaller than tubes, a transistorized machine would run inherently faster, because electrons had a shorter distance to move.

The University of Manchester demonstrated its prototype transistorized computer on November 16, 1953. The machine made use of the "point-contact" transistor, a piece of germanium that was in contact with two wires held in very close proximity to each other—the two "points." The Manchester machine had 92 point-contact transistors and 550 diodes. The system had a word size of 48 bits. (Many of today's microprocessors can operate on words that are 8, 16, 32, or 64 bits.) A few months later, Jean H. Felker at Bell Labs created the TRADIC (transistor digital computer) for the US Air Force, with 700 point-contact transistors and more than 10,000 diodes.

This point-contact transistor was soon replaced by the bipolar junction transistor, so named because it is formed by a junction involving two kinds of semiconductors. Manchester updated its prototype in 1955 with a new design that used 250 of these junction transistors. Called the *Metrovick 950*, that computer was manufactured by Metropolitan-Vickers, a British electrical engineering company.

In 1956, the Advanced Development Group at MIT Lincoln Lab used more than 3,000 transistors to build the TX-0 (Transistorized eXperimental computer zero), a transistorized version of the Whirlwind and the forerunner to Digital Equipment Corporation's (DEC®) PDP-1.

SEE ALSO Silicon Transistor (1947), Whirlwind (1949), PDP-1 (1959)

Close-up of the prototype of the Manchester transistorized computer.

Artificial Intelligence Coined

John McCarthy (1927–2011), **Marvin Minsky** (1927–2016), **Nathaniel Rochester** (1919–2001), **Claude E. Shannon** (1916–2001)

Artificial intelligence (AI) is the science of computers doing things that normally require human intelligence to accomplish. The term was coined in 1955 by John McCarthy, Marvin Minsky, Nathaniel Rochester, and Claude Shannon in their proposal for the "Dartmouth Summer Research Project on Artificial Intelligence," a two-month, 10-person institute that was held at Dartmouth College during the summer of 1956.

Today we consider the authors of the proposal to be the "founding fathers" of AI. Their primary interest was to lay the groundwork for a future generation of machines that would use abstraction to let them mirror the way humans think. So the founding fathers set off a myriad of different research projects, including attempts to understand written language, solve logic problems, describe visual scenes, and pretty much replicate anything that a human brain could do.

The term *artificial intelligence* has gone in and out of vogue over the years, with people interpreting the concept in different ways. Computer scientists defined the term as describing academic pursuits such as computer vision, robotics, and planning, whereas the public—and popular culture—has tended to focus on science-fiction applications such as machine cognition and self-awareness. On *Star Trek* ("The Ultimate Computer," 1968), the AI-based M5 computer could run a starship without a human crew—and then quickly went berserk and started destroying other starships during a training exercise. The *Terminator* movies presented Skynet as a global AI network bent on destroying all of humanity.

Only recently has AI come to be accepted in the public lexicon as a legitimate technology with practical applications. The reason is the success of narrowly focused AI systems that have outperformed humans at tasks that require exceptional human intelligence. Today AI is divided into many subfields, including machine learning, natural language processing, neural networks, deep learning, and others. For their work on AI, Minsky was awarded the A.M. Turing award in 1969, and McCarty in 1971.

SEE ALSO *Rossum's Universal Robots* (1920), *Metropolis* (1927), Isaac Asimov's Three Laws of Robotics (1942), HAL 9000 Computer (1968), Japan's Fifth Generation Computer Systems (1981), Home-Cleaning Robot (2002), Artificial General Intelligence (AGI) (~2050), The Limits of Computation? (~9999)

Artificial intelligence allows computers to do things that normally require human intelligence, such as recognizing patterns, classifying objects, and learning.

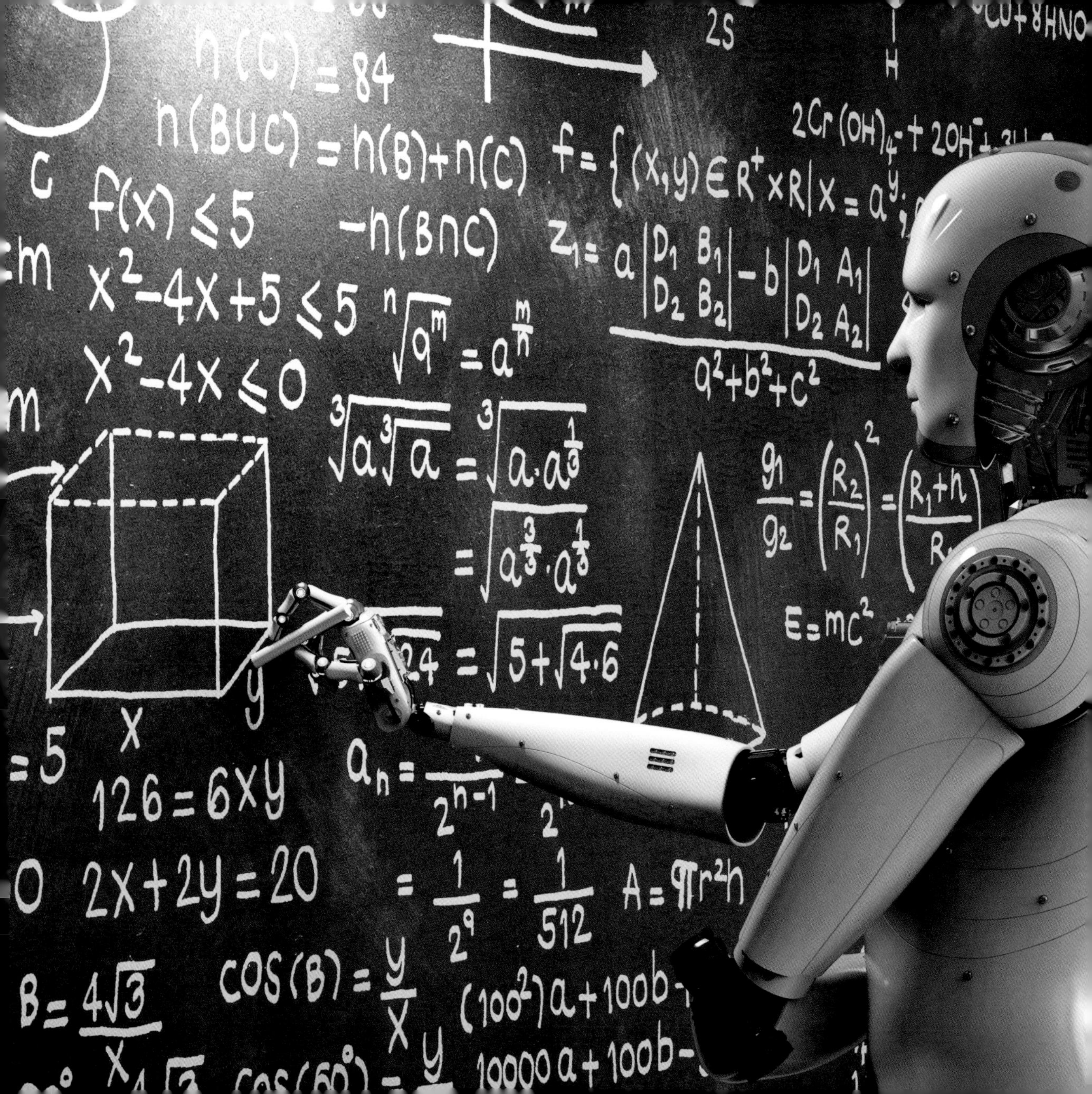
n(C) = 84
n(BUC) = n(B)+n(C)
−n(BnC)
f(x) ≤ 5
x²−4x+5 ≤ 5
x²−4x ≤ 0
E=mc²
126 = 6xy
2x+2y=20
A=πr²h
cos(B) = y/x

Computer Proves Mathematical Theorem

Allen Newell (1927–1992), **John Clifford Shaw** (1922–1991), **Herbert Simon** (1916–2001)

"Kind of crude, but it works, boy, it works!" So said Allen Newell to Herb Simon on Christmas Day 1955, about the program that the two of them had written with the help of computer programmer John Clifford Shaw. The program, Logic Theorist, had been given the basic definitions and axioms of mathematics and programmed to randomly combine symbols into successively complex mathematical statements and then check each one for validity. If the program discovered a true statement, it added that statement to its list of truths and kept going.

As Newell said, the approach was kind of crude, but it worked. As Logic Theorist ran, it discovered more and more mathematical truths. Over time, the program proved 38 of the 52 theorems in *Principia Mathematica*, the classic text on mathematics written by Alfred Whitehead (1861–1947) and Bertrand Russell (1872–1970).

Simon sent a letter to Lord Russell telling him of the program's contributions. In one case, it turned out that Logic Theorist had found a proof to one of Russell's theorems that was more elegant than the one published in the text. Russell wryly wrote back: "I am delighted to know that 'Principia Mathematica' can now be done by machinery. I wish Whitehead and I had known of this possibility before we wasted 10 years doing it by hand."

For Newell, Shaw, and Simon, their program kindled expectations that many of the secrets of thought and intelligence would be cracked in only a few years.

When they wrote Logic Theorist, Newell and Shaw were both computer researchers at RAND Corporation, a research and development think tank. Simon, a political scientist and economist at Carnegie Mellon, was working for RAND as a consultant. Newell ended up moving to Carnegie Mellon, where he and Simon started one of the first AI laboratories. Together Newell and Simon shared the 1975 A.M. Turing Award for their work on AI.

SEE ALSO Algorithm Influences Prison Sentence (2013)

To write Principia Mathematica, *Whitehead and Russell painstakingly derived most of modern mathematics from a small number of axioms and rules of inference. Using a similar approach, the program Logic Theorist was also able to find and prove mathematical truths.*

PRINCIPIA MATHEMATICA

BY

ALFRED NORTH WHITEHEAD, Sc.D., F.R.S.

Fellow and late Lecturer of Trinity College, Cambridge

AND

BERTRAND RUSSELL, M.A., F.R.S.

Lecturer and late Fellow of Trinity College, Cambridge

VOLUME III

Cambridge

at the University Press

1913

EV.

First Disk Storage Unit

Reynold B. Johnson (1906–1998)

Faster than tape but slower than main memory, magnetic disk drives have been an important part of computing since they were invented by IBM and publicly demonstrated on September 14, 1956.

The IBM 305 RAMAC (Random Access Method of Accounting and Control) was designed to store accounting and inventory files that had previously been stored as boxes of IBM punch cards or on tape. To do this, the RAMAC shipped with the IBM 350 disk storage unit, a new device that stored data on 50 spinning disks, each 24 inches (61 centimeters) in diameter and revolving at 1,200 revolutions per minute (RPM). Arranged in 100-character blocks that could be randomly accessed, read, and rewritten, the RAMAC made it possible for a computer with only a few kilobytes of main memory to rapidly access 5 million characters—the equivalent of 64,000 punch cards.

Unlike modern drives, which have a head for every disk, the RAMAC had a single head that moved up and down to select the disk, and then in and out to select the specific block where data would be read or written. The average access time was six-tenths of a second.

The RAMAC also came equipped with a rotating drum memory that spun at 6,000 RPM and stored 3,200 characters on 32 tracks of 100 characters each.

Over the 60 years that followed, the capacity of disk-drive systems increased from 3 megabytes to 10 terabytes—a factor of 3 million—thanks to improvements in electronics, magnetic coatings, drive heads, and mechanical head-positioning systems. But the time it takes for the disk to reposition its head to read the data, something called the *seek time*, only dropped from an average of 600 milliseconds to 4.16 milliseconds, a factor of just 144. That's because reducing seek times depended on improving mechanical systems, which, unlike electronics, are subject to constraints resulting from friction and momentum: in all the years since the RAMAC was introduced, rotation rates have only increased from 1,200 RPM to 10,000 RPM for even the most expensive hard drives.

SEE ALSO Magnetic Tape Used for Computers (1951), Floppy Disk (1970), Flash Memory (1980)

The RAMAC actuator and disk stack, with fifty 24-inch (61-centimeter) disks spinning at 1,200 RPM, held 5 million characters of information.

DISK → ROTATION

The Byte

Werner Buchholz (b. 1922), Louis G. Dooley (dates unavailable)

Designers of the early binary computers faced a fundamental question: how should the computers' storage be organized? The computers stored information in bits, but computer users didn't want to write programs that manipulated bits; they wanted to solve math problems, crack codes, and generally work with larger units of information. The memory of decimal computers such as ENIAC and the UNIVAC I was organized in groups of 10 alphanumeric digits, called *words*. The binary computers also organized their memory into words, but these groups of bits were called *bytes*.

It appears that the word *byte* was coined simultaneously in 1956 by Werner Buchholz at IBM, working on the IBM STRETCH (the world's first supercomputer), and by Louis G. Dooley and others at MIT Lincoln Lab working on the SAGE air-defense system. In both cases, they used the word byte to describe the inputs and outputs of machine instructions that could operate on less than a full word. The STRETCH had 60-bit words and used 8-bit bytes to represent characters for its input/output system; the SAGE had instructions that could operate on 4-bit bytes.

Over the next 20 years, the definition of a byte was somewhat fluid. IBM used 8-bit bytes with its System/360 architecture, and 8-bit groups were the standard for AT&T's long-distance digital telephone lines. DEC, on the other hand, successfully marketed a series of computers with 18-bit and 36-bit words, including the PDP-7 and the PDP-10, which both utilized 9-bit bytes.

This lack of consistency resulted in the early internet standards avoiding the word byte entirely. Instead, the word *octet* is used to describe a group of 8 bits sent over a computer network, a usage that survives to this day in internet standards.

Nevertheless, by the 1980s, the acceptance of 8-bit bytes was almost universal—largely a result of the microcomputer revolution, because micros used 8-bit bytes almost exclusively. In part, that's because 8 bits is an even power of 2, which makes it somewhat easier to design computer hardware with 8-bit bytes than with 9-bit bytes.

Today the era of 9-bit bytes is all but forgotten. And what about collections of 4 bits? Today these are called a *nibble* (sometimes spelled *nybble*).

SEE ALSO ENIAC (1943), The Bit (1948), Digital Long Distance (1962)

Today's computers most frequently use bytes consisting of 8 bits, represented by 1s and 0s.

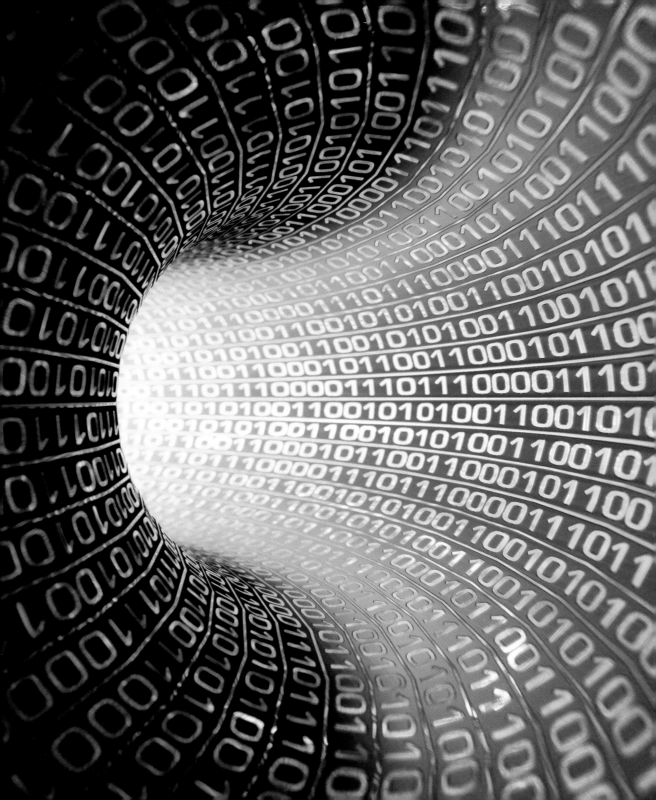

Robby the Robot

Robby the Robot made his world debut as a fictional character in the movie *Forbidden Planet*, the same year that nonstick frying pans went on the market and the hovercraft was invented. In the 1950s, the impact and potential application of technology—both good and bad—was showing up in varying forms in popular culture, as well as diffusing into the home as commodities in different forms. Known for his clever personality and unique visual presence, Robby was symbolic of deeper themes and anxieties the public had toward advancing technology. Believable as a real character and as a robot, Robby quickly became a lasting symbol of the friendly, helpful promise of robots—even though Robby was actually a person walking around inside a 6-foot, 11-inch vacuum-form plastic tube.

In the movie *Forbidden Planet*, Robby was the creation of Dr. Morbius, who built him using the blueprints of an alien race called the Krell that lived a millennium before. The Krell once lived on the planet Altair IV, now home to Dr. Morbius and his daughter, the only survivors from an expedition of scientists sent to the planet two decades earlier. Despite the threatening undertones of the marketing poster for *Forbidden Planet*, showing Robby carrying an injured woman, Dr. Morbius programmed Robby to obey Isaac Asimov's Three Laws of Robotics, which dictated that robots protect and obey human beings.

Following his motion picture debut, Robby appeared in dozens of films and TV shows, including *The Invisible Boy*, *Lost in Space*, *The Twilight Zone*, *Mork & Mindy*, and a 2006 AT&T commercial in which he appears alongside other well-known robots, including Rosie from *The Jetsons* and KITT from *Knight Rider*.

Robby was truly an advanced machine, with the ability to converse fluently in 187 languages and cook Dr. Morbius's food by reproducing molecules in any shape and quantity. Robby—like *R.U.R.* in 1920 and *Metropolis*'s Maria in 1927, helped envision for computer scientists and the general public what a computer's technical and practical potential could be and the role such machines could play in human society. Robby was as much an inspiration to computer scientists and budding inventors as he was an entertainer for the general public.

SEE ALSO Isaac Asimov's Three Laws of Robotics (1942), Unimate: First Mass-Produced Robot (1961)

Robby the Robot, from the film Forbidden Planet, *1956.*

FORTRAN

John Warner Backus (1924–2007)

Machine code is native language that each computer speaks. It's a primitive numeric code, made from simple instructions that computers execute with great speed. And yet, simply adding two numbers and printing the result in machine code requires a long series of tedious instructions: transfer the contents of a memory location into one of the central processing unit's registers, transfer the contents of a second memory location to a second register, add the two registers together, store the results in a third memory location, save the memory location containing the number somewhere else, and, finally, call a function to print the information stored at the memory location. Writing this machine code is grueling work. To make things worse, every early computer had its own machine code with its own peculiarities.

American computer scientist John Warner Backus had a better idea. Instead of programmers laboriously translating the mathematical functions they wanted to solve into machine code, why not have the computer do it? He proposed his vision of a computer program that automatically translated formulas into machine code to his management at IBM in 1953, assembled a team in 1954 to create the IBM Mathematical Formula Translating System, and delivered the first FORTRAN compiler to IBM customers in April 1957.

FORTRAN, which stands for FORmula TRANslation, dramatically simplified the practice of writing programs. It was easier to write, easier to read, and resulted in more reliable code. For example, to compute the length of a triangle's hypotenuse, a programmer could simply write C=SQRT(A * A + B * B) instead of a sequence of perhaps 20 machine code instructions. For his work on FORTRAN and subsequent work involving compiler theory, Backus was awarded the A.M. Turing Award in 1977.

FORTRAN was actually the second computer language that Backus invented. The first one, Speedcoding, was an interpreted code that was easier to write but ran 10 to 20 times slower than machine code. IBM's customers were not willing to pay such a performance price for mere programmer productivity. FORTRAN gained acceptance, in part, because it frequently produced machine code that ran faster than the code produced by the typical human coder.

SEE ALSO COBOL Computer Language (1960)

A programmer's quick reference for the FORTRAN programming language, showing the different arguments for the FORMAT statement.

FORMAT Arguments

Input Blank is equivalent to + in FORTRAN input Form.

FORMAT Argument	FORTRAN Input Form	Internal Value

Ew.d — Decimal with Exponent — Real
$\pm x \cdots x . y \cdots y$ E±ee

E10.3	b0.238E+03	238.
	or bb 2.38E + 02	238.
	or 23.8bb E + 01	238.

The exponents E3, E +3, E03, E + 03, +3, and +03 are equivalent

Fw.d — Decimal without Exponent — Real
$\pm x \cdots x . y \cdots y$

nPFw.d — nP means (input value) $* 10^{-n}$ — 1

F10.4	bb238.5000	238.5
2PF10.4	23850.0bbb	238.5

Iw — Decimal Integer — Integer
$\pm x \cdots x$

I5	bb238	238

Ow — Octal Integer — Binary Number / Any Type Name — 2
$\pm x \cdots x$

O7	b +34670	0034670

wHh···h — Alphameric $g \cdots g$ — Alphameric in FORMAT statement (g's replace h's) — 3

5HbBETA	ALPHA	5HALPHA

Aw — Alphameric $h \cdots h$ — Alphameric in Variable or Array — 4

If w ≤ 6 ; input is left justified and blank filled. — 5
w > 6 ; input is rightmost 6 characters. — 5

A8	ANbERROR	bERROR
A4	bPHI	bPHIbb

wX — (any characters) — w characters are skipped — 6

Lw (w = 1 recommended) — Single Character b or 0 or F / 1 or T — Logical .FALSE. / .TRUE. — 7

9a

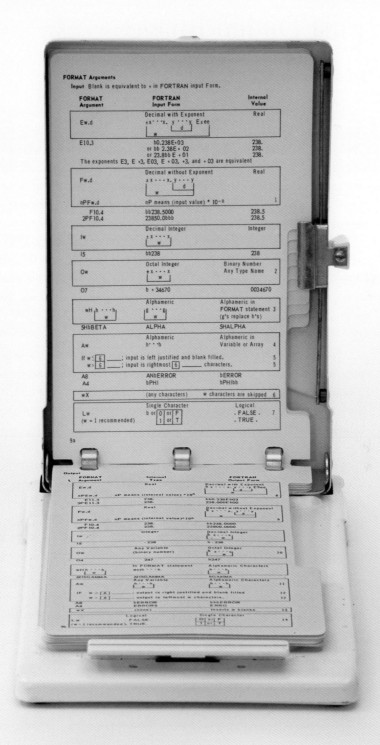

First Digital Image

Russell Kirsch (b. 1929)

Working at the US National Institute of Standards and Technology (NIST), Russell Kirsch supported a team of mathematicians using the Standards Eastern Automatic Computer (SEAC) to model thermonuclear weapons, predict weather, and perform other typical functions of government computers in the 1950s.

In 1957, NIST set about answering a question: What would happen if computers could look at pictures? And with that, Kirsch created a scanner that made the first digital photograph.

Kirsch's scanner consisted of a rotating drum and an optical sensor that could be independently moved along the drum's axis. The sensor assembly started at one end of the drum and moved a little bit after each full revolution. The sensor could only detect the presence or absence of light. To make a grayscale image, Kirsch made multiple scans, each with successively darker filters in front of the sensor, and then electronically combined the results.

For the first picture, Kirsch affixed a 2-inch (5-centimeter) square photo of his three-month-old son to the drum and engaged the machine. The resulting image was a matrix of 176 digitized rows, each with 176 cells of grayscale. Kirsch had invented raster graphics, the idea of displaying a picture or graphic with a matrix of picture elements (now called *pixels*).

The SEAC scanner opened up entire new areas for computer research and applications. Kirsch's approach of storing images as a grid of numbers dominates all computer applications involving images, including satellite imagery, medical imagery, and even the simple two-dimensional color display on the screen of the modern cell phone. An alternative approach for computer graphics based on vectors of light drawn on a screen (called *vector graphics*) vied with raster graphics in the 1960s and 1970s, but ultimately lost out due to its expense.

Kirsch went on to create the Kirsch operator, an algorithm that detects the edges of objects in a digital photograph. Later, he researched ways to use computers in the visual arts, both for analyzing existing artwork and for remixing existing concepts to create new art, all under computer control.

SEE ALSO Fax Machine Patented (1843)

Russell Kirsch's three-month-old son, the first digital photograph, foreshadowing the millions of digital baby photos that would be shared in the future.

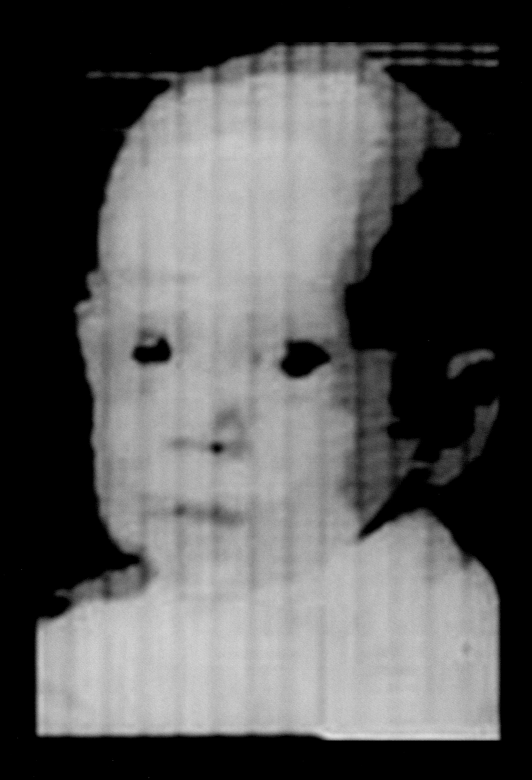

The Bell 101 Modem

A modem (short for *modulator/demodulator*) converts digital information into an analog signal (a process known as *modulation*) so that the signal can be transmitted, and then, on the receiving side, converts the analog signal back into digital bits (the *demodulation*). From 1958 until the late 1990s, acoustic modems that interfaced with the analog telephone network were the primary way that computers communicated with remote users.

The first acoustic modem was probably SIGSALY, a voice-encryption system developed by the Allies during World War II to let Winston Churchill speak directly with Franklin Roosevelt. That modem might have been developed by the Air Force Cambridge Research Center (AFCRC), which developed a digital device for sending radar images over the telephone lines.

Then, in 1958, AT&T released the Bell 101 modem for use with SAGE (Semi-Automatic Ground Environment), a US air defense system. The modem allowed communications over ordinary phone lines at 110 bits per second (bit/s). The following year, AT&T made the device available for commercial customers. The Bell 101 was superseded in 1962 with the Bell 103 modem that could send and receive data at 300 bit/s.

The Bell modems connected directly to ordinary telephone lines, but AT&T, which at the time provided both long-distance and local telephone service, prohibited its customers from attaching equipment manufactured by other companies. Then, in 1968, the US Federal Communications Commission (FCC) ruled that AT&T could not prohibit devices from connecting to telephone lines if they used an acoustic coupler. Within a few years, companies like Novation® and Hayes Microcomputer Products® were offering Bell-compatible 300-baud modems.

A 300-baud modem can deliver text at 30 characters per second or 250 words per minute. In 1979, AT&T introduced the Bell 212 modem, which could send and receive information four times faster. Hayes released the Smartmodem 1200, which was compatible with the Bell 212 but cost much less, in 1982 for $699. Two years later, the International Telegraph and Telephone Consultative Committee (CCITT) released v.22bis, a worldwide standard for 2400-baud modems. Those modems set the ground for the first dial-up time-sharing services.

SEE ALSO SAGE Computer Operational (1958), Telebit Modems Break 9600 bps (1984)

The Bell 101 Dataset (1958) was the first commercial modem able to transmit digital data.

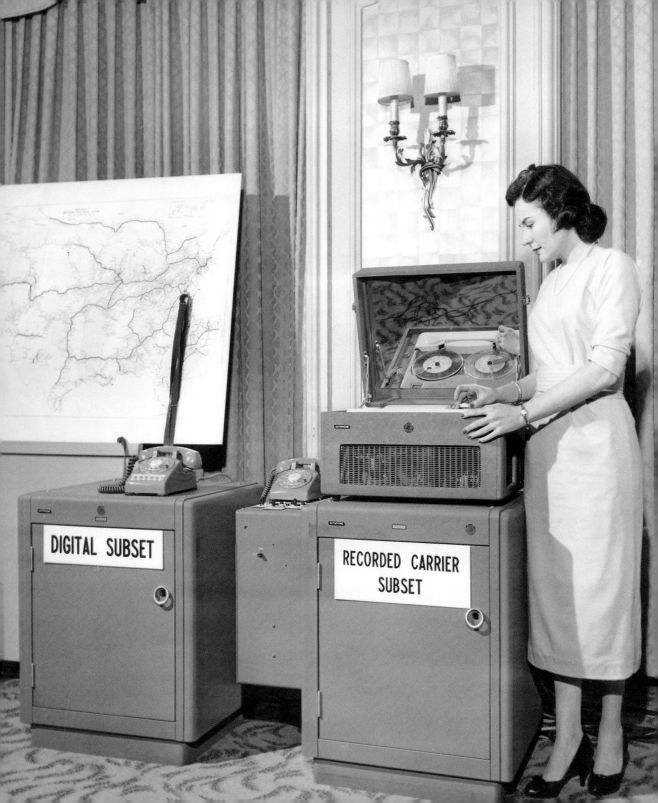

SAGE Computer Operational

Jay Forrester (1918–2016)

The Semi-Automatic Ground Environment (SAGE) was a network of computers that protected the United States from a surprise Soviet attack from the time it was operational in 1958 until it was decommissioned in 1984. The system, developed at MIT Lincoln Laboratory, remains the largest computer project of all time: $10 billion in 1954 dollars (around $90 billion in 2018), meaning SAGE cost roughly three times more than the Manhattan Project did.

SAGE's task was huge: a networked system of 24 computers that watched the airspace inside and surrounding the United States, tracking every airplane that moved and flagging those that had not filed flight plans. When unknown objects appeared, SAGE would determine which interceptor missile to launch and calculate the location of the interception. Besides protecting the nation against Soviet bombers, SAGE assisted in numerous rescues of small aircraft that went down at sea.

MIT professor Jay Forrester chose IBM as the institute's partner for SAGE; the system was responsible for 80 percent of IBM's revenue during its construction and made IBM a corporate giant. The network itself consisted of many pairs of computers (a primary and a backup), with each pair built in a concrete-block house that was an acre in area and four stories tall. The original IBM AN/FSQ-7 computers were the largest computers ever built at the time, each having 60,000 vacuum tubes and 256 kibibytes of magnetic core RAM, and weighing 250 tons. The pair of computers controlled 150 video consoles equipped with light guns that operators used to select their targets.

As the SAGE computers were removed from service and replaced with cheaper, more modern systems, the old systems started showing up in movies and in television shows. Between 1996 and 2016, AN/FSQ-7 components starred on screen more than 80 times (according to the website Starring the Computer), including as the "Bat Computer" in the 1960s *Batman* TV series, as a 22nd-century computer in Woody Allen's 1973 movie *Sleeper*, and in the 1983 movie *WarGames*—one of the few instances in which the AN/FSQ-7 system played itself on screen.

SEE ALSO Core Memory (1951), *WarGames* (1983)

Captain Charbonneau sits at SAGE's situation display console at the Experimental Direction Center at Lincoln Labs.

IBM 1401

The IBM 1401 was IBM's second-generation computer. Built from transistors and designed primarily for keeping business records, the 1401's low cost and flexibility made it the most successful computer on the planet.

The computer could store data on both magnetic tape and disk, read punch cards at the astounding rate of 800 per minute, punch new cards at 250 per minute, and print 600 lines per minute on a "chain printer" that had just 26 upper-case letters, 10 numbers, and 12 special characters (& , . ¤ - $ * / % # @ ≠).

The 1401 was programmed in Autocode, or in IBM's Symbolic Programming System, which today we would call an *assembler*. To avoid binary rounding errors, the system worked with decimal numbers. The 8-bit memory used 6 bits for encoding a number or letter, a parity check bit (for detecting hardware errors), and an 8th bit that marked the end of a number or text. "No space is wasted by filling in fixed-length words," boasted the model's manual, written at a time when every bit was precious.

The 1401's processor consisted of individual printed circuit boards, also called *cards*, that each contained a few transistors and other discrete components. Systems could be purchased with 1,400, 2,000, 4,000, 8,000, 12,000, or 16,000 characters of 8-bit core storage. The small configurations were particularly attractive: small businesses could easily computerize by renting a 1401 for just $2,500 per month. (For comparison, IBM's 701 business computer had rented for $15,000 per month in 1953.) Large organizations that already had a mainframe could rent a 1401 to transfer data from slower-speed punch cards to tape, feed the tape into the mainframe, perform their calculations, transfer the results back to tape, and use the 1401 to print the results.

The 1401 was a runaway success. By the end of 1961, there were more than 2,000 systems installed in the United States, representing a quarter of all installed computers. By the time IBM was ready to replace the 1401 with the radically different System/360 line, one-third of the computers on the planet would be IBM 1401s.

SEE ALSO IBM System/360 (1964)

IBM engineers Chester Siminitz and C. Fred Woidt review data on the IBM 1401 computer equipped with the IBM 1009 data transmission unit, which converts binary-coded decimal signals from the computer into a special transmission code.

PDP-1

Ben Gurley (1926–1963)

In 1957, Digital Equipment Corporation was founded. The idea was to sell ready-to-use electronic logic "modules" that would make it easy for other labs to experiment with the new technology. The company made a profit in its first year and hired Ben Gurley, a brilliant designer from MIT, to create Digital's first computer.

Except that it couldn't be called a computer. After all, computers were big and expensive, and Digital's financial backers didn't want to compete with IBM. Besides, Digital had a vision for something different: an interactive computer that was small, affordable, exciting, and fun. So instead, the machine was called a *Programmed Data Processor*. Using Digital's electronics modules as building blocks, Gurley and his small team designed and built the machine in three and a half months, with the first customer shipment in December 1959.

The PDP-1 was a completely different machine from the batch-oriented systems sold by IBM and Sperry Rand. Those machines cost $10,000 a month to rent; the PDP-1 cost between $85,000 and $120,000 to purchase. And the PDP-1 was interactive, with options including a large graphics display, a small high-resolution display, a light pen, a real-time clock, a multiplexed analog-to-digital converter (for interfacing with laboratory equipment), and audio output. It was slower and smaller than the computers of the day, but it was very usable, even friendly. People called it a *minicomputer*.

Digital gave one of its first PDP-1s to MIT, specifically for use by students. "They learned more about the computer and how to do things with it than probably anybody had before that, because you had dozens of bright people spending all hours of the day studying this," Ken Olsen (1926–2011), Digital's cofounder, said in his 1988 oral history interview with the Smithsonian Institution.

On the strength of the PDP-1 and the computers that followed, Digital became the second-largest computer company in the world, and the largest private employer in Massachusetts. Sadly, Gurley never saw any of it: he was murdered by a disgruntled former employee, shot by rifle as he had dinner with his wife and five children in 1963.

SEE ALSO Core Memory (1951), *Spacewar!* (1962), AltaVista Web Search Engine (1995)

A photo of the PDP-1 computer, a class of machine that would eventually be referred to as a "minicomputer."

Quicksort

Charles Antony Richard Hoare (b. 1934)

Sorting a list of names or numbers is a common task in many computer programs.

The most obvious way to sort is to give the program a list of randomly shuffled numbers and have it cycle through the list, comparing every pair in order, putting the lower-valued number first. After n passes, where n is the number of elements in the list, the list will be sorted. This is called *bubble sort*, and it is a horribly inefficient sorting method.

In 1959, Tony Hoare came up with a better approach for sorting numbers. Called Quicksort, the algorithm involves partitioning a list of elements into two parts, sorting each, and then applying an algorithm recursively to each partition.

Hoare developed Quicksort in 1959 to sort words in a Russian–English dictionary while he was visiting the Soviet Union as part of an exchange program. On his return, he discovered that the algorithm was faster than other sorting algorithms, so he sent it to the algorithms department of *Communications of the* ACM (Association for Computing Machinery), a professional magazine. At the time, ACM was trying to stimulate the development of the field by publishing as many algorithms as possible; it published Quicksort as "Algorithm 64" in July 1961.

Quicksort was a groundbreaking algorithm because it was so simple (fewer than 10 lines of code) and yet dramatically more efficient than other, more obvious ways to sort a list of elements. In the five decades since, there have been minor tweaks to the algorithm, but Quicksort is still widely used in computer programs, and implementations are built into most computer languages.

Hoare went on to develop techniques for analyzing and reasoning about the correctness of computer programs. He became a professor at Queen's University Belfast in 1968, and then moved to the University of Oxford in 1977. He won the A.M. Turing Award for his contributions to algorithms in 1980 and became a Fellow of the Royal Society in 1982.

Looking back over his career, in 2009 he told the editor of *Communications of the ACM*, "I think Quicksort is the only really interesting algorithm that I've ever developed."

SEE ALSO Ada Lovelace Writes a Computer Program (1843), Software Engineering (1968)

The Quicksort algorithm, pictured in part on the computer screen, offered a surprisingly efficient way to sort lists of numbers.

Airline Reservation System

R. Blair Smith (dates unavailable), **C. R. Smith** (1899–1990)

As the era of the jet engine took off and the popularity of air travel increased, so did the complexity of synchronizing passenger demand with flight availability. Flight reservations were booked manually, with index cards, files, and a giant Lazy Susan with half a dozen humans sitting around it. Booking a flight took on average 90 minutes.

American Airlines® (AA) was not new to the challenge of automation and had researched solutions, resulting in early systems called the *Electromechanical Reservisor* and the *Magnetronic Reservisor*.

In 1953, IBM sales representative R. Blair Smith was on an American flight from Los Angeles to New York when he happened to be seated at the back of the plane next to C. R. Smith, the unassuming president of the airline. The IBM salesman realized that the challenge faced by American Airlines (and in fact the entire airline industry) was grounded in the same kinds of challenges that IBM had faced when creating the SAGE system for the US Air Force. Smith and Smith got to talking and by the end of the flight, the spark that would eventually become the Semi-Automated Business Research Environment reservation system—SABRE®—had been lit.

The initial SABRE was run on two IBM 7090® mainframes located in a computer center in Briarcliff Manor, New York. By the time it was fully operational in 1964, SABRE was processing 84,000 telephone transactions a day and was one of the largest civilian data-processing systems in the world. The system had more than 1,500 data terminals, spread across the country, that remotely connected to the mainframe to query and complete reservation and booking requests. The time required to make a reservation went from 90 minutes down to several seconds.

SABRE not only revolutionized the travel industry, prompting other airlines to develop their own proprietary systems or sign up for SABRE, but also laid the groundwork for the concepts and methods that would enable the launch of what would become a juggernaut e-commerce industry in the following decades. In 1996, after SABRE had been spun off as its own company, it founded a company called *Travelocity*®, one of the earliest consumer-facing online booking systems.

SEE ALSO SAGE Computer Operational (1958)

The SABRE airline reservation system, developed jointly by American Airlines and IBM, is still in use today.

COBOL Computer Language

Mary K. Hawes (dates unavailable), **Grace Hopper** (1906–1992)

In 1959, the US Department of Defense (DOD) operated 225 computers and had another 175 on order. These computers were rapidly replacing paper filing systems, utilizing dozens of different programming languages to track people, supplies, and money. Realizing that the government could not afford the skyrocketing cost of software development, the DOD funded Mary K. Hawes, a computer scientist from the Burroughs Corporation, to create the Conference/Committee on Data Systems Languages (CODASYL) to design a Common Business Language (CBL). The plan was to create a new language that would be easier to use, allowing computers to be programmed with English-like sentences.

Numerous committees were created to design the language, and many manufacturers announced their intention to support it. The plan was to have a short-range committee come up with an interim, stopgap measure, and then for others to refine the language at a more careful, leisurely pace. But the task was huge, and the project soon got bogged down in competing designs.

Meanwhile, members of the short-range committee took FLOW-MATIC, a computer language developed within the DOD by computer scientist Grace Hopper, made a few modifications, and distributed it as the Common Business Oriented Language (COBOL). Calling themselves alternately the Short-Range Committee and the PDQ (Pretty Darn Quick) Committee, the team put together a language specification between August and December 1959; a year later, on December 7, 1960, a COBOL program could run on the RCA 501 computer and on a Remington Rand UNIVAC computer.

COBOL soon came to dominate the world of business computing. And while it has gone through many revisions, COBOL remains in use today, driving the back-office systems of many banks and payroll systems.

COBOL is also one of the first examples of free software. At a time when users rented their computers and corporations jealously guarded their intellectual property rights, the designers and users of COBOL insisted it should be available to all.

SEE ALSO *FORTRAN (1957), BASIC Computer Language (1964), C Programming Language (1972)*

Grace Hopper explains the operation of the COBOL compiler.

Recommended Standard 232

For more than three decades, the Electronics Industries Association's Recommended Standard 232 was the communications protocol that connected the wired world. Systems of all kinds came equipped with RS-232 connectors on their back sides that allowed them to transmit bytes of data as a series of serialized bits sent down a single *transmit data* wire; the same connector had a second *receive data* wire that could receive bytes from something at the other end.

Standardized in 1960, by the mid-1970s the RS-232 protocol was "spoken" by practically every computer on the planet. The original purpose was to enable worldwide communication of terminals and computers using the telephone network. In the machine room, RS-232 connected the computer with a dial-up telephone modem. Getting online meant making a phone call and having the two modems communicate using audio tones.

The original RS-232 connector had 25 pins. In addition to the data pins, one pin indicated that the phone was ringing, another pin that the carrier tone was present, and two pins indicated that each side was ready to accept data; two others indicated if each side had data to transmit. The 25-pin standard allowed the extra pins to be used as a second data channel. In practice, that channel was rarely used, so early PCs had 9-pin RS-232 connectors, necessitating 9-to-25 pin converters. At many universities, it was common for terminals to be connected with just three wires, with the others "looped back." This setup was less reliable, but it allowed schools to use cheaper telephone cable to connect their machines.

Early terminals and modems ran RS-232 at 110, 300, or 1200 bits per second; in 1981, the IBM PC was introduced with a new chip from the National Semiconductor® company: the 8250 UART (universal asynchronous receiver/transmitter). The 8250 had a programmable bit-rate generator that could run RS-232 up to 115,200 bits per second.

The introduction of the Universal Serial Bus (USB) in 1996 marked the beginning of RS-232's decline. Today few PCs have RS-232 connectors, although many motherboards still have the necessary hardware. Meanwhile, RS-232 is still widely used for communicating with embedded computers, such as computerized door locks.

SEE ALSO The Bell 101 Modem (1958), Universal Serial Bus (USB) (1996)

First developed in 1960, RS-232 ports are still incorporated into modern circuit boards.

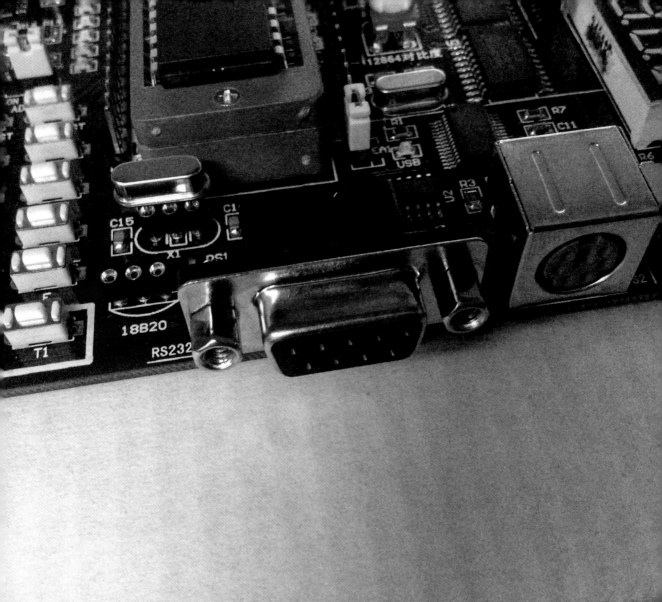

ANITA Electronic Calculator

Norbert "Norman" Kitz (dates unavailable)

The ANITA Mk VII and ANITA Mk 8 were the world's first electronic calculators to be sold commercially. Created by Norbert Kitz, who had previously worked on the pilot ACE (Automatic Computing Engine) at the British National Physical Laboratory in the late 1940s, ANITA's user interface was modeled on the mechanical calculators of the era, but inside it was all electronics: tubes, resistors, diodes, lots of wire, and one transistor (apparently as a voltage regulator).

Like a mechanical calculator, the ANITAs had 10 buttons for each decimal position, rather than the single 10-digit keypad that is standard today. Numbers were displayed on 13 "nixie tubes," neon tubes (still made today) that each contain 10 individually controllable wires, each one in the shape of a different number.

ANITA was manufactured by London's Bell Punch Co., a company that had been established in 1878 to sell accounting products to England's early train lines. Over the years, Bell Punch developed a number of mechanical counting devices for the railroads. The calculator was most likely named after the inventor's wife, although the company variously claimed that ANITA stood for "A New Inspiration to Arithmetic" and "A New Inspiration to Accounting."

The ANITA Mk VII and ANITA Mk 8 were introduced within a week of each other, the Mk 8 in England, and the Mk VII a week later on the continent, mostly in Germany, Holland, and Belgium. Only a thousand Mk VII machines were made, due to a design flaw resulting from the use of Dekatron cold-cathode vacuum tubes. Those tubes were damaged from being repeatedly turned on and off—a common occurrence in a desktop calculator—and over time the Mk VII units developed errors. The Mk 8 was manufactured with more reliable ring counters made from thyratron tubes.

Priced at £355 (about $1,000), the Mk 8 sold for roughly the same price as mechanical calculators. It was silent, however, a fact that Bell Punch highlighted in its advertisements. The company had no competition until 1964, when companies in the United States, Italy, and Japan simultaneously introduced transistor-based calculators. At that point, Bell Punch was producing more than 10,000 ANITA calculators each year.

SEE ALSO Curta Calculator (1948), HP-35 Calculator (1972)

The Anita Mk 8 calculator's rows of numeric push-buttons were designed to mimic the user interface of mechanical calculators.

Unimate:
First Mass-Produced Robot

George Devol (1912–2011), **Joseph F. Engelberger** (1925–2015)

After seeing a picture of assembly-line workers in a technical journal, American inventor George Devol wondered if there could be a tool to replace the repetitive, mind-numbing tasks people had to perform. This question led him to design something akin to a mechanical arm, which he patented in 1961, called the *Programmed Article Transfer device*.

Devol had a fortuitous introduction to engineer and businessman Joseph Engelberger at a cocktail party in 1956. Engelberger, fascinated by Isaac Asimov's robot stories, immediately recognized the business potential of Devol's "robot" device. As business partners, they had to perfect the device and convince others to buy it. Engelberger's sales strategy was to identify jobs that Unimate (the name suggested by Devol's wife, Evelyn) could do that were dangerous or difficult for humans to perform. General Motors® (GM) was the first to buy into the idea, and in 1959 the Unimate #001 prototype was installed in an assembly line in Trenton, New Jersey. Unimate's job was to pick up hot door handles that had just been made from molten steel and drop them into cooling liquid before they were sent down the line for human workers to finish polishing. The Unimate would go on to spawn new industries and revolutionize production and manufacturing plants around the world.

The Unimate weighed 4,000 pounds (1,814 kilograms) and was controlled by a series of hydraulics. Memory was stored on a magnetic drum, and pressure sensors inside the arm enabled it to adjust the strength of its grip as needed. Unimate "learned" a job by first having a person manually move its parts in the sequence of steps desired to complete the task. The movements were recorded by its computer and then simply repeated over again.

In 1966, Unimate was featured on *The Tonight Show Starring Johnny Carson*, where it demonstrated how it could knock a golf ball into a hole, pour a can of beer, and conduct the *Tonight Show* orchestra. An early model of this robot can be found at the Smithsonian's National Museum of American History; in 2003, the Unimate was inducted into the Carnegie Mellon Robot Hall of Fame.

SEE ALSO Isaac Asimov's Three Laws of Robotics (1942)

This 52-inch (1.3-meter) -thick, oil-filled glass window protects a nuclear engineer from radiation while he operates a robotic arm at NASA's Plum Brook Station in Sandusky, Ohio, 1961.

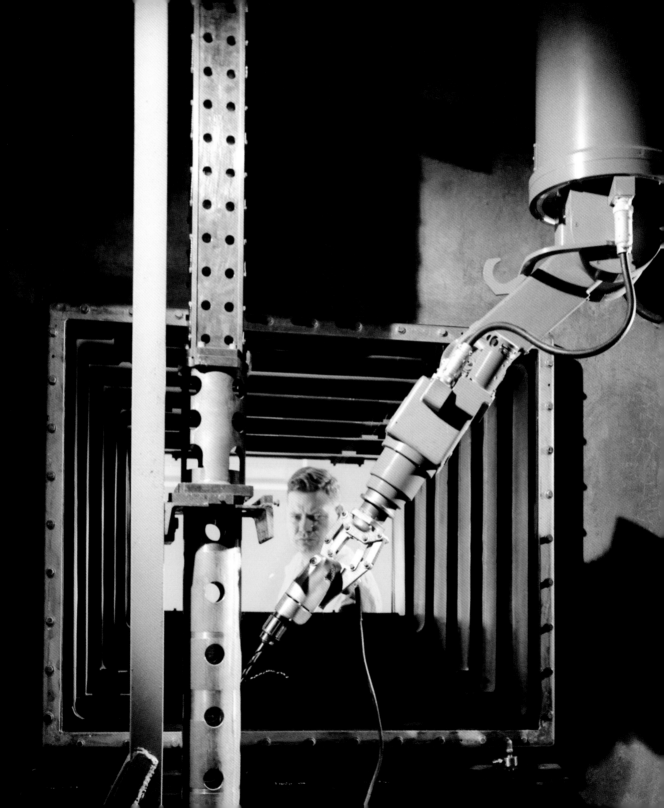

Time-Sharing

John Warner Backus (1924–2007), **Fernando J. Corbató** (b. 1926)

A computer's CPU can run only one program at a time. Although it was possible to sit down at the early computers and use them interactively, such personal use was generally regarded as a waste of fantastically expensive computing resources. That's why batch processing became the standard way that most computers ran in the 1950s: it was more efficient to load many programs onto a tape and run them in rapid succession, and then make the printouts available to the much slower humans in due time.

But while batch processing was efficient for the computer, it was lousy for humans. Tiny programming bugs resulting from a single mistyped letter might not be discovered for many hours—typically not until the next day—when the results of the batch run were made available.

Researchers at MIT realized that a single CPU could be shared between several people at the same time if the CPU switched between different programs, running each for perhaps a 10th of a second. From the users' point of view, the computer would appear to be running slower, but for the users, this system still would be more efficient, because they would find out about their bugs in seconds, rather than hours.

John Backus first proposed this method in 1954 at an MIT summer session sponsored by the Office of Naval Research, but it couldn't be demonstrated until IBM delivered its 7090 computer to MIT—a computer that was large enough to hold several programs in memory at the same time.

MIT professor Fernando J. Corbató demonstrated his Experimental Time-Sharing System in November 1961. The system time-shared between four users. The operating system had 18 commands, including *login, logout, edit* (an interactive text editor), *listf* (list files), and *mad* (an early programming language). Later, this became the Compatible Time-Sharing System (CTSS), so named because it could support both interactive time-sharing and batch processing at the same time. Corbató was awarded the 1990 A.M. Turing award for his work on CTSS and Multics.

Time-sharing soon became the dominant way of interactive computing and remained so until the PC revolution of the 1980s.

SEE ALSO Utility Computing (1969), UNIX (1969)

Photograph of Fernando Corbató at MIT in the 1960s.

Spacewar!

Steve Russell (b. 1937), **Martin Graetz** (dates unavailable),
Wayne Wiitanen (dates unavailable)

It was obvious to Steve Russell and friends at MIT that the best way to demonstrate the power of the new PDP-1 machine would be a multiuser video game in which players tried to shoot down one another's spaceships. So Russell, along with his friends Martin Graetz and Wayne Wiitanen, came up with *Spacewar!*, inspired in part by American and Japanese science-fiction and pulp novels.

The basic program took about six weeks to develop and featured two spaceships, the *Needle* and the *Wedge*, both orbiting around the gravitational well of a sun, with a starfield as background. The program required more than 1,000 calculations per second to compute the spaceships' motion and location, plot the relative positions of the stars and sun, and apply the player inputs. Players could launch torpedoes using a toggle switch on the computer or by pressing a button on a control pad. Because the game followed Newtonian physics, the ships remained in motion even when the players were not accelerating them. Part of the challenge then was to shoot down the opponent's ship without colliding with a star. *Spacewar!* featured hyperspace, gravity-assist powers, and forced cooldowns between firings, so it required some strategy to win rather than just aiming weapons and firing at the other player as quickly as possible. There was even a lone asteroid that the players could fire upon.

First unveiled to the public at MIT's 1962 Science Open House, *Spacewar!* could soon be found on most of the PDP-1 research computers in the country. Considered one of the 10 most important video games of all time by the *New York Times* in 2007, much of *Spacewar!*'s success is evident in the decades after its creation—Stewart Brand (b. 1938) and *Rolling Stone* sponsored a *Spacewar!* tournament in 1972, reporting on it with the excitement of a physical sporting event. In 1977, *BYTE* magazine published a version of *Spacewar!* in assembly language that could run on the Altair 8800, and *Spacewar!* was the inspiration for one of the first-ever arcade games, *Computer Space*, in 1971, designed by the same person—Nolan Bushnell (b. 1943)—who would go on to launch *Pong*® and the Atari Corporation®. And that lone asteroid in *Spacewar!* became the inspiration for the video game *Asteroids*®, which would become Atari's most successful game.

SEE ALSO PDP-1 (1959), *Pong* (1972), First Personal Computer (1974), *BYTE* Magazine (1975)

Dan Edwards (left) and Peter Samson (right) play Spacewar!—*one of the earliest digital-computer video games—on the PDP-1 Type 30 display.*

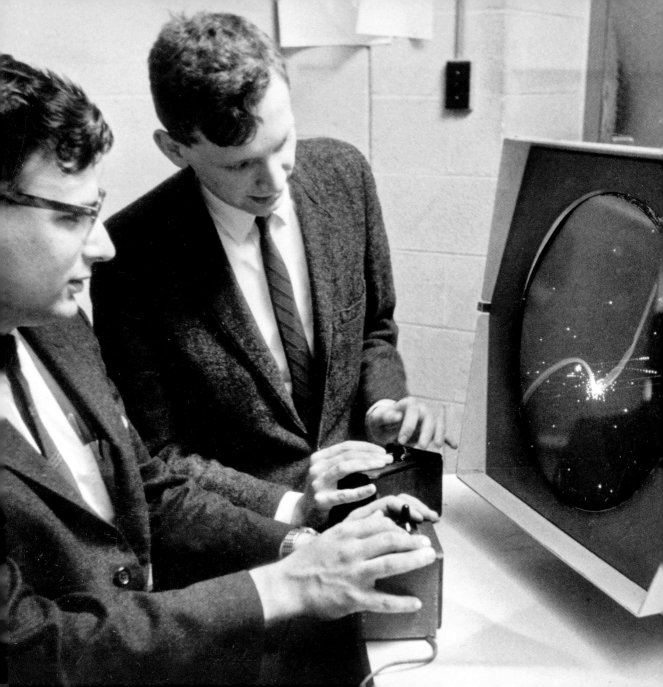

Virtual Memory

Fritz-Rudolf Güntsch (1925–2012), Tom Kilburn (1921–2001)

Memory was one of the key limitations of early computers. To deal with this limitation, programmers would split code and data into many different pieces, bring them into memory for processing, and then swap them back out to the auxiliary storage device when the computer's memory was needed for something else. All of this movement was controlled by the programmer, and it took a lot of effort to get everything correct.

Working on his doctoral thesis at Technical University Berlin, Fritz-Rudolf Güntsch proposed having the computer automatically move data from storage into memory when it was referenced, and automatically save it back when the memory was needed for something else. To make this work, the computer's data would be assigned to different portions of a larger *virtual memory address space*, which the computer would map, or assign, to much smaller pages of *physical memory* as needed.

Even by the standards of the day, the machine that Güntsch designed was tiny: it had just six blocks of core memory, each 100 words in size, that it used to create a virtual address space of 1,000 blocks. But he accurately described the idea.

Meanwhile, at the University of Manchester, a team led by Tom Kilburn was working to build what was a massive and fast machine by the day's standards. For that machine, they designed and implemented a virtual memory system with 16,384 words of core memory and an auxiliary storage of 98,304 words. Each word had 48 bits, large enough to hold a floating-point number or two integers. Originally called *MUSE*, for "microsecond engine," and later renamed *Atlas*, the transistorized computer took six years to create, with the British electronics firms Ferranti and Plessey joining in.

Today virtual memory is a standard part of every modern operating system: even cell phones use virtual memory. Seymour Cray (1925–1996), known for creating the world's fastest computers in the 1960s, '70s, '80s, and '90s, famously eschewed virtual memory because moving data between auxiliary storage and main memory takes precious time and slows down a computer. "You can't fake what you don't have," Cray often said.

When it comes to virtual memory, it turns out that you can.

SEE ALSO Floating-Point Numbers (1914), Time-Sharing (1961)

Console of the Atlas 1 computer, used to provide information on the operation of the machine. The computer was installed at the Atlas Computer Laboratory in Britain and was mainly used for particle physics research.

Digital Long Distance

Picture this: It's Mother's Day around 1960. All over the country, sons and daughters who live far away from their mothers are calling to wish them a happy day and say thank you for all they do. Except many of them can't, because they can't get their calls to go through. All they hear when they dial is a busy signal or an automated voice saying to try again later. That was because there was a relatively small number of copper wire pairs crisscrossing the country as part of the telecommunications network, and each pair could carry just a single conversation.

With the introduction of AT&T's digital T1 carrier service, the capacity of each pair of copper wires dramatically increased. Rather than one conversation per pair of twisted wires, two pairs could carry 24 conversations simultaneously. The T1 service did this by converting all the analog voice data to digital format and sequencing or organizing that data to travel together on a copy pair and then get accurately separated for delivery to the intended residence or phone line. In essence, there was suddenly more than 10 times the capacity on each copper pair. (For technical reasons, the T1 required a copper pair to carry data in each direction.) The first T1 was installed in Chicago, where the city had run out of space in places to add more buried cable under the city streets.

The digital long-distance service required three things: the T1 digital communication protocol, a technology called a *multiplexer* to combine the 24 conversations into a single data stream, and a converter that changed analog data to digital and digital back to analog.

The T1 created the possibility of connecting two computers with a high-speed digital network ordered from the phone company. The evolution and maturation of the specifications and standards surrounding the T1 carrier service, popularly referred to as a *T1 line*, was fundamental to a lot of other innovation occurring, including both the early internet and the eventual computerization of the local telephone network with the invention of the 5ESS switch.

SEE ALSO Computerization of the Local Telephone Network (1983)

Engineers at Bell Telephone replace the T1 interface deep within a telephone switch.

Sketchpad

Ivan Sutherland (b. 1938)

Sketchpad is generally accepted as the first interactive computer graphics program. Created as part of Ivan Sutherland's doctoral thesis at MIT, Sketchpad helped launch a new era in human–computer interaction. The program used a light pen, but instead of pointing to dots on a cathode ray tube (CRT), Sketchpad used the pen to let users draw shapes on a computer screen. It was a revolutionary concept to input data into a computer graphically, rather than as code made up of numbers and letters. One of the earliest examples of a graphical user interface (GUI), Sketchpad also introduced the idea of using the computer for art and design rather than for strictly technical or scientific work.

Sketchpad relied on a variety of knobs and toggle switches as input mechanisms to control the size and ratio of the lines. It ran on the Lincoln experimental TX-2 computer, which at the time had more memory than any commercial machines on the market. Sketchpad was the foundation for numerous developments in the field of computer graphics. Computer-aided design (CAD) software, human–computer interaction (HCI), and object-oriented programming (OOP) can all trace their roots in some way back to Sketchpad.

Sketchpad influenced Douglas Engelbart's design of the oN-Line System at the Stanford Research Institute (SRI). Sutherland built upon his own work in numerous ways, including the invention of the first head-mounted display in 1967, a milestone in what would become known as *virtual reality*.

It would be decades before readily available PCs had similar capabilities. But when they did, designers would finally escape the practical limitations of drawing and drafting on paper. For many, the revolution came in the 1980s with AutoCAD and similar programs, which allowed drawings to be made with greater accuracy and complexity and without the time-consuming burden of erasing work because of mistakes or a change in design direction.

Sutherland won the A.M. Turing Award in 1988 and the Kyoto Prize in 2012 for his work on Sketchpad.

SEE ALSO "As We May Think" (1945), Head-Mounted Display (1967), Mother of All Demos (1968), VPL Research, Inc. (1984)

Ivan Sutherland's Sketchpad allowed users to create graphic images directly on a display screen using a light pen.

ASCII

Bob Bemer (1920–2004)

A character code assigns a numeric value to every printable character. But no code is more fundamentally correct than any other. By the end of the 1950s, there were more than 60 different coding standards in use—IBM itself had nine different character sets on its various computers—making it difficult to move information in digital form from one kind of system to another.

The clear solution was for the industry to get together and decide, once and for all, the mapping between codes in a computer and human-readable characters. The project started in 1960 when Bob Bemer, an engineer at IBM, started lobbying for a common code. Bemer submitted his proposal in May 1961 to the American Standards Association (ASA), which responded by forming a subcommittee, which, after two years of work, published the American Standard Code for Information Interchange, better known as *ASCII*.

At first, ASCII was designed to support the teletypes and teleprinters of the day without consideration for future developments. So the original code had only capital letters, numbers, a few symbols, and special control characters for governing the movement of a teleprinter. This included *carriage return* (which returned the printhead to the left), *line feed* (which advanced the paper), and *bell* (which rang the teleprinter's bell). In 1967, ASCII was expanded to include lowercase letters and a few more symbols.

Because ASCII was only a 7-bit code, it could represent only $2^7 = 128$ different characters, so there wasn't room for accented characters such as É. Symbols that didn't make the cut, such as ¢, ⁋ and ‰, stopped appearing on keyboards in the United States in the 1980s, as computer keyboards started to replace typewriters in homes and offices.

With the growing popularity of 8-bit machines, different manufacturers used codes 129–255 in different ways to represent different characters. Meanwhile, in Asia, a variety of incompatible techniques were devised to encode Chinese, Japanese, and Korean. Once again, the computer industry seemed intent on creating another tower of Babel. Things weren't resolved until many years later, when Unicode was adopted.

SEE ALSO Baudot Code (1874), Unicode (1992)

ASCII table of 7-bit ASCII characters.

ASCII TABLE

Decimal	Hexadecimal	Binary	Octal	Char	Decimal	Hexadecimal	Binary	Octal	Char	Decimal	Hexadecimal	Binary	Octal	Char
0	0	0	0	[NULL]	48	30	110000	60	0	96	60	1100000	140	`
1	1	1	1	[START OF HEADING]	49	31	110001	61	1	97	61	1100001	141	a
2	2	10	2	[START OF TEXT]	50	32	110010	62	2	98	62	1100010	142	b
3	3	11	3	[END OF TEXT]	51	33	110011	63	3	99	63	1100011	143	c
4	4	100	4	[END OF TRANSMISSION]	52	34	110100	64	4	100	64	1100100	144	d
5	5	101	5	[ENQUIRY]	53	35	110101	65	5	101	65	1100101	145	e
6	6	110	6	[ACKNOWLEDGE]	54	36	110110	66	6	102	66	1100110	146	f
7	7	111	7	[BELL]	55	37	110111	67	7	103	67	1100111	147	g
8	8	1000	10	[BACKSPACE]	56	38	111000	70	8	104	68	1101000	150	h
9	9	1001	11	[HORIZONTAL TAB]	57	39	111001	71	9	105	69	1101001	151	i
10	A	1010	12	[LINE FEED]	58	3A	111010	72	:	106	6A	1101010	152	j
11	B	1011	13	[VERTICAL TAB]	59	3B	111011	73	;	107	6B	1101011	153	k
12	C	1100	14	[FORM FEED]	60	3C	111100	74	<	108	6C	1101100	154	l
13	D	1101	15	[CARRIAGE RETURN]	61	3D	111101	75	=	109	6D	1101101	155	m
14	E	1110	16	[SHIFT OUT]	62	3E	111110	76	>	110	6E	1101110	156	n
15	F	1111	17	[SHIFT IN]	63	3F	111111	77	?	111	6F	1101111	157	o
16	10	10000	20	[DATA LINK ESCAPE]	64	40	1000000	100	@	112	70	1110000	160	p
17	11	10001	21	[DEVICE CONTROL 1]	65	41	1000001	101	A	113	71	1110001	161	q
18	12	10010	22	[DEVICE CONTROL 2]	66	42	1000010	102	B	114	72	1110010	162	r
19	13	10011	23	[DEVICE CONTROL 3]	67	43	1000011	103	C	115	73	1110011	163	s
20	14	10100	24	[DEVICE CONTROL 4]	68	44	1000100	104	D	116	74	1110100	164	t
21	15	10101	25	[NEGATIVE ACKNOWLEDGE]	69	45	1000101	105	E	117	75	1110101	165	u
22	16	10110	26	[SYNCHRONOUS IDLE]	70	46	1000110	106	F	118	76	1110110	166	v
23	17	10111	27	[ENG OF TRANS. BLOCK]	71	47	1000111	107	G	119	77	1110111	167	w
24	18	11000	30	[CANCEL]	72	48	1001000	110	H	120	78	1111000	170	x
25	19	11001	31	[END OF MEDIUM]	73	49	1001001	111	I	121	79	1111001	171	y
26	1A	11010	32	[SUBSTITUTE]	74	4A	1001010	112	J	122	7A	1111010	172	z
27	1B	11011	33	[ESCAPE]	75	4B	1001011	113	K	123	7B	1111011	173	{
28	1C	11100	34	[FILE SEPARATOR]	76	4C	1001100	114	L	124	7C	1111100	174		
29	1D	11101	35	[GROUP SEPARATOR]	77	4D	1001101	115	M	125	7D	1111101	175	}
30	1E	11110	36	[RECORD SEPARATOR]	78	4E	1001110	116	N	126	7E	1111110	176	~
31	1F	11111	37	[UNIT SEPARATOR]	79	4F	1001111	117	O	127	7F	1111111	177	[DEL]
32	20	100000	40	[SPACE]	80	50	1010000	120	P					
33	21	100001	41	!	81	51	1010001	121	Q					
34	22	100010	42	"	82	52	1010010	122	R					
35	23	100011	43	#	83	53	1010011	123	S					
36	24	100100	44	$	84	54	1010100	124	T					
37	25	100101	45	%	85	55	1010101	125	U					
38	26	100110	46	&	86	56	1010110	126	V					
39	27	100111	47	'	87	57	1010111	127	W					
40	28	101000	50	(	88	58	1011000	130	X					
41	29	101001	51	)	89	59	1011001	131	Y					
42	2A	101010	52	*	90	5A	1011010	132	Z					
43	2B	101011	53	+	91	5B	1011011	133	[					
44	2C	101100	54	,	92	5C	1011100	134	\					
45	2D	101101	55	-	93	5D	1011101	135	]					
46	2E	101110	56	.	94	5E	1011110	136	^					
47	2F	101111	57	/	95	5F	1011111	137	_					

RAND Tablet

The RAND Corporation was founded in 1946 as a research project backed by the US Army Air Forces. In 1948, it became a nonprofit research and policy institute covering a wide variety of multidisciplinary topics for both government and nongovernment clients. In 1964, the company's research team created the first digitizing tablet: a flat device that was placed on a table that could detect and capture the movements of a pen-like object called a *stylus*, allowing a user to easily enter drawings, measurements, or even their signature into a computer.

Called the *RAND Tablet*, the device measured 24¼ inches wide, 20¼ inches deep, and just 1 inch high (approximately 60 × 50 × 2½ centimeters). It had a pen-like instrument connected by a wire to the tablet's base with which the user could draw or write freehand on its horizontal surface. As users made contact with the tablet's surface using the stylus, the text or images appeared in real time on a connected display, giving the user a sensory experience that felt as if they were somehow drawing directly on the screen.

The tablet was part of a study RAND conducted that examined how a person and a computer could more efficiently interact and exchange information. In this case, the tablet demonstrated how communication could occur by leveraging a human's physical dexterity and the kinetic movements used to express oneself in written form. RAND described the tablet as a "live pad of paper."

The RAND Tablet was a seminal example of reimagining the traditional shape and interface of a computer—a large machine with a screen and a keyboard—into a configuration designed to function as an analog communication device with a natural means for self-expression. The graphic elements of the device were informed by the groundbreaking research that went into Ivan Sutherland's Sketchpad, which is also credited with contributing to the development of the first computer-aided design (CAD) software.

Because of their flexibility and familiarity, tablets became a popular method for graphical computer input. The RAND Tablet is also the ancestor of signature pads used by banks and even the tiny graphical digitizer on the screen of the PalmPilot® and the modern smartphone.

SEE ALSO Sketchpad (1963), Touchscreen (1965), AutoCAD (1982), PalmPilot (1997)

Tom Ellis, one of the inventors of the RAND tablet, with a pen in his hand, writing on the tablet.

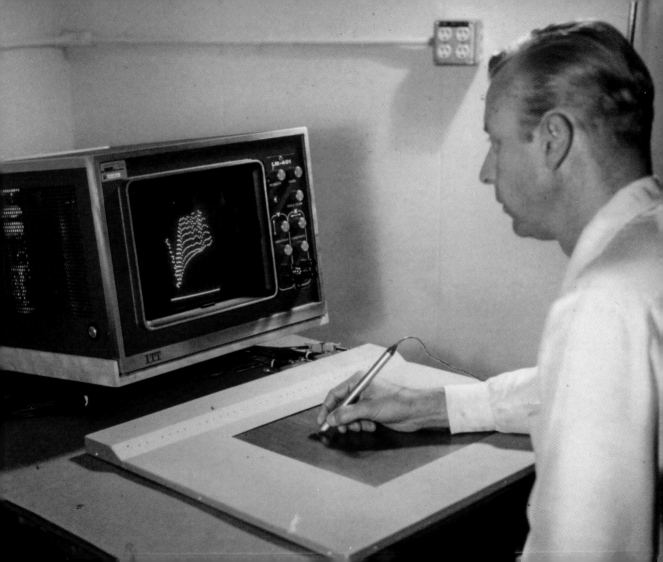

Teletype Model 33 ASR

Press a key on a teleprinter and a letter would appear on a printout and be sent down a wire at 110 bits per second; another wire would receive characters from the remote computer and print them in lines across a long paper roll. Teleprinters existed before computers: they were used to send telegrams, because typing was faster than keying out a message in Morse code. But with the dawn of interactive computing, the dominant purpose of teleprinters soon became interacting with computers.

The letters *ASR* stood for "automatic send and receive," which described the purpose of the paper tape reader and punch located to the left of the keyboard. For example, a student could use the keyboard to type a program into a computer, and type *LIST* to print the program on the paper. To make a permanent copy of a program, the student could press a clear button that enabled the punch and type *LIST*; as the program printed, a copy would be punched to paper tape. To load the program, the student would then feed the tape into the reader and press another button: the entire tape would be read in, as if the student were typing the program on the ASR's keyboard.

In 1964, Model 33 from Teletype Corporation was the first teleprinter to implement the newly adopted ASCII. The machines soon became common fixtures in computer rooms all over the world, where they remained until the late 1970s, replaced by cathode-ray tube terminals (CRTs) and line printers. On the surplus market, a flood of Model 33 ASRs were repurposed as printers for hobbyists.

Teletype Corporation produced more than 600,000 Model 33s. Their spirit lives on today in the UNIX® operating system, which uses "/dev/tty" as the name of the programmer's console—a "teletype device" that faithfully implements the Model 33's control codes for *carriage return*, *line feed*, *back space*, and *bell*.

SEE ALSO ASCII (1963), UNIX (1969)

The ASR-33 Teletype, on display at the Musée Bolo, at the École Polytechnique Fédérale in Lausanne, Switzerland.

IBM System/360

Gene Amdahl (1922–2015), **Fred Brooks** (b. 1931)

In 1950, IBM had made just four different kinds of electronic computers—and two of them were prototypes. Ten years later, IBM was manufacturing more than a dozen different computer families, with more than 10,000 systems delivered to customers.

IBM defined the catastrophe of success: it had five separate product lines, each with its own design, instruction set, and operating system. Hardware and software created for one line could not run on any of the others. IBM's solution was a radical, bet-the-company gamble called *System/360*. The idea was to create a single computer architecture that would allow software to run across the whole range of IBM computer systems, from the smallest to the largest.

Work on System/360 started in secret in 1959 and involved IBM laboratories and plants throughout the world. Fred Brooks was the project manager; Gene Amdahl was the project's chief architect. They oversaw teams that created six different-but-compatible computer lines that could support 19 combinations of speed and memory, with the fastest running roughly 50 times faster than the slowest. It had no tubes: System/360 was based entirely on the company's integrated circuits. Memory size ranged from 8,000 to 8,000,000 characters of magnetic core.

System/360 was announced on April 7, 1964, simultaneously in 165 cities: more than 100,000 people attended. The first customer system shipped in 1965.

Originally estimated to cost just $675 million, the company ending up spending $750 million on engineering alone and then another $4.5 billion on factories, equipment, and the machines themselves, which were typically rented to customers at a cost of between $5,330 per month for a System 30 to $115,000 per month for the largest systems, called *mainframes*. Just one year after the first system shipped, System/360 had generated $1 billion in pretax profit; annual profits more than doubled by 1970, when IBM replaced System/360 with upward-compatible System/370.

Brooks was awarded the 1999 A.M. Turing award for his work "on computer architecture, operating systems, and software engineering."

SEE ALSO *The Mythical Man-Month* (1975)

The IBM System/360 Model 30 computer. The first Model 30 was purchased by McDonnell Aircraft Corporation and was the least expensive of the System/360 models shipped in 1965.

BASIC Computer Language

John Kemeny (1926–1992), **Thomas Kurtz** (b. 1928)

Dartmouth professors John Kemeny and Thomas Kurtz created the Beginner's All-purpose Symbolic Instruction Code (BASIC) so that ordinary students—not computer nerds—could learn how to program and use computers to solve challenging problems.

Unlike other programming languages at the time, Dartmouth's BASIC used simple, easy-to-understand commands. For example, commands typed on the teleprinter would be immediately executed. So typing *PRINT* 2+2 would print 4. To create a program, the student would simply preface each line with a line number. So typing *10 PRINT* 2+2 would create a program that, when run (by typing *RUN*, no less), would cause the number 4 to be printed when the computer reached line 10. BASIC was designed to be so simple that students would find programs self-evident just by reading them.

By 1968, BASIC was being used in more than 80 courses at Dartmouth—including courses as diverse as Latin, statistics, and psychology. And it had spread to four other colleges and 23 secondary schools, for a total of more than 8,000 users.

BASIC soon escaped Dartmouth. Versions were written for computers manufactured by Digital, Data General®, and Hewlett-Packard (HP®). In 1973, David Ahl (b. 1939), an engineer at Digital, convinced his company to publish a book called *101 BASIC Computer Games*. In 1976, *Dr. Dobb's Journal* published a listing of Tiny BASIC, a version of BASIC that could run on a computer with less than 3 kilobytes of RAM—ideal for the Altair 8800 computer. The free Tiny BASIC program was soon joined by Altair BASIC, a commercial program written by Bill Gates and Paul Allen. Altair BASIC sold for a whopping $150. The steep price encouraged many hobbyists to obtain and run unauthorized copies, inadvertently creating the first widespread case of software piracy.

For the following decade, practically every microcomputer sold on the planet came with some version of BASIC installed in read-only memory (ROM). (ROM is like RAM, except the contents can't be changed; it's frequently used for software distributed by the computer manufacturer.) For millions of people, it was their first programming language. Today, Visual Basic for Applications is built into Microsoft Access, Excel, Word, Outlook, and PowerPoint.

SEE ALSO First Personal Computer (1974), *Dr. Dobb's Journal* (1976), Microsoft and the Clones (1982)

BASIC textbook created by General Electric.

Now YOU can program a computer in

BASIC

A new dimension in data processing

GENERAL ⊗ ELECTRIC

First Liquid-Crystal Display

George Heilmeier (1936–2014)

The liquid-crystal state of matter was discovered at the Karl-Ferdinands-Universität in Prague in 1888 by Friedrich Reinitzer and further investigated in the 1900s and 1930s. But otherwise, the curious ability of some liquids to have crystalline properties—specifically the ability to change the polarization of light—remained a chemical curiosity, and not the subject of practical exploration or exploitation.

Then, in the early 1960s, engineers at RCA's David Sarnoff Laboratories in Princeton, New Jersey, were looking for a new kind of display that could replace the vacuum tubes used in color TVs. RCA® physical chemist Richard Williams turned his attention to liquid crystals and discovered that certain chemicals, heated to 243° Fahrenheit (117° Celsius), would change their appearance from transparent to opaque when placed in a high-voltage electric field.

Williams soon gave up on the idea of using these crystals for a display, but a young engineer at the company named George Heilmeier saw their promise. Over the following years, Heilmeier and his newly formed research group discovered materials that exhibited the liquid-crystal effect at room temperature and when exposed to very small electric fields.

Heilmeier's group built the first liquid-crystal display in 1965, using a bit of liquid crystal sandwiched between a polarizer and a reflective surface, separately controlling the seven segments of a single-digit display; next, the group created an LCD that displayed a tiny TV test pattern.

In 1971, RCA sold its computing division to Sperry Rand, taking a $490 million write-off. Doubtful that liquid-crystal displays would ever make money, RCA sold that technology in 1976 to Timex® for its digital watches. Today, LCD screens are used for desktop computers, portable phones, televisions, projectors, and more.

Heilmeier left RCA for a fellowship at the White House, and in 1975 he became director of the Defense Advanced Research Projects Agency (DARPA), where he continued to oversee the development of technology on behalf of the US government.

SEE ALSO First LED (1927), E Ink (1997)

Today, liquid-crystal displays (LCDs) are commonly used in television screens.

Fiber Optics

Narinder Singh Kapany (b. 1926), **Jun-ichi Nishizawa** (b. 1926),
Manfred Börner (1929–1996), **Robert Maurer** (b. 1924),
Donald Keck (b. 1941), **Peter Schultz** (b. 1942),
Frank Zimar (dates unavailable)

Fiber-optic transmission takes information in the form of 1s and 0s, encodes it as pulses of light, and shoots it through a tiny cylindrical glass pipe no wider than a human hair. After moving through the glass at the speed of light, the pulses are converted back into their original electronic form. For computer data, that would be 1s and 0s. Voice must go through an additional step of being digitized to 1s and 0s before being sent, and being reconstructed into analog waves on the receiving end.

The idea of bending or controlling light and using it to solve everyday problems, including faster transmission of information, was not new. Early examples of optical communication include the heliograph from the 1800s, which used mirrors to produce flashes of sunlight coded for letters or numbers. Later in the century, Alexander Graham Bell (1847–1922) and his assistant Thomas Augustus Watson (1854–1934) invented the photophone, which used modulated beams of light against a selenium receiver to carry spoken words.

Many contributed to the modern era of fiber-optic communications, including Indian American physicist Narinder Singh Kapany and Jun-ichi Nishizawa of Japan's Tohoku University. In 1965, Manfred Börner, a German based in Ulm, created the first working fiber-optic data transmission system. But it wasn't until the 1970s, when four scientists at Corning Glass Works®—Robert Maurer, Donald Keck, Peter Schultz, and Frank Zimar—developed a kind of glass that could carry the light from light-emitting diodes (LEDs) and semiconductor lasers over dozens of miles without significant loss of power, that the technology was mature enough to be adopted as a general purpose communication system.

Compared to wire communications such as traditional copper or a T1 line, a single fiber optic can carry more than a thousand billion times more information in the same amount of time.

SEE ALSO Digital Long Distance (1962)

Fiber-optic cables, shown here, shoot information encoded as pulses of light through tiny cylindrical glass pipes.

DENDRAL

Joshua Lederberg (1925–2008), **Bruce G. Buchanan** (dates unavailable), **Edward Feigenbaum** (b. 1936), **Carl Djerassi** (1923–2015)

DENDRAL was an early influential computer research project in the development of modern AI systems. It helped to shift the focus of AI research from developing general intelligence to creating systems tailored for specific areas. It did this by representing experts' knowledge of chemistry in a way that could be used by a computer, allowing the system of code and data to solve narrowly defined chemistry problems and draw conclusions the same way that a human expert might, thus earning it the name "expert system."

DENDRAL started in 1965, when geneticist Joshua Lederberg was looking for a computer-based research platform to further his understanding of organic compounds in support of his exobiology research—a branch of astrobiology that seeks to understand the evolution of life on other planets. Lederberg enlisted the partnerships of Stanford assistant professor Edward Feigenbaum, one of the founders of the school's computer science department, Stanford chemist Carl Djerassi, and virtuoso AI programmer Bruce Buchanan to develop a system that could suggest chemical structures and the mass spectra that might comprise them. The project unfolded over the course of approximately 15 years, evolving a program designed to model scientific reasoning and explain experimental chemistry into a system that chemists could use to generate hypotheses and, eventually, to learn new things about chemistry.

By the end it resulted in two main components—Heuristic DENDRAL and Meta-DENDRAL. Heuristic DENDRAL aggregated existing data from different sources (such as the experts' core knowledge base of chemistry) and produced the sets of chemical structures and their potentially corresponding mass spectra. Meta-DENDRAL was the learning side of the house. This program took the output of Heuristic DENDRAL and produced sets of hypotheses that might explain the correlation between chemical structures and the combinations of mass spectra that might be associated with them. For his work on DENDRAL, Edward Feigenbaum was awarded the 1994 A.M. Turing Award.

SEE ALSO AI Medical Diagnosis (1975)

Joshua Lederberg in front of exobiology equipment at Stanford.

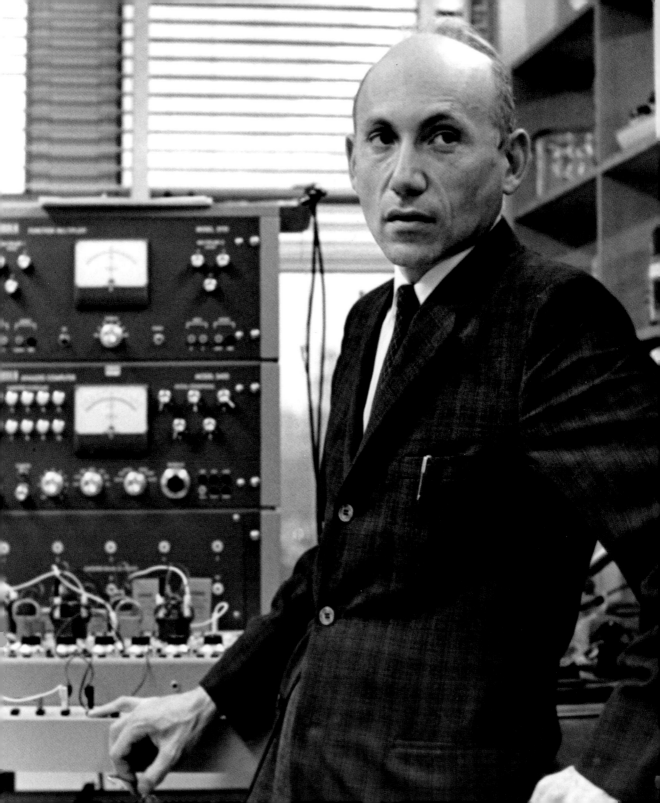

ELIZA

Joseph Weizenbaum (1923–2008)

ELIZA, named after the character Eliza Doolittle in the play *Pygmalion* by George Bernard Shaw, was the first program that could converse in English. It would take a line of text that a human being typed on a teleprinter, transform the text—for example, changing *you* to *I* and *me* to *you*— and then send back the text to the human typist. Just like a parrot, the computer had no idea what it was saying.

MIT professor Joseph Weizenbaum modeled his "language analysis program" on Rogerian psychotherapy, a type of nondirective therapy that does not involve interpretation of a client's statements. Following this technique, ELIZA turned "I am unhappy" into "DO YOU THINK COMING HERE WILL HELP YOU NOT TO BE UNHAPPY?" Most of these transformations were triggered by simple keywords; when the program was stumped, it displayed a preprogrammed question.

Weizenbaum wasn't prepared for what happened next. The program became immensely popular. People started conversing with the program as if it were intelligent— even people who knew better. Weizenbaum's secretary, who knew that he had written the program, asked him to leave the room so that she could use the system without being watched. She was horrified when she discovered that Weizenbaum had logs of her interactions with the computer. A visitor to MIT who found it left running thought that he was using the teleprinter to talk to another professor—and then got angry when the "professor" answered his questions with other questions. Later, some people claimed that a computerized psychotherapist might be better in some cases than human ones— after all, they were always available and didn't bill by the hour.

At its core, ELIZA knew nothing about the language that it used to communicate. In the final analysis, that lack of understanding didn't matter.

"A large part of whatever elegance may be credited to ELIZA lies in the fact that ELIZA maintains the illusion of understanding with so little machinery," Weizenbaum wrote in a 1966 article about the program. The power of the program was not its ability to understand what a person typed, but to conceal its lack of understanding.

SEE ALSO The Turing Test (1951), AI Medical Diagnosis (1975)

ELIZA *was named after the character Eliza Doolittle, who was given speech lessons in George Bernard Shaw's* Pygmalion *and later its musical adaptation,* My Fair Lady. *Here Eliza Doolittle is portrayed by Mrs. Patrick Campbell in 1914.*

Touchscreen

E. A. Johnson (dates unavailable), **Dr. Sam Hurst** (1927–2010),
Nimish Mehta (dates unavailable)

In 1955, the MIT Whirlwind project created a light pen that let a user indicate a point on a computer screen. But being able to point more naturally with your own finger? That was the invention of E. A. Johnson at the Royal Radar Establishment (RRE).

Johnson was a researcher working on the RRE's air traffic control system. In 1965, he published an article in *Electronic Letters* titled "Touch Display—A Novel Input/ Output Device for Computers" that described a touch-sensitive screen. Two years later, he expanded on the idea in an article in *Ergonomics*, showing how a touchscreen could be used to interact with graphs and pictures.

Johnson invented what is called today a *capacitive touchscreen*. It uses a layer in the screen to store an electric charge. When a user touches the screen, some of the charge is transferred to the user. This sends a signal to the device's operating system where the touch occurred on the screen.

A few years later, Sam Hurst, a researcher at Oak Ridge National Laboratory in Tennessee, invented a similar kind of transparent, touch-sensitive film, but one based on changes in resistance that result when two transparent layers of material are pressed together. Unlike capacitive screens, touchscreens can be used with a stylus or a finger. They are generally cheaper but less accurate than capacitive screens. A third kind of screen relies on changes to ultrasonic waves sent over the surface of the touchscreen to measure where the screen is touched.

Early touchscreens could sense only one touch at a time. Nimish Mehta developed the first multitouch device at the University of Toronto in 1982. The technology continued to develop along with potential commercial applications, finally reaching the masses in the 2000s. During this period, it became a popular tool for design collaboration.

Today touchscreens are the primary way that people interact with smartphones and tablets, which in 2016 became the primary way that the world's population accesses information on the World Wide Web.

SEE ALSO Trackball (1946), The Mouse (1967), PalmPilot (1997)

Touchscreens came to be used with many applications. Pictured here is an operator using a touchscreen on the shared computer-based education system PLATO (Programmed Logic Automated Teaching Operation).

Star Trek Premieres

Gene Roddenberry (1921–1991), **William Shatner** (b. 1931), **Leonard Nimoy** (1931–2015), **Nichelle Nichols** (b. 1932)

On September 8, 1966, the National Broadcasting Company (NBC®) premiered *Star Trek*, a new science-fiction television show by Gene Roddenberry. The show ran from 1966 to 1969 and would grow into the most influential science-fiction franchise in history, eventually spinning off 7 different TV shows, 13 movies, and more contributions to popular culture—*"Beam me up, Scotty!"*—than can be accurately measured.

At a time when war and fears of out-of-control technology were everyday fears, *Star Trek* showcased a vision of an alternative future in which humanity had evolved and overcome poverty and material needs, harnessing technology to overcome myriad challenges.

Computers and robots play important roles in many *Star Trek* episodes. The *Enterprise* is equipped with a ship's computer that maintains massive information stores, offers universal accessibility throughout the ship, and responds to human commands by voice. Meanwhile, the crew encounters societies that make war by computer and civilizations ruled by computers, as well as robots that devour planets, send people back through time, or manifest in humanoid form.

The original series documented many firsts, including showing the first women in positions of authority, the first televised interracial kiss—Captain Kirk (William Shatner) and Lieutenant Uhura (Nichelle Nichols)—and the first series of storylines that used the veil of science fiction to draw analogies to current events involving war and discrimination that TV censors of the day would have otherwise banned. Alien beings and a multiethnic bridge crew, including the alien Mr. Spock (Leonard Nimoy), were often depicted working together to solve problems, drawing subtle comparisons to moral and ethical issues about the Vietnam War and women's rights that were in the public's consciousness at the time.

To this day, *Star Trek*'s vision of the future—"To boldly go where no man has gone before"—continues to inspire progress among scientists, engineers, artists, teachers, philosophers, and others.

SEE ALSO *Artificial Intelligence* Coined (1955), HAL 9000 Computer (1968)

Dr. McCoy (DeForest Kelley), Captain Kirk (William Shatner), and Mr. Spock (Leonard Nimoy) on the transporter platform aboard Star Trek's USS Enterprise.

Dynamic RAM

Robert H. Dennard (b. 1932)

From the beginning, small memories limited what computers could do. Called *random access memory* (RAM) because any location could be read or written at any time, memory was fantastically expensive. And because programmers frequently did not have enough memory to work with, they had to split their programs and data into segments, copying one segment into memory, processing it, and then saving the results on tape.

Enter semiconductors. Smaller and cheaper than core memory, semiconductors were the obvious next technology to use for RAM. IBM assigned electrical engineer Robert Dennard the task of designing its next-generation electronic memory system. Dennard's original approach used six transistors to create an electronic switch, called a *flip-flop*, to store each bit. But halfway into the project in 1966, Dennard realized that he could make the devices even cheaper by storing each bit in a capacitor—a device that stores charge—and using a single transistor to alternatively store the charge and read it back when the data was needed. However, there was a catch: capacitors leak their charge away. The solution was to refresh the bit, perhaps a thousand times every second, by reading the bit and writing it back. Because the charge would be constantly in motion, Dennard described his invention as *dynamic random access memory* (DRAM).

IBM was committed to finishing the design that used six transistors—a design that today is called *static random access memory* (SRAM)—so Dennard pursued the DRAM as a side project. Finally, in 1967, IBM filed for a patent for DRAM, which was granted in 1968.

The company that ended up commercializing DRAM wasn't IBM, but Intel, using a less-efficient three-transistor design licensed from Honeywell, an American technology company. Released in 1970, the Intel 1103 was the first commercially available DRAM. It stored 1024 bits and offered both a better price and better performance than magnetic core memories.

Since then, the storage of DRAM has increased as the size of transistors has shrunk. By the early 1990s, manufacturers were putting a million bits of storage on a chip; by the 2000s, a billion bits. Today the rate of progress has slowed, and modern DRAM chips can hold "only" 4 billion to 32 billion bits.

SEE ALSO Atanasoff-Berry Computer (1942), Core Memory (1951)

Dynamic RAM memory chips assembled on a dual inline memory module (DIMM), a format widely used in modern laptops.

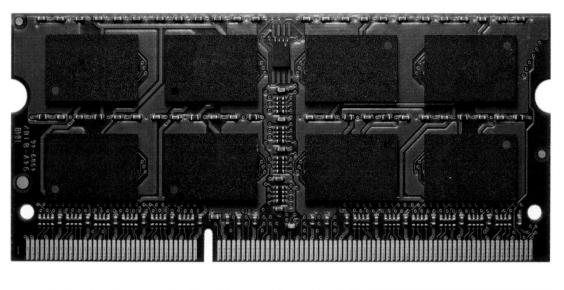

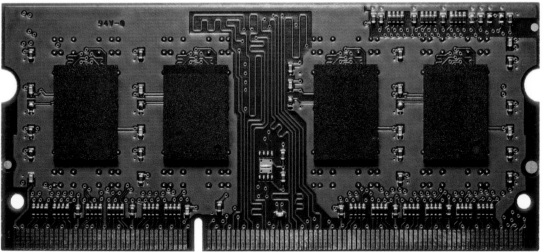

Object-Oriented Programming

Ole-Johan Dahl (1931–2002), **Kristen Nygaard** (1926–2002)

The first programs did important but repetitive tasks, such as printing artillery tables, performing calculations for nuclear weapons, and cracking codes. These programs consisted of loops that executed the same mathematical functions over and over, each time with slightly different parameters. Early business computers performed similar iterative computations on business ledgers and other records, repeatedly reading data from disk, processing it, and saving the results.

At the Norwegian Computing Center, professors Ole-Johan Dahl and Kristen Nygaard wanted to use their computer to simulate physical systems, in particular ship simulations, and here they found the programming languages developed to enable simple, repetitive tasks to be lacking. So they developed a new way to program and a new computer language, which they called *SIMULA 67*.

The key idea of SIMULA was for data representing physical objects to be bundled together with the computer code for acting on that data. For example, a simulation of traffic might have a data type called *CAR* that might have variables for the car's location and speed. A special function might handle the car's behavior when it encountered a traffic light. SIMULA refers to each of these data types as a "class." A different class called *TRUCK* might represent trucks. Another key idea was inheritance, which allows classes with shared characteristics to be arranged in a hierarchy. So a TRUCK and a CAR might both inherit from an abstract class called a *VEHICLE*, which itself might inherit from another abstract class called an *OBJECT*.

Today, SIMULA's style of programming is called *object-oriented programming*, and SIMULA 67 is recognized as the first object-oriented language. It turns out that the ideas of SIMULA were good for a whole lot more than just writing simulations: practically every modern computer language is object-oriented, including C++, Java, Python, and Go, and today object-oriented programming is the dominant way that software is written.

SEE ALSO Programming for Children (1967), C Programming Language (1972)

Programmers create object-oriented programs by designing classes of objects that represent physical objects, processes, or arrangements of data. They then connect the objects with code.

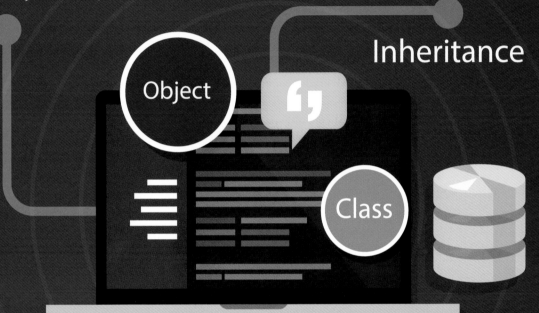

Polymorphism

Object

Inheritance

Class

Encapsulation

Abstraction

First Cash Machine

James Goodfellow (b. 1937), **John Shepherd-Barron** (1925–2010),
Donald Wetzel (b. 1921), **Luther George Simjian** (1905–1997)

There was no one inventor of the automated teller machine (ATM), although there have been plenty of people who have tried to lay claim to it. Rather, it was the result of a series of innovations by multiple people over years, leading to the pivotal 1967 unveiling of two competing but different machines in the UK that were introduced within a month of each other.

James Goodfellow's machine, introduced by Westminster Bank in London, used a plastic card and a personal identification number (PIN) for customer verification and access. The John Shepherd-Barron machine, introduced by Barclays, used mildly radioactive checks infused with carbon-14 that the ATM relied on to match against a customer's identification number. While Shepherd-Barron's machine beat Goodfellow's to market by a month (and thus received much of the "first-of" glory), it was Goodfellow's PIN design that stuck and ended up seeing mass commercialization and licensing by manufacturers. He also had the first patent for an ATM that used a PIN for authentication.

In the United States, the ATM was pioneered by Donald Wetzel, who worked for a technology company called *Docutel*. The first US ATM was installed by Chemical Bank at its Rockville Centre, New York, branch in 1969. The bank's advertisement for its new darling read, "On September 2, our bank will open at 9:00 and never close again."

As the computer technologies that constituted the ATM and the customer features it offered evolved—magnetic stripes, improved security, free-standing machines, the ability to accept deposits as well as dole out cash, and so on—the definition of what exactly defined an ATM frequently depended on who was making the claim, especially for people claiming to be "first." For example, in 1939, the Armenian American Luther George Simjian had an idea for a hole-in-the-wall machine that would enable financial transactions. He would eventually create the Bankograph, which let people make utility-bill payments and get receipts. Sadly, the Bankograph was a failure. "It seems the only people using the machines were a small number of prostitutes and gamblers who didn't want to deal with tellers face to face," wrote Simjian in his autobiography.

SEE ALSO Data Encryption Standard (1974), Digital Money (1990), Bitcoin (2008)

British actor Reg Varney poses at the unveiling of the world's first ATM at Barclays Bank in Enfield, Middlesex, just north of London, on June 27, 1967.

Head-Mounted Display

Ivan E. Sutherland (b. 1938)

Ivan Sutherland's head-mounted display (HMD), better known as "The Sword of Damocles," was an early immersive video display and the first to use a computer instead of a camera to simulate a physical space. It consisted of six parts: a general-purpose computer, matrix multiplier, vector generator, special electronics called a *clipping divider* that eliminated lines hidden behind the observer or outside the field of view, headset, and head-position sensor. It had to be suspended from the ceiling by a cable because it was too heavy for a person to comfortably support on his or her own. Like the fabled sword that dangled above Damocles as he sat in his ruler's throne, you did not want this thing to fall on your head.

There was a screen in front of the eyes that presented a slightly different image to each eye, which produced a sense of depth. Sensors told the computer when you moved your head, and the computer updated the screen accordingly.

The idea for the display occurred to Sutherland during a visit to Bell Helicopter in Texas. Bell was testing how to mount an infrared camera on the bottom of a helicopter for night landings. The idea was to have the camera's position follow the position of the pilot's head; this way, if the pilot looked to the right, a head-mounted display would show what was to the right. During the visit, Sutherland saw an operator wearing the helmet watch two people on the roof throwing a ball to each other. Suddenly one of the people tossed the ball in the direction of the camera. The person wearing the helmet jumped back as if the ball could hit him: he had become immersed in the reality of the rooftop. It was in that moment that Sutherland hit upon the idea of getting rid of the camera, and having the image for each eye generated by a computer.

The key concept behind Sutherland's device was to enable users to observe an image that moved in accordance with their own head movements. The hardware to achieve this did not exist, so Sutherland and his team designed and built what they needed. It would be decades before the technology would be readily accessible.

SEE ALSO Sketchpad (1963), RAND Tablet (1964), VPL Research, Inc. (1984)

Ivan Sutherland's head-mounted display allowed a person to change the direction of their gaze and even move around the room; when the person moved, each eye's display updated accordingly.

Programming for Children

Seymour Papert (1928–2016), **Wally Feurzeig** (1927–2013), **Cynthia Solomon** (dates unavailable)

Seymour Papert, Wally Feurzeig, and Cynthia Solomon thought that programming could be a powerful tool for teaching children about thinking, planning, and abstract thought. So while working at the research company Bolt, Beranek, and Newman (now BBN Technologies®) in Cambridge, Massachusetts, they designed a computer language for children, with a few commands that could be combined to enable complex tasks. Programming in this language, called *Logo*, let children create instructions for the computer much in the way that they might string together many simple words to convey complicated thoughts. Some of the early programs that children wrote were math quizzes and chatbots.

Although programs in early versions of Logo could do little more than communicate with the user through text, the inventors soon made it possible for the machine to control a mechanical robot called a *turtle*, which the program could instruct to move forward, move backward, or turn. The original turtle was a robot named Irving; it crawled around on a piece of paper in response to commands from a machine nearby. Irving had a pen that it could drop and drag to make a drawing. Eventually, Irving became a virtual turtle on a computer screen, making it possible to share Logo with many more people.

Children were taught that the turtle understood certain words and numbers that they could use to make the turtle move around. To draw a square, for example, a child would learn through trial and error to type *FORWARD 40* (or some other number) to go forward 40 steps, then type the command *RIGHT 90* to turn to the right, and then *FORWARD 40* and *RIGHT 90* three more times to complete the shape. The child would get immediate visual feedback on the screen. Kids soon learned that they could "teach" the turtle a new word such as *SQUARE* and associate individual steps with that word, so that they did not have to type multiple commands to create an object. Often referred to as "turtle graphics," teaching games and different flavors of Logo evolved over time and proved that, yes, children can program.

SEE ALSO BASIC Computer Language (1964), Object-Oriented Programming (1967), Nintendo Entertainment System (1983)

Seymour Papert shows two children the inner workings of a robotic turtle controlled by programs written in the Logo programming language.

The Mouse

Douglas C. Engelbart (1925–2013), **Bill English** (dates unavailable)

Douglas Engelbart, a pioneer in the field of human–computer interaction, is generally credited with inventing the mouse in the early 1960s, producing the first prototype in 1964 and introducing it to the public in 1968 during what has come to be known as the "Mother of All Demos." He patented it with Bill English, a coworker at the Stanford Research Institute (SRI). Engelbart's idea contributed to the already-evolving pursuit of corresponding a user's hand motion to a specific location in two-dimensional space as visualized on a computer screen. In the 1967 patent, Engelbart described the mouse as an "X-Y position indicator for a display system."

The early origins of the mouse can be seen in Ralph Benjamin's roller ball, recognized as the origin of the modern trackball. Unlike the trackball, which was stationary and required users to move their hands and fingers over it, the mouse necessitated the entire device to be repositioned. Engelbart's mouse consisted of a wooden box with three buttons on top and two wheels underneath positioned at right angles to control horizontal and vertical movement. As the wheels rolled, the distance and location were captured as binary code in the computer, which was translated into visual output on the screen. Engelbart invented the mouse to control the navigation on a groundbreaking computer collaboration system he designed called the *oN-Line System* (NLS).

Xerox Corporation's Palo Alto Research Center (PARC) further evolved the mouse after Bill English left SRI and went to work for PARC in 1971. Xerox was the first to commercialize the mouse, selling it as part of the Star 8010 Dandelion computer system. It was Steve Jobs, however, who finally brought the mouse mass commercial success by simplifying its design and packaging it with the famous Apple Macintosh® computer.

Engelbart's mouse was part of a larger vision he had to make computers more interactive and accessible to everyone.

SEE ALSO "As We May Think" (1945), Trackball (1946), Mother of All Demos (1968), Xerox Alto (1973), Macintosh (1984)

Replica of Engelbart's mouse prototype, built by engineer Bill English, 1964.

Carterfone Decision

Thomas Carter (1924–1991)

When a Texas oil worker wanted to communicate from a remote location such as an oil field, he used a long-distance radio to talk to other people because there were no telephone lines in such isolated places. Meanwhile, family, friends, and colleagues had phones, but they didn't have radios.

Enter Thomas Carter, Texas entrepreneur and inventor who created the Carterfone, a device that linked two-way radios to the telephone network, enabling those in far-flung places to stay in contact with others.

The Carterfone worked by acoustically (as opposed to electrically) connecting a radio to the public telephone network. Once a station operator was in contact with parties at both ends—the radio operator and the person on the other end of the telephone call—the operator would place the telephone handset into a cradle, which aligned a small speaker with the handset's microphone and a microphone with the handset's speaker. A voice-operated switch in the Carterfone would then automatically turn on the radio transmitter when the person on the telephone spoke. When the person stopped talking, the Carterfone would stop transmitting. The device's microphone would then pick up any sound received by the radio receiver and send it down the phone line. This enabled both parties to hear each other and converse.

Even though the Carterfone did not electrically connect to the phone system, it violated the phone company's rules. In 1968, AT&T controlled the US telecommunications system, with Western Electric® as the manufacturer producing all the equipment. No one owned their phones: they were leased. AT&T's rules prohibited users from attaching third-party gear to its network. So Carter filed suit against AT&T, and—to the surprise of many—the Federal Communications Commission (FCC) ruled in Carter's favor.

The FCC's landmark decision is a reminder that regulation is sometimes needed to protect and enable the innovations that lead to technology advances. Without the FCC's Carterfone ruling, innovations such as the fax machine, answering machine, and modem would not have had the regulatory space to enter the market and evolve, paving the way for what would become the internet and the dynamic communications ecosystem that exists today.

SEE ALSO The Bell 101 Modem (1958), Telebit Modems Break 9600 bps (1984)

The original Carterfone, which connected mobile radios to the telephone network.

The Original "Carterfone"

This original Carterfone, manufactured by Carter Electronics in 1959, served a need for mobile radio users to interconnect with the public telephone network. Use of the Carterfone was challenged by the telephone companies in 1966, and a lengthy struggle began that ultimately led to the Federal Communications Commission.

On June 26, 1968, the FCC handed down the landmark Carterfone Decision. The resolution of Tom Carter's struggle for acceptance of the concept of interconnection permitted the creation of a multi-billion dollar industry that today serves all areas of communications needs—data, voice, message. The historic Carterfone Decision allowed an open, competitive market to exist for communications equipment and facilities to the benefit of the communications user.

This original Carterfone is one of the few remaining devices in existence, and has been preserved to commemorate the historic legal milestone it represents.

Carterfone Communications Corporation
Dallas, Texas

Software Engineering

Peter Naur (1928–2016), **Brian Randell** (b. 1936)

Few of computing's pioneers anticipated how complex software would become, in part because they were mostly building hardware. There is a logic to computer hardware, no matter whether it is built with relays, tubes, or transistors. Hardware executes one instruction at a time, and the operation of each instruction can be individually analyzed and proven correct. Hardware design errors were relatively easy to detect and fix.

Software was entirely different because it is dynamic. The programmer needs not only to specify the correct instructions, but also to specify them in the correct order, and that order can be different with different data. Some programmers can catch some errors when they are writing code, but that process is slow and error prone itself, because many errors aren't evident until the program runs.

Next, there is the issue of architecture and design. There are many different ways to code a program, each with a different set of tradeoffs. Sometimes the simplest, most elegant program is the most efficient. Other times, it is highly inefficient.

By the mid-1960s, there was growing consensus that the whole software thing was quickly spinning out of control. The problem wasn't just bugs—it was that experts couldn't predict, in advance, how long it would take to write a program that performed new tasks that had not been previously computerized. Another problem was that software, once written, couldn't be easily extended to perform new features.

In October 1968, the North American Treaty Organization (NATO) Science Committee held a weeklong Working Conference on Software Engineering in Garmisch, Germany, to address the growing software crisis. Fifty experts from around the world attended. The conference is widely credited with popularizing the phrase *software engineering* and establishing it as a subject of serious academic study.

The resulting January 1969 report, edited by computer scientists Peter Naur and Brian Randell, identified five key areas where work was needed: the relationship between software and computer hardware, software design, software production, software distribution, and software service.

In the years since, many organizations have identified the need for software engineering, yet few have mastered the ability to do it well.

SEE ALSO Church-Turing Thesis (1936), Actual Bug Found (1947)

This relationship-entity diagram shows how different tables in an online-portfolio management system are structured and queried.

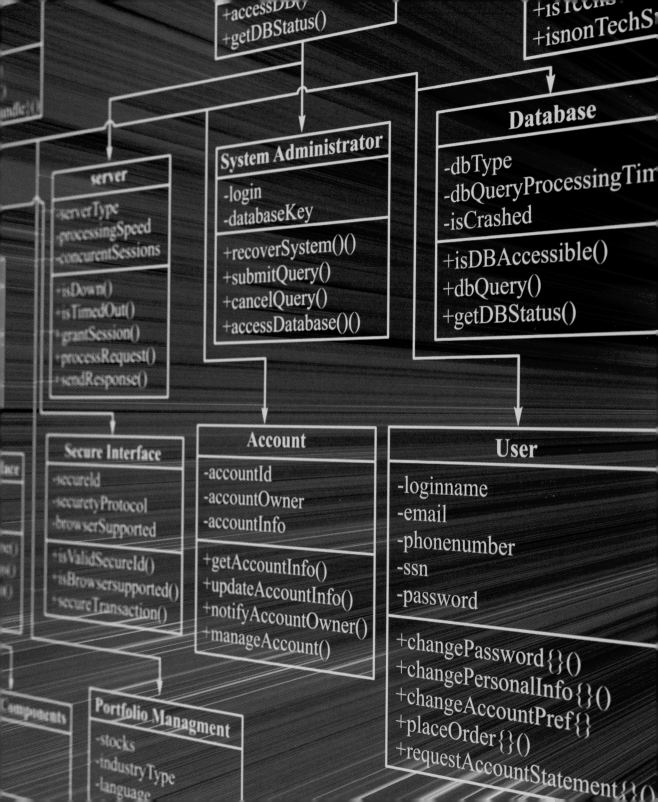

HAL 9000 Computer

Stanley Kubrick (1928–1999), **Arthur C. Clarke** (1917–2008),
Douglas Rain (b. 1928)

The HAL 9000 (**H**euristically programmed **AL**gorithmic computer) is a self-aware,
artificially intelligent computer that controls the *Discovery* spacecraft in the movie *2001: A
Space Odyssey*. HAL accompanies a six-member human crew on an interplanetary mission
to Jupiter, four of whom are in suspended animation during the movie and never wake up.

In the movie, astronauts David Bowman and Frank Poole discuss the idea of
shutting HAL off when he appears to make a mistake. "No 9000 computer has ever
made a mistake or distorted information. We are, by all practical definition of the words,
foolproof and incapable of error," HAL had previously stated in a television interview.
So if HAL *was* making mistakes, the astronauts surmised, then the machine was no
longer reliable and had to be shut off.

Of course, as a sentient, self-preserving being programmed to continue the mission
if the crew becomes incapacitated, HAL decides that it is the humans who are in error,
and decides therefore to kill them.

Visually, HAL is represented in the movie as a red television camera. HAL's skills
include human-like reasoning and conversation, artificial vision, facial recognition,
emotional interpretation, opinions on highly subjective topics including art appreciation,
and understanding of human interactions. Later, the surviving crew learns that HAL
can lip-read.

Directed by Stanley Kubrick and written by science-fiction author Arthur C. Clarke,
with technical support by AI pioneer Marvin Minsky, many film critics consider *2001: A
Space Odyssey* to be among the best movies ever made. The movie broke new cinematic
ground with its cutting-edge special effects and the authenticity of its space exploration
narrative. HAL "was the first computer to become a famous personality and become
part of public mythology," said Clarke in a 1992 interview with the *Chicago Tribune*.
The movie provoked questions about where technology was heading and the potential
consequences of developing an artificial entity that could surpass human intelligence.

SEE ALSO *Rossum's Universal Robots* (1920), *Star Trek* Premieres (1966)

A movie still from 2001: A Space Odyssey, *showing HAL 9000 depicted as a red television camera.*

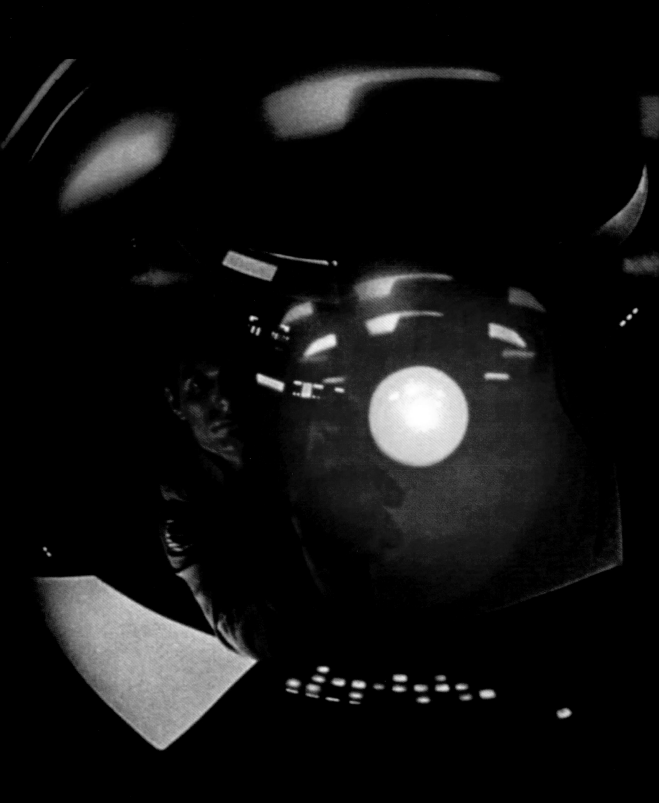

First Spacecraft Guided by Computer

Margaret Hamilton (b. 1936)

On May 25, 1961, President John F. Kennedy promised to send an American to the moon before the end of the decade. Soon after, MIT's Instrumentation Laboratory signed a contract with the National Aeronautics and Space Administration (NASA) to design and develop the flight and guidance systems.

The system that MIT produced was called the *Apollo Guidance Computer* (AGC). It had a real-time operating system and 32,768 bits of core memory, operated at 43 kilohertz (0.043 megahertz), and included integrated circuits rather than transistors. Amazingly, with a millionth the storage and computing power of a modern cell phone, the AGC got the Apollo 11 astronauts to the moon and back safely—roughly half a million miles in total.

Margaret Hamilton, a 24-year-old, led the programmers who created the onboard flight software for both the Apollo command-modules and lunar modules. Her work helped accelerate advancements for an entire industry and, in a quirky turn of events, likely saved the lives of the Apollo 8 astronauts. As the story goes, Hamilton's young daughter Lauren was playing with the command module simulator and crashed it by hitting a key that ran program P01, which was supposed to be run only before the launch. NASA would not let Margaret add code to the system to prevent this from happening in-flight, because the astronauts' extensive training was supposed to prevent such a mishap. So Hamilton added a programming note instead.

Of course, during the Apollo 8 mission, an astronaut accidentally ran program P01, wiping the computer's memory. Already aware of what caused the loss of data, Hamilton and her team spent the next nine hours solving the problem.

Today the entire AGC program is available online, showing, among other things, the programming team's sense of humor. One block of code reads "INITIALIZE LANDING RADAR . . . OFF TO SEE THE WIZARD."

SEE ALSO Actual Bug Found (1947)

The sextant of the Apollo Guidance and Navigation system was used to make precise measurements of the stars from the command module, which were then entered into the Apollo Guidance Computer.

Cyberspace Coined — and Re-Coined

Susanne Ussing (1940–1998), **Carsten Hoff** (b. 1934), **William Gibson** (b. 1948)

According to *Oxford Dictionaries Online*, the term *cyberspace* refers to "the notional environment in which communication over computer networks occurs." This definition originates from author William Gibson's use of the word in his 1982 short story, "Burning Chrome," in which he described cyberspace as "a consensual hallucination experienced daily by billions of legitimate operators, in every nation" and "a graphic representation of data abstracted from the banks of every computer in the human system."

Widespread use of the word really took off after Gibson's immensely popular techno-futuristic novel, *Neuromancer* (1984). The media seized on the term and grew it into a general expression for describing the rapid changes occurring in technology, including the phenomenon of social activity taking place in a nonphysical environment.

But while the modern definition of cyberspace is traced to 1982, the word itself made a brief appearance in 1968, when Danish visual artists Susanne Ussing and Carsten Hoff named themselves *Atelier Cyberspace* and incorporated the word into a series of physical artworks titled *Sensory Spaces*, as reported in 2015 in the Norwegian arts magazine *Kunstkritikk*.

Ussing and Hoff's inspiration for putting the words *cyber* and *space* together evolved in part from their appreciation of, and curiosity about, the work of American mathematician and philosopher Norbert Wiener and his concept of cybernetics. Wiener defined cybernetics in his 1948 book of that name as "the scientific study of control and communication in the animal and the machine." Of particular interest to Ussing and Hoff was a 1968 exhibition they saw in London titled *Cybernetic Serendipity*. Hoff described the exhibition as being about "art's potential for employing modern technology, particularly information technology." For Ussing, the definition of cyberspace was meant to convey "open-ended systems where things could grow and evolve as required," she told *Kunstkritikk*.

SEE ALSO *The Shockwave Rider* (1975)

The term cyberspace *dates back to 1968, when Danish artists Susanne Ussing and Carsten Hoff incorporated the word into a series of artworks.*

Mother of All Demos

Douglas C. Engelbart (1925–2013)

On December 9, 1968, a team at the Stanford Research Institute (SRI) publicly demonstrated a computer system that enabled knowledge sharing, content creation, and personal collaboration, the likes of which no one had ever seen before. It has since become known as the "Mother of All Demos," showcasing the application of tools and concepts now recognized as hypertext, word processing, real-time editing, file sharing, teleconferencing, multiple window views, and graphical navigation using a device called a *mouse*. The person behind the design of the system called the *oN-Line System* was Douglas Engelbart, a pioneering inventor and engineer in the field of human-computer interaction.

The demonstration was given at a joint conference of the ACM and the IEEE. It was the culmination of years of work on a project produced at SRI's Augmentation Research Center (ARC), jointly sponsored by DARPA, NASA, and the Air Force's Rome Air Development Center. Engelbart conducted the collaborative demo in San Francisco, with the rest of his team in their lab at Menlo Park, connected by modem and microwave link.

Like others of his generation, Engelbart was hugely influenced by Vannevar Bush's essay "As We May Think," which laid out the blueprint for an interactive computing machine that would improve the quality of human performance in intellectual tasks and enable people to make greater use of the staggering amount of knowledge at their disposal. Today you can find Engelbart's heavily annotated copy of the Bush essay in the Engelbart archives, showing how he mined it for ideas that he later implemented.

The NLS was a mechanism to implement a broader goal that Engelbart pursued throughout his life: to scale cooperation and collaboration among people in order to solve the most challenging problems the world faced and enable humanity to reach its full potential. The work earned Engelbart the 1997 A.M. Turing award.

The Mother of All Demos inspired a generation of technologists and subsequent inventions, including the Xerox Alto in 1973, a PC that would influence the design and interface of the Apple Macintosh. Of note, Stewart Brand—the creator of the *Whole Earth Catalogue*, a publication focused on shared community and collective goals—operated the video camera during the demonstration in Menlo Park.

SEE ALSO "As We May Think" (1945), The Mouse (1967), Xerox Alto (1973)

Doug Engelbart's demonstration of the first graphical user interface set in motion what became the personal computer revolution.

Dot Matrix Printer

Rudolf Hell (1901–2002), **Fritz Karl Preikschat** (1910–1994)

Dot matrix printers use a cluster of closely spaced dots to form individual letters. Each dot is controllable by the printer, making it possible for the printers to produce text in any typestyle and at any size; they can also construct elaborate graphics. In contrast, other contemporary printers, such as computerized electric typewriters and typeball and "daisy wheel" printers, all stamped letters wholly formed from a die.

To create the dots, small metal pins or rods are mechanically pushed forward against an ink-soaked ribbon or fabric that makes physical contact with the paper. Tiny electromagnets called *solenoids* power the forward motion of the pins against a guide plate with tiny holes to help steer the pins to the appropriate place. The print quality of dot matrix printers depends heavily on the number of pins used to transfer the image, which is typically 7 to 24, for maximum resolution around 240 dots per inch (dpi). Speeds range from 50 to 500 characters per second (cps).

The birth of the modern dot matrix printer is generally understood to have occurred in Japan in 1968 with the introduction of the Shinshu Seiki company's EP-101 (later EPSON®) and OKI Data Corporation's Wiredot printer that same year. Earlier devices, such as Rudolf Hell's 1929 Hellschreiber teletypewriter, sent the raw collection of dots from one machine to the other and are more properly thought of as *facsimile machines*.

Dot matrix printers were as popular with businesses as they were in the home office. Because they made their print with mechanical pressure, they could easily print on multipart forms, simultaneously creating two or more copies, making them fixtures at car-rental kiosks, where they were commonly used to print rental contracts.

Electrostatic discharge printers (which print on silver paper), thermal printers (typically used for credit card receipts), and inkjet printers are all fundamentally dot matrix printers, but with different kinds of mechanisms to transfer the dots to the paper. Even 3-D printers can be thought of as a special kind of dot matrix printer that prints a single dot of material at a time.

In 2013, the Wiredot printer received the Information Processing Technology Heritage award from the Information Processing Society of Japan.

SEE ALSO Laser Printer (1971), 3-D Printing (1983)

Wiredot printer, manufactured by the OKI Electric Industry Co., Ltd., 1968.

Interface Message Processor (IMP)

In 1968, the former Bolt, Beranek, and Newman (BBN) company won a million-dollar contract with the US Department of Defense's Advanced Research Projects Agency (ARPA) to build the computers that would create the first computer network. The design called for a host computer to connect to a small computer known as an *Interface Message Processor* (IMP) at each location, and then for the IMPs to connect to one another using long-distance data links.

The IMP was based on a Honeywell DDP-516 minicomputer, with a 0.96-microsecond cycle time, a 16-bit word length, 12 kibibytes of memory, 16 interrupt channels, 16 data channels, and a relative-time clock to "facilitate program event timing," according to the manual.

The computers of the day had not yet standardized on using 8-bit bytes for their memory systems, so the IMP had a bit-serial, asynchronous interface, allowing it to be used by computers with 8-, 12-, 16-, 24-, 36-, and 60-bit words. Programmers had to write software to connect each kind of computer to the IMPs. But each IMP could run the same program. That interconnection protocol was contained in BBN Report 1822, "Specification for the Interconnection of a Host and an IMP." A second document, "RFC 1: Host Software," specified the requirements for the computers.

Messages were sent in variable-length packets that could be up to 8159 bits in length and included the 24-bit address of the destination host, a 24-bit checksum (an error-detection code that was computed in hardware), and a variable-length data payload. Packets that were received reliably were acknowledged; those with errors were dropped and eventually retransmitted by the sender.

BBN also created a Terminal IMP, or TIP, which allowed teletypes, modems, and eventually video display terminals to connect directly to a computer network, and to directly access remote computers.

IMPs were created for the Advanced Research Projects Agency (ARPA)'s ARPANET network, and they formed its backbone. ARPA was renamed the Defense Advanced Research Projects Agency (DARPA) in 1972. IMPs continued to operate the network until DARPA formally decommissioned the ARPANET in 1989. Some IMPs were transferred to the Department of Defense's MILNET (military network), while others were taken apart, put on display, or given to museums.

SEE ALSO ARPANET/Internet (1969), *Network Working Group Request for Comments: 1* (1969)

The front panel of the first Interface Message Processor (IMP), used at the UCLA Boelter 3420 lab to transmit the first message on the internet.

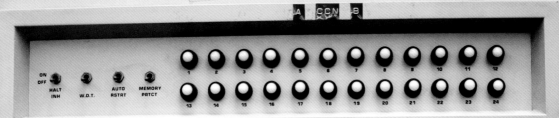

INTERFACE MESSAGE PROCESSOR

Developed for
the Advanced Research Projects Agency
by Bolt Beranek and Newman Inc.

1969

ARPANET/Internet

Leonard Kleinrock (b. 1934), **J. C. R. Licklider** (1915–1990),
Thomas Marill (dates unavailable), **Lawrence G. Roberts** (b. 1937),
Ivan Sutherland (b. 1938)

The internet as we know it came into being in the fall of 1969, when three computers in California and one in Utah were connected and started exchanging messages.

What makes the internet successful is the concept of *packet switching*. The telegraph and telephone networks had allowed remote access to computers for more than a decade, but those systems required that each conversation have its own wire. With packet switching, each communications stream is split up into a number of packets, each with a header that contains the address where it is heading and a payload of information it is carrying there. The network's function is to route these packets of data, much as a post office routes paper mail.

Leonard Kleinrock at MIT published the first paper on packet-switching theory in July 1961. The next year, J. C. R. Licklider, also at MIT, wrote a memo describing his vision for a "Galactic Network," as he playfully called it. Motivated by Licklider's vision, in 1965, Ivan Sutherland and Thomas Marill connected a TX-2 computer at MIT Lincoln Lab in Lincoln, Massachusetts, with a Q-32 computer at the System Development Corporation (SDC) in Santa Monica, California. The two groups wrote software that allowed remote login and file transfer.

The following year, Lawrence G. Roberts took a job at the US Advanced Research Projects Agency (ARPA) with the intention of making Licklider's vision of computer networking a reality. He increased the design speed of the network to 50 kilobits per second (kbps), and then requested teams of scientists and engineers who could design and build—basically invent—the packet-switch network hardware.

In 1969, the first pieces of hardware were delivered to the University of California, Los Angeles (UCLA); the Stanford Research Institute (SRI); the University of California, Santa Barbara; and the University of Utah. The first successful connection was on October 29, in which the string *LOGIN* was sent from the machine at UCLA to the machine at SRI. It took three more years for the invention of network email.

SEE ALSO Interface Message Processor (IMP) (1968), *Network Working Group Request for Comments: 1* (1969)

Map of the ARPANET, May 1973.

ARPA NETWORK, LOGICAL MAP, MAY 1973

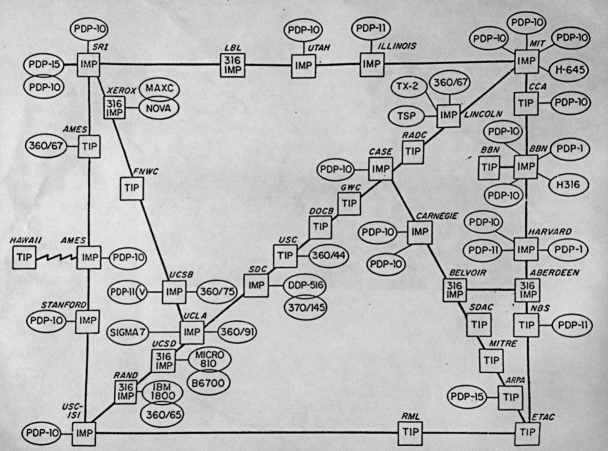

Digital Imaging

Bruce E. Bayer (1929–2012), **Willard S. Boyle** (1924–2011),
George E. Smith (b. 1930)

The first photograph was sent by wire in 1907, and the first photograph was digitized with a scanner in 1957. But capturing an image directly from light to digital form required the invention of light-sensitive semiconductors that could be packed into a rectangular array.

That invention was the *charge-coupled device* (CCD), developed by Willard S. Boyle and George E. Smith at Bell Laboratories in New Jersey. The device makes use of the photoelectric effect, by which certain kinds of materials eject electrons when stimulated by light. The CCD collects these electrons in an array of capacitors, with each capacitor receiving an electric charge proportional to the amount of light to which it is exposed. The electrons then cascade across the CCD to its edge, where their voltage is measured and digitized.

In a digital camera, the light-sensing capacitors are arranged in a two-dimensional array, so that the entire image is captured at once. But in satellites and fax machines, the CCD is arranged as a one-dimensional strip, and the image (the Earth below, or a piece of paper) sweeps by the sensor.

The first CCDs captured only black-and-white photos; color requires capturing three separate images through three colored filters—red, green, and blue—and electronically combining them by overlaying the image sensor with an array of tiny red, green, and blue filters. In practice, the filters are arranged in an *R, G, B, G* pattern, called a *Bayer color mask* after its inventor, Bruce Bayer at Kodak. This RGBG pattern produces twice as many green pixels as red or blue, which works out well, because people are most sensitive to green light. As a result, the sensor provides greater spatial resolution and dynamic range, without sacrificing too much color resolution and accuracy.

For their work on the CCD, Boyle and Smith shared the 2009 Nobel Prize in Physics.

SEE ALSO Fax Machine Patented (1843), First Digital Image (1957)

The technology inside smartphones' cameras can be traced back to the invention of the charge-coupled device (CCD).

Network Working Group Request for Comments: 1

Steve Crocker (b. 1944), Jon Postel (1943–1998)

Once the ARPANET equipment started being installed, a group of graduate students and staff members at the first four sites began meeting in person to work out the technical details of what they would do with the world's first computer network.

The group expected that someone from ARPA or a senior professor at one of the labs would take charge. In the meantime, they decided to write down the ideas they had come up with and elaborate on their existing questions. As Steve Crocker, a graduate student working on the project, told *WIRED* magazine many years later, to avoid the "presumption of authority," the group utilized an informal approach regarding these records: "You could ask questions without answers. . . . And then to emphasize the informal nature, I hit upon this silly little idea of calling every one of them a 'Request for Comments'—no matter whether it was really a request" or if it was a statement about the network that someone in the group had written down.

Network Working Group Request for Comments: 1, dated April 7, 1969, was a summary of the software that would run on the interface message processors (IMPs), the requirement for software that sent messages between hosts, and the nature of the host software itself. Key among the requirements for the system was that it needed to be simple to use, mirror existing software, and have error checking, so that information sent over the network would be reliable. Because there wasn't an organization running the network, the six researchers involved called themselves the *Network Working Group* (NWG).

RFC 3, *Documentation Conventions*, written later that month, established the written specifications for the RFC series, such as the idea that notes should be "timely rather than polished" and could be as brief as a single sentence.

There are now more than 8,000 RFCs, documenting all kinds of things, including network protocols, data formats, organizational rules, etiquette, and even April Fools' jokes.

SEE ALSO Interface Message Processor (IMP) (1968)

Published April 1, 1978, RFC 748 jokes about a fictional option that would prevent remote computers from randomly crashing and losing data.

Network Working Group   M. Crispin
Request for Comments 748   SU-AI
NIC 44125   1 April 1978

TELNET RANDOMLY-LOSE Option

1. Command name and code.

    RANDOMLY-LOSE   256

2. Command meanings.

    IAC WILL RANDOMLY-LOSE

        The sender of this command REQUESTS permission to, or confirms
        that it will, randomly lose.

    IAC WON'T RANDOMLY-LOSE

        The sender of this command REFUSES to randomly lose.

    IAC DO RANDOMLY-LOSE

        The sender of this command REQUESTS that the receiver, or grants
        the receiver permission to, randomly lose.

    IAC DON'T RANDOMLY-LOSE

        The command sender DEMANDS that the receiver not randomly lose.

3. Default.

    WON'T RANDOMLY-LOSE

    DON'T RANDOMLY-LOSE

    i.e., random lossage will not happen.

4. Motivation for the option.

    Several hosts appear to provide random lossage, such as system
    crashes, lost data, incorrectly functioning programs, etc., as part
    of their services.   These services are often undocumented and are in
    general quite confusing to the novice user.   A general means is
    needed to allow the user to disable these features.

Utility Computing

Fernando Corbató (b. 1926), **Jerome Saltzer** (b. 1939)

Fresh from the success of the Compatible Time-Sharing System, MIT partnered with Bell Labs and General Electric (GE®) to create a computing "utility" that would make information resources available to anybody who wanted to pay for them, providing the same kinds of reliability, scalability, and flexibility guarantees for computation that electric, gas, and water utilities did for those commodities earlier in the century. The project was called *Multics*, short for Multiplexed Information and Computing Service.

The goal was to build a thoroughly modern operating system that was reliable, easy to use, and secure. The project was a technical success. For example, most modern concepts of computer security originated in the Multics design. But it was ultimately a commercial failure.

The Multics team started by creating a detailed design specification. The entire specification was more than 3,000 typewritten pages long and described more than 1,500 distinct software modules. No operating system had ever before been specified in such exacting advance detail; it let the developers catch many bugs before a line of code was written. The system became operational in 1969. By 1972, the system was supporting 55 users at a time, including system programmers who continued to develop the underlying operating system.

Multics was ahead of its time: its vision of a utility computer would not become the dominant paradigm until the 2000s, with the rise of cloud computing.

Bell Labs pulled out of Multics in 1969 because of the cost. The Bell programmers, Ken Thompson (b. 1943) and Dennis Ritchie (1941–2011), took their best ideas from the project and created a simplified version of Multics they called *UNIX*—a pun meant to indicate that their system was much simpler.

Some governments and the Ford Motor Company® bought Multics systems because of their security, but the system was never widely used. The last Multics was shut down in 2000; in 2006, Groupe Bull, which had bought Multics years before, released the operating system's code as open-source software. Today you can download Multics and run it in a simulator.

SEE ALSO Time-Sharing (1961), UNIX (1969)

Careful design evidenced by detailed documentation and in-depth academic papers was a hallmark of the Multics project. By first writing and analyzing the design on paper, many bugs were anticipated and eliminated before a line of code was written.

Perceptrons

Seymour Papert (1928–2016), Marvin Minsky (1927–2016)

By the late 1940s, some computer scientists thought that the way to achieve human-level problem solving would be to create artificial neurons, borrowing the model for how the human brain works, and wire them in some kind of network. An early demonstration was the Stochastic Neural Analog Reinforcement Calculator (SNARC), a network of 40 artificial neurons that learned how to solve a maze, created in 1951 by Marvin Minsky, then a first-year graduate student at Princeton University.

Following SNARC, researchers throughout the world took up the idea of artificial neural networks. The most significant effort was at the Cornell Aeronautical Laboratory, where in 1958, Frank Rosenblatt (1928–1971) built a massive machine that "learned" how to recognize images.

Minsky, though, gave up on neural networks in the 1950s and instead pursued symbolic artificial intelligence, an approach that aims to mirror higher-level human thought by representing knowledge with symbols and rules. He moved to MIT and was joined in 1967 by Seymour Papert, an expert in the field of learning.

Annoyed by the attention (and perhaps funding) that neural networks continued to attract throughout the 1960s, Papert and Minksy wrote the book *Perceptrons: An Introduction to Computational Geometry*, published in 1969, in which they mathematically proved that there were fundamental limits to the artificial neural networks approach. *Perceptrons* was so persuasive that researchers around the world (and at many funding agencies) simply gave up on neural networks and moved on to other ideas; the book was credited with singlehandedly destroying the field of artificial neural networks.

Papert and Minsky, however, had only proved limits for a very specific kind of artificial neural network, one that had just a single layer of neurons. A few researchers who stayed with the idea eventually figured out how to efficiently train multistage neural networks, and by the 1990s, computers were finally fast enough that neural networks with multiple hidden layers were solving complex problems that could not be solved symbolically. Today, neural networks are the dominant approach used in AI.

SEE ALSO Watson Wins *Jeopardy!* (2011), Google Releases TensorFlow (2015), Computer Beats Master at Go (2016)

The cover of the book Perceptrons, *designed by Muriel Cooper, shows a problem that is difficult to solve with a neural network: it is hard to tell if there is an unobstructed path between any two given points.*

Marvin L. Minsky and Seymour A. Papert

Reissue of the 1988 Expanded Edition with a new foreword by Léon Bottou

Perceptrons

An Introduction to Computational Geometry

UNIX

Ken Thompson (b. 1943), **Dennis Ritchie** (1941–2011),
Malcolm Douglas McIlroy (b. 1932)

After Bell Labs decided to pull out of the Multics project, Bell computer scientists Ken Thompson, Dennis Ritchie, Malcolm McIlroy, and others decided to build a modern, streamlined operating system with a fraction of the resources that MIT and Honeywell were throwing into Multics.

Bell Labs had a five-year-old PDP-7 computer from DEC that wasn't being otherwise used, so in 1969, Ken Thompson wrote an operating system for it that would implement the core Multics ideas, including a hierarchical file system with a root directory that could contain both files and other directories, a "shell" program that allowed the user to type commands that could support any number of directories and files within other directories, and the ability to expand the system with user-written commands.

By 1972, the system had a name—UNICS—a play of the word *eunuch* (a castrated man) and Multics. UNICS was a castrated Multics! Perhaps a bit childish, the name was changed to UNIX (nobody quite remembers by whom), and the new name stuck.

UNIX was rewritten into the newly invented C programming language in 1972. Although the minimalist operating system lacked many features found on more sophisticated operating systems, UNIX offered just enough features to let researchers and businesses develop their own software.

UNIX got a huge boost in 1983 when the University of California, Berkeley, added support for the internet's transmission control protocol/internet protocol (TCP/IP) networking protocol to the operating system. Now any school or business that wanted to get on the internet could do so with a network connection and a computer running the Berkeley Standard Distribution (BSD) of UNIX, which also came with an email server, a mail client, and even games. Soon workstations were being created specifically for the purpose of running the operating system.

UNIX survives today as the operating system in the Apple Macintosh and the iPhone®.

SEE ALSO Utility Computing (1969), C Programming Language (1972), IPv4 Flag Day (1983), Linux Kernel (1991)

Photograph of Ken Thompson (sitting) and Dennis Ritchie (standing) at the PDP-11.

Fair Credit Reporting Act

Alan Westin (1929–2013)

In March 1970, a professor from Columbia University testified before the US Congress about shadowy American businesses that were maintaining secret databases on American citizens. These files, said Alan Westin, "may include 'facts, statistics, inaccuracies and rumors' . . . about virtually every phase of a person's life: his marital troubles, jobs, school history, childhood, sex life, and political activities."

The files were used by American banks, department stores, and other firms to determine who should be given credit to buy a house, a car, or even a furniture set. The databanks, Westin explained, were also used by companies evaluating job applicants and underwriting insurance. And they couldn't be outlawed: without credit and the ability to pay for major purchases with installments, many people couldn't otherwise afford such things.

Westin was well known to the US Congress: he had testified on multiple occasions before congressional committees investigating the credit-reporting industry, and he had published a book, *Privacy and Freedom* (1967), in which he argued that freedom in the information age required that individuals have control over how their data are used by governments and businesses. Westin defined privacy as "the claim of individuals, groups, or institutions to determine for themselves when, how, and to what extent information about them is communicated to others." And he coined the phrase *data shadow* to describe the trail of information that people leave behind in the modern world.

On October 26, 1970, Congress enacted the Fair Credit Reporting Act (FCRA), which gave Americans, for the first time, the right to see the consumer files that businesses used to decide who should get credit and insurance. The FCRA also gave consumers the right to force the credit bureaus to investigate a claim that the consumer felt was inaccurate, and the ability to insert a statement in the file, telling his or her side of the story.

The FCRA was one of the first laws in the world regulating what private businesses could do with data that they collect—the beginning of what is now called *data protection*, an idea that eventually spread worldwide.

Today there are privacy commissioners in almost every developed country. The passage of the European Union's General Data Protection Regulation (GDPR) marked the most far-reaching privacy law on the planet.

SEE ALSO Relational Database (1970)

Columbia professor Alan Westin was concerned about American businesses keeping secret databases on American citizens.

Relational Database

Edgar F. Codd (1923–2003)

Storing large amounts of data was one of the early uses for computers, but it wasn't immediately obvious how the data should be organized. At IBM's San Jose research laboratory, computer scientist Edgar Codd devised an approach for organizing and arranging data that was more efficient than other models. Instead of grouping together data belonging to the same entity, his approach created large tables of data that had the same conceptual types, with identifying numbers (IDs) defining the relationships between records in different the tables.

For example, an insurance company might have one table of customers, with each customer having a CUSTOMER ID and a name. Then there might be another table of insurance policies, with each having a POLICY ID, a CUSTOMER ID, and a POLICY TYPE ID. A third table might link the POLICY TYPE ID and the details of the policy. In this example, to find the insurance policies for a customer, the computer would first find the CUSTOMER ID, and then find all of the policies that had the same CUSTOMER ID. To get the details of each policy, the system would take the POLICY ID, look it up in the table of policies to get the POLICY TYPE ID, and then search the table of policy types to find the details.

Codd's groundbreaking research showed that organizing data in this fashion made it more efficient to store, faster to access, and easier to program. Most importantly, he showed that it was possible to create a general-purpose database engine for storing data on the computer's hard drive, freeing programmers from the task and allowing them to concentrate on their applications. Once the database was developed and deployed, improvements to the underlying software benefitted all of the applications that relied on it. For his work, Codd was awarded the 1981 A.M. Turing Award.

Today the operating systems of both Apple's iPhone and Google's Android create a relational database on every smartphone for every app that's installed, making Codd's invention one of the dominant ways of storing data.

SEE ALSO First Disk Storage Unit (1956)

Edgar Codd created large tables of data that had the same conceptual types.

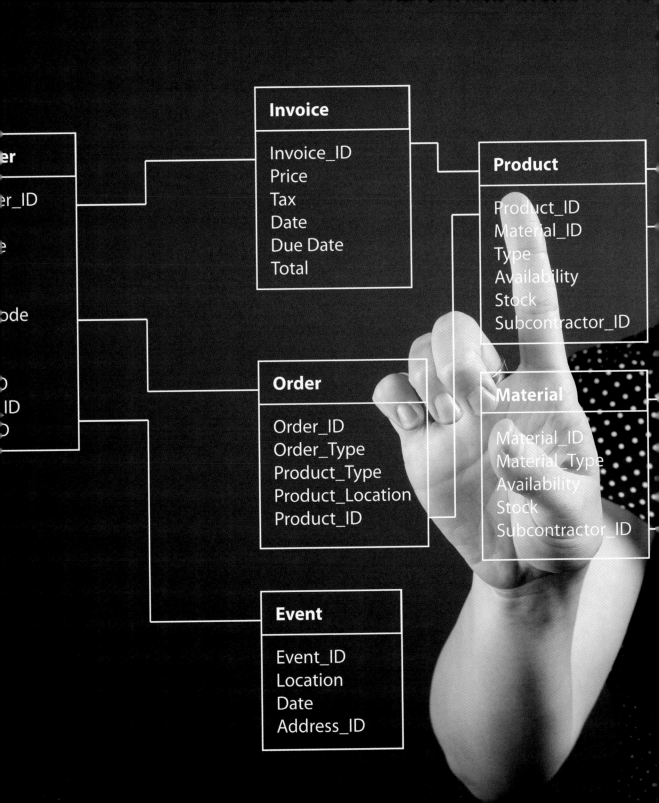

Floppy Disk

The floppy disk provided computer users with a reliable, compact system for storing data. The system consisted of a single spinning magnetic disk located inside a cardboard or plastic envelope. Inside the envelope was special fabric designed to clean the spinning media and trap dust that might otherwise contaminate the magnetic surface. The other part of the system was the floppy disk drive, an electromechanical contraption that included a circular clamp that grabbed the magnetic disk, a motor that spun it, and a read/write head assembly mounted on a radial track and moved with a stepper motor that could be used to read and write concentric magnetic rings on the media.

Floppy disks stored less data and were slower than the larger disk and drum storage systems, but they were dramatically cheaper. Even better, one disk could be removed from the drive and replaced with another, creating a system with limitless storage (providing that the user was willing to purchase another box of floppy disks).

The first IBM 8-inch floppy released in 1971 could store just 80 kilobytes (KB) of information—the same amount of data as a thousand punched cards. Within a few years, 8-inch floppies could store 200–300 KB. Shugart Associates, founded by one of the developers of the 8-inch floppy, produced the world's first 5¼-inch drive and disk in 1976. The smaller drive was adopted by both IBM for the PC and Apple for the Apple II, although the disks were not compatible because the companies developed different approaches for storing bits on the magnetic surfaces. The 3½-inch floppy (with a hard-plastic shell) was introduced in 1983 with a capacity of 360 KB; Apple took the same hardware and introduced a version with an incompatible format for the Macintosh computer (storing 400 KB) in 1984. Double-sided drives were soon introduced that could store 400 or 800 KB, followed by double-sided, high-density drives that could store 1.44 megabytes (MB).

SEE ALSO First Disk Storage Unit (1956), CD-ROM (1988), DVD (1995)

Floppy disks offered an inexpensive, limitless storage system. Pictured are examples of an 8-inch, 5.25-inch, and 3.5-inch floppy disk.

Laser Printer

Gary Starkweather (b. 1938)

In 1967, Gary Starkweather was an engineer at Xerox's research and engineering facility in Webster, New York. Xerox copiers worked by illuminating an original document with a bright light and focusing the reflected light on a photosensitive rotating drum, which then passed through toner and pressed against a piece of plain paper. The toner stuck only to the parts of the paper where the drum had not been illuminated. The paper was then heated so that the toner would melt and fuse in place. Voilà! Instant copies.

Starkweather's big idea was to remove the bright lights and the original, and instead scan the xerographic drum with a laser that was modulated by a computer, allowing the computer to precisely control what appeared on the page. The result would be a Xerox copy without an original, and a fundamentally new kind of computer output device.

Starkweather's management was resistant to the idea and directed him to work on something else. Instead, he spent three months building his prototype in secret. When Xerox's New York management office still wasn't interested, Starkweather transferred to the company's Palo Alto Research Center (PARC), which was filled with engineers who were building the machines of the future. He arrived at PARC in January 1971 and had a working printer nine months later.

PARC had an easier time making prototypes than convincing Xerox to sell them. The XGP (Xerox Graphics Printer) was the first laser printer to make it out of PARC. The XGP printed at 180 dots per inch on a continuous roll of 8½-inch paper that the printer then cut into individual pages. One XGP was lent to Stanford, just down the street, while another was installed across the country on the ninth floor of the MIT Artificial Intelligence Laboratory; however, none were ever sold.

In 1976, PARC created the Dover, which could print two pages per second at 300 dots per inch. Like the XGP, the Dover was also an experimental machine. That same year, IBM released the first commercial laser printer, the IBM 3800. Not wanting to fall behind, Xerox released the Xerox 9700 laser printer the following year, in 1977.

SEE ALSO Xerox Alto (1973), PostScript (1982), Desktop Publishing (1985)

Dover laser printer displayed at the Computer History Museum.

NP-Completeness

Stephen A. Cook (b. 1939), **Richard Karp** (b. 1935),
Leonid Anatolievich Levin (b. 1948)

The Church-Turing thesis answers the question about what fundamentally can and cannot be computed. But it ignores the question of efficiency—whether a calculation would require an hour or a million years to complete.

For even the first computers, some calculations were blindingly fast: ENIAC could add 10-digit numbers together in a few milliseconds. Other problems grew easier as scientists developed efficient methods for solving them. Cracking the German codes during World War II was impossible at first—but eventually tricks were discovered that let the Allies crack each day's messages in just a few hours.

Other tasks didn't get significantly more efficient, no matter how many programmers attacked the problem. Examples include creating a class schedule for a university so that no classrooms, professors, or students are double-booked, or devising a plan for a traveling salesman to visit 50 cities without running out of time or money. For these complex problems, solutions were hard to find, but once found, their correctness was easy to verify.

In 1971, Stephen Cook, an associate professor of computer science at the University of Toronto, showed that, surprisingly, any specific traveling salesman problem could be transformed into the university scheduling problem, so that a solution for the second could be used to solve the first. Two years later, Richard Karp showed the same kind of equivalency between 21 classes of problems.

Today computer scientists say that such problems are NP-complete: we still don't know how to efficiently solve the university scheduling or the traveling salesman problems, but if we can ever get a solution for one of them, we will instantly have a solution for them all. Annoyingly, we don't know if efficient solutions to NP-complete problems are even possible. Answering that question is one of today's great mysteries of computer science.

For their work, Cook was awarded the A.M. Turing Award in 1982, and Karp in 1985. In the USSR, Leonid Levin had independently developed a theory of NP-completeness using a different line of mathematical reasoning. For this reason, the theory of NP-completeness is called the *Cook-Levin theorem*.

SEE ALSO Church-Turing Thesis (1936)

The traveling salesman problem, which seeks to find the most efficient route for a salesperson traveling between cities, is an example of a problem computer scientists call NP-complete.

@Mail

Ray Tomlinson (1941–2016)

Many early time-sharing systems allowed users to leave messages for one another. For example, early PDP-10 systems had a program called *SNDMSG* that let a user compose a message and then append it to another user's mailbox. When the second user logged in, he or she could see the message that the first user had left by running another program, READMAIL. The messages composed and displayed with these programs, however, were confined to a single computer.

In 1971, Ray Tomlinson, an engineer at BBN, developed a program called *CPYNET* for sending files between computers. Shortly thereafter, Tomlinson realized that he could combine aspects of SNDMSG and CPYNET. A user could author a message with CPYNET and use a modified version of the CPYNET protocol to deliver the message to another computer, where it would be appended to the specified user's mailbox. Shortly thereafter, Tomlinson sent a message between two computers in BBN's Cambridge, Massachusetts, laboratory. Tomlinson didn't keep that first message, but he told numerous journalists that it was probably the keyboard pattern "QWERTYUIOP."

Beyond sending the first network mail message, Tomlinson is also credited with choosing the @ symbol to separate the name of the mailbox from the name of the destination host. Instead of its original meaning to denote prices—e.g., 2 eggs @ 35¢ = 70¢—the sign soon represented the middle part of a network mail address.

The original ARPANET had been designed to support remote computer access through virtual terminals and file transport; mail was not part of the initial design. Nevertheless, mail soon became ARPANET's "killer app"—the reason that many research laboratories spent time and money to get on the network. No doubt this was helped along by ARPA's tendency to use email for official business: researchers who found it easy to exchange email with ARPA program managers also found it easier to get funding.

Despite the adoption of the @ sign for mail, the *From:*, *Date:*, and *Subject:* mail headers would not be standardized until ARPANET standard RFC 561, *Standardizing Network Mail Headers*, was adopted in 1973. The *To:*, *Cc:*, and *Bcc:* headers would have to wait for RFC 680, *Message Transmission Protocol*, adopted in 1975.

SEE ALSO Time-Sharing (1961), ARPANET/Internet (1969)

BBN engineer Ray Tomlinson chose the @ symbol to separate the name of a user's mailbox from the name of the destination host.

First Microprocessor

Federico Faggin (b. 1941), **Ted Hoff** (b. 1937), **Stanley Mazor** (b. 1941)

Intel was founded in July 1968 with $2.5 million in funding from a venture capitalist to build integrated circuits. The company's first chip was the 1101 memory chip; it stored 256 bits of memory and did not sell well. The second chip, the 1103, stored 1024 bits and was hugely successful, taking the company public in 1971.

That same year, Intel introduced the 4004, the world's first general-purpose computer on a chip. It was designed by Federico Faggin, Ted Hoff, and Stanley Mazor and had 2,300 transistors.

Modern computers are based on von Neumann architecture, credited to mathematician and physicist John von Neumann in the 1940s. The computer has an *arithmetic logic unit* that can perform basic math (addition and subtraction), a few fast memory cells called *registers*, logic that can fetch data and instructions from memory, and other logic that can store data back to memory. A special register called the *program counter* (PC) points to a specific location in the computer's memory bank. The computer operates by reading an instruction from the memory location specified by the PC, executing it, incrementing the PC so that it points to the next memory location, and then repeating. The 4004 marked the first time all of these functions were combined on a single piece of silicon. The 4004 had 16 4-bit registers and 45 instructions, kept its program in 4,096 bytes of read-only memory, and could address another 1,280 words of 4-bit RAM.

The 4004's word size was just 4 bits—enough to store the numbers "0" through "9" coded as binary. Not surprisingly, the 4004 was designed for use in a calculator. But because it was general purpose, the chip could be used in other applications. For example, one company used the 4004 to control a pinball machine.

Five months later, Intel brought out an 8-bit version of the device called the *8008*, which could address up to 16,384 bytes of memory and intermix programs and data. In April 1974, the company released the 8080, which could address up to 65,536 bytes of memory. Despite the differences, all versions used the same basic assembler code—a code that would later be shared with Intel's Pentium® and Core® processors.

SEE ALSO EDVAC *First Draft* Report (1945), First Personal Computer (1974)

Federico Faggin stands before an enlarged blueprint of the Intel 4004, which he designed and which became the world's first microprocessor.

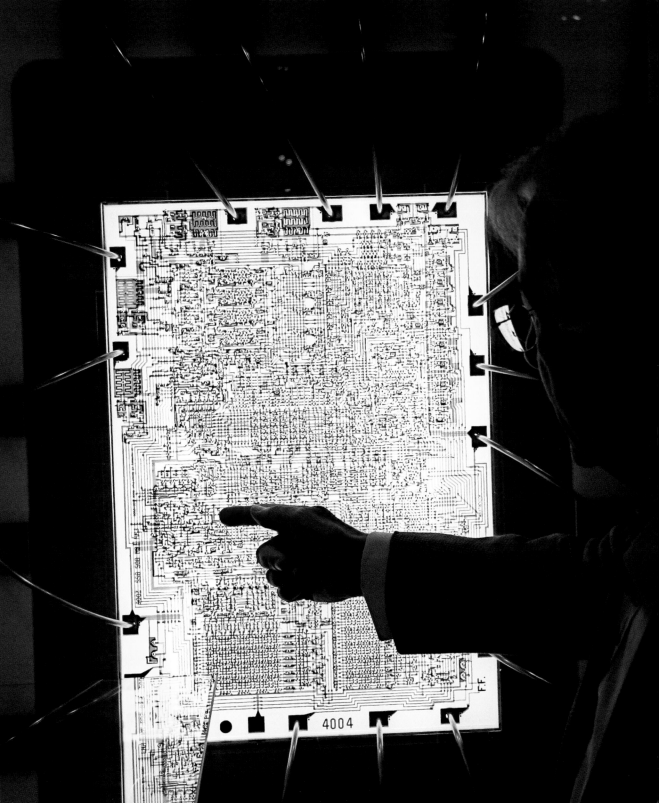

First Wireless Network

Norman Abramson (b. 1932), **Robert Metcalfe** (b. 1946)

By 1968 it was clear that the voice telephone network would not be adequate to serve the emerging requirements of networked computing. ALOHA was designed to explore the possibility of using wireless communications as a potentially superior alternative to wired.

A team of researchers at the University of Hawaii (UH), led by Norman Abramson, set out with the goal of linking computers on the main campus in Manoa Valley (near Honolulu) with terminals at a college in Hilo, Hawaii—and five community colleges on the islands of Oahu, Kauai, Maui, and Hawaii. If successful, the project would allow students at the schools to use the computers without having to travel to Hilo.

At the time, point-to-point microwave channels were well understood—and expensive. Such channels were also wasteful, because the nature of terminal communication was such that it was sporadic and frequently idle and could tolerate only small delays. Soon the group hit on the idea of sharing a single high-speed wireless channel between all the senders: if a sender didn't get an acknowledgement, it would retransmit its packet until it did.

The first packet was transmitted in June 1971 from a terminal attached by an RS232 interface to a new device called a *terminal control unit* (TCU). With the TCU, the terminal could be used anywhere within 100 miles of the UH campus. Soon the group built more TCUs, networking the islands together with ALOHANET, the world's first wireless computer network.

The ALOHANET system was interconnected to the ARPANET on December 17, 1972, over a single 56-kilobits-per-second satellite communications channel.

Electrical engineer Robert Metcalfe realized that the same broadcast architecture could be run over a piece of coaxial cable. He improved the basic protocol by having the radios listen for traffic before sending their packet, an approach called *carrier sense multiple access* (CSMA), and the Ethernet was born.

Versions of the ALOHANET protocol made their way into many other wireless networks, including early cellular systems. But the lasting contribution of the project was the impact on Ethernet protocols and, eventually, today's Wi-Fi standards.

SEE ALSO IPv4 Flag Day (1983)

The campus at the University of Hawaii, Manoa, where computers were linked with terminals at other colleges throughout Hawaii.

C Programming Language

Dennis Ritchie (1941–2011)

When a program runs on a computer, the computer's CPU executes a sequence of low-level machine instructions that invoke operations such as fetching data from memory, adding numbers together, and storing the result back into memory. Humans write in high-level languages that are translated into machine code by special-purpose programs called *compilers*. This translation lets programmers create programs that are vastly more complex and powerful than if they had to write directly in machine code, or in assembly language, which was trivially translated into machine code.

Created at Bell Laboratories by Dennis Ritchie, C was designed for writing operating systems. Key aspects of the language include the ability to precisely control the layout of data in the computer's memory and to intermix high-level instructions and machine code, and the fact that C runs faster than other high-level languages.

Still widely used today, C also lets programmers write at a high level of abstraction. The language comes with a library of built-in functions that implement complex behaviors, such as reading and writing data files, and performing advanced mathematical operations. What's more, programmers can create their own functions and use them as if they were built-in functions, making it easy for even beginning programmers to extend the language. As C gained popularity, programmers started to share their libraries, naturally creating an open-source culture.

The original UNIX operating system was written in assembly language for an old Digital Equipment Corporation PDP-7 computer and then rewritten into the assembly language for the PDP-11. In 1973, Version 2 of UNIX was rewritten into C, making it the third operating system ever to be written in a high-level language. Rewriting it in C made UNIX much easier to maintain and extend.

Today C is one of the most popular computer languages in the world. It's also inspired many other languages, including C++, C#, Java, PHP, Perl, and others.

SEE ALSO UNIX (1969)

This fragment of source code, containing a comment and a complex instruction, appeared in the UNIX operating system kernel. The instruction suspends one program and starts another.

```
/*
 * If the new process paused because it was
 * swapped out, set the stack level to the last call
 * to savu(u_ssav).  This means that the return
 * which is executed immediately after the call to aretu
 * actually returns from the last routine which did
 * the savu.
 *
 * You are not expected to understand this.
 */
if(rp->p_flag&SSWAP) {
        rp->p_flag =& ~SSWAP;
        aretu(u.u_ssav);
}
```

Cray Research

Seymour Cray (1925–1996)

For more than two decades, Seymour Cray's last name was synonymous with the word *supercomputer*. "The Cray" wasn't just a cornerstone of high-performance computing; it was part of popular culture. Cray's computer appeared in movies like *TRON* and *Sneakers*.

Seymour Cray was the lead designer of the CDC 6600, the first computer to be called a supercomputer, because it ran more than 10 times faster than any other system when it was released in 1964. It was system engineering that made the difference. "Anyone can build a fast CPU. The trick is to build a fast system," Cray was widely quoted as saying.

Among those tricks: the 6600 was the first computer that could dynamically evaluate and execute CPU instructions out of order; doing so sped up execution without altering the result of a calculation. It also had multiple execution units that could function at the same time—something called *instruction-level parallelism*. And instead of a large number of complex instructions, the system had a small number of instructions, each designed to run fast—an approach that inspired the so-called Reduced Instruction Set Computers (RISCs) of the 1980s.

Cray designed three generations of computers, each successively more ambitious but more risky. With a $300,000 investment from his friend William Norris, the CEO of Control Data Corporation (CDC), Cray founded Cray Research, forerunner of Cray Inc., in 1972. Four years later, the first Cray-1 supercomputer was given on a six-month loan to Los Alamos National Laboratory.

Sometimes called "the world's most expensive loveseat" because of its iconic design of a central column surrounded by a semicircular-padded base, the Cray-1 was an instant success. Its curious shape came from a legitimate design decision: the curved backplane meant that no wire within the computer needed to be more than 4 feet (1.2 meters) long, which minimized the propagation delay of signals within the machine. With the Cray-1 priced at more than $8 million, Cray Research sold more than 80 systems, establishing the company as the world's leading manufacturer of high-performance computers. They were used primarily for scientific computing, including weather predictions, and even by Apple to design the cases of Macintosh computers.

SEE ALSO RISC (1980), *TRON* (1982)

The Cray–1 supercomputer at the National Magnetic Fusion Energy Computer Center (NMFECC)—in California, 1983.

Game of Life

John H. Conway (b. 1937)

Mathematician John Conway's *Game of Life* is a digital grid of square cells. Each cell has eight "neighbors," or other cells touching it (horizontally, vertically, or diagonally). Cells can be alive, indicated by a stone in the square, or dead. On each turn, the computer examines every cell. A live cell with zero or one live neighbors dies on the next turn (or generation), presumably from loneliness (or underpopulation). A cell with four or more live neighbors also dies, this time from overpopulation. A cell with two or three live neighbors survives to the next generation. A dead cell with three live neighbors will become alive—a birth. As time progresses, the pattern of cells changes and evolves, usually (but not always) reaching stable patterns. Beyond setting up the initial pattern and starting the game, there is nothing for a human to do, which is why *Life* is sometimes referred to as a *zero-player game*.

Conway developed *Life* after hearing John von Neumann ask if it was possible for a machine to replicate itself. In the simplified world of *Life*, Conway showed that a machine could.

This game was significant to computer science because it was the first time a program had been created that could copy itself independently of any human coding activity (outside of starting the program, of course). *Life* launched a new field of modeling and simulation research, wherein the cycles and evolutions in nature—whether environmental, human, or even organizational—could be observed and studied as emergent and evolutionary behavior in a dramatically simplified environment. These research activities and the questions they purported to answer would become known as *simulation programs*.

Life's popularity grew significantly after Martin Gardner (1914–2010) mentioned it in the October 1970 issue of *Scientific American*. The game's rules were easy to implement, and the resulting complexity from relatively simple initial configurations was completely unexpected. Today there are many versions of *Life* available on the internet that can be played in a web browser.

SEE ALSO First International Meeting on Synthetic Biology (2004), Computer Beats Master at Go (2016)

John Conway's Game of Life *on an LED matrix showing an assortment of gliders, oscillators, and reflecting patterns.*

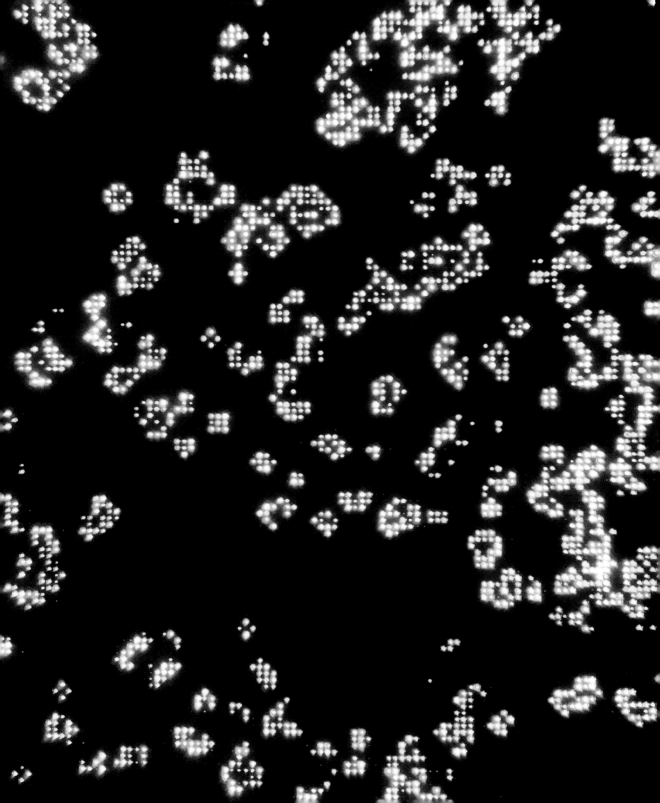

HP-35 Calculator

Bill Hewlett (1913–2001), David Packard (1912–1996)

Founded in a Palo Alto garage by Bill Hewlett and his Stanford classmate David Packard in 1939 with just $538, Hewlett-Packard (HP) was a respected manufacturer of test equipment for the electronics industry by 1950. The company introduced its first computer in 1966, and a programmable desktop calculator with a screen, printer, and magnetic card storage in 1968 that cost $4,900.

In 1968, Bill Hewlett decided that HP should create a portable electronic calculator that would fit in a shirt pocket. Hewlett was positive such a product would be successful, even though HP's marketing department did not see a need for it, arguing that the company's desktop calculators were selling just fine. But the project went ahead per Hewlett's wishes and in 1972 was introduced to the public.

Hewlett's idea was to build a portable calculator, powered by three AA batteries, using new integrated circuits from Intel and two other vendors, and have it show results on a single-line display made from light-emitting diodes. Inside, the calculator was powered by a microprocessor running at 200 kilohertz with a program that was just 768 instructions long. As the device's manual stated, "Its ten digit accuracy exceeds the precision to which most of the physical constants of the universe are known."

Just as much thought and care went into the design of the physical shell. The plastic keys were manufactured with a special two-step process so that the numbers could not rub off. Designed to survive a 3-foot drop onto concrete, the calculator was praised for its strength and reliability. Reportedly, HP's sales representatives would drop and even throw the calculators to demonstrate how tough they were.

From its launch, the $395 calculator took the market by storm—even at a time when the average month's rent in the United States was just $165. The company needed to sell 10,000 units to break even; it sold 100,000 in the first year. HP's calculator single-handedly destroyed the market for slide rules.

Originally called just "the calculator," it was renamed the HP-35 (because it had 35 keys) in 1973, after HP started selling a portable business calculator, the HP-80. Confusingly, the HP-80 also had 35 keys.

SEE ALSO Slide Rule (1621), Thomas Arithmometer (1851), First LED (1927), Curta Calculator (1948), ANITA Electronic Calculator (1961)

The HP-35 calculator, the world's first handheld scientific calculator. Note the π (pi) button.

Pong

Allan Alcorn (b. 1948), Nolan Bushnell (b. 1943)

In the late 1960s the idea of putting a computer into an arcade hall to make money was still quite novel. Nolan Bushnell, a huge fan of the highly influential *Spacewar!* game, had tried it with *Computer Space*—generally accepted as the first coin-operated video game. While *Computer Space* never achieved commercial success, Bushnell's next attempt—*Pong*—would be the first product from Atari, a new company he cofounded with Ted Dabney (1937–2018).

Pong is often referred to as the game that helped launch the video game industry, bringing into focus a new kind of entertainment that would revolutionize electronic "play" and drive advancements in other fields, including AI, which benefitted from the innovations in computer graphics technologies that run modern games.

As legend goes, Bushnell wanted to create a game that was blazingly simple to understand by anyone. He was familiar with the Magnavox® Odyssey, the first home console game, and its version of table tennis. However that influenced his thinking (which resulted in a series of protracted lawsuits), he asked new employee, Allan Alcorn, to design an arcade version with similar game mechanics. Alcorn figured out that he could build such a game purely with digital circuitry—no programming involved. He took a black-and-white television set and placed it inside a wooden cabinet, soldering the circuiting to boards as needed. It was good enough to develop a prototype and test out, so Bushnell and Dabney installed a coin collector in the case and charged 25 cents per game. It was "launched" in 1972 in Andy Capp's Tavern in Silicon Valley as a test run.

The game consisted of a screen interface divided in half as two sides of a playing court. On either side were two vertical sticks or paddles that players moved up and down as a ball bounced between the sides of the screen. Scores increased each time a player's opponent failed to volley the ball back to the opposing player. The game was an instant hit. Two weeks later, the bar's owner called up and told the engineers to come and fix their contraption: nobody could play it anymore, because the coin box was jammed with quarters.

SEE ALSO PDP-1 (1959), *Spacewar!* (1962)

The Pong arcade game and coin box placed in Andy Capp's Tavern in Silicon Valley in 1972. Each player controls their paddle by turning the respective knob.

First Cell Phone Call

Martin Cooper (b. 1928)

On April 3, 1973, Motorola® employee Martin Cooper did something no one else had ever done before: he made a phone call while he walked down the street. It was the first time a call had been made on a handheld cellular telephone, and its key developer decided not to call his mom but—whom else?—his chief rival at Bell Labs to rub it in. With a journalist and photographer in tow to publicize the event, and pedestrians watching slack-jawed, those first words were: "Joel, this is Marty. I'm calling you from a cell phone, a real, handheld, portable cell phone."

The call was made on Sixth Avenue in New York City between Fifty-Third and Fifty-Fourth Streets. Cooper's only concern was whether the phone would work when he turned it on.

It took Cooper's team just five months to build the prototype using existing technology from their research labs. Without the advent of large-scale integrated circuits, Motorola engineers had to jam thousands of inductors, resistors, capacitors, and ceramic filters in a device that would be lightweight enough to carry. The prototype weighed 2.5 pounds, stood 11 inches tall, and cost $1 million in today's dollars to produce.

Up until then, the industry (with AT&T in the lead) had focused on placing mobile technology in the car, not in people's hands. Cooper and his team believed that AT&T's vision was too limited. As Cooper explained to the BBC years later in a retrospective interview, he wanted to create "something that would represent an individual, so you could assign a number not to a place, not to a desk, not to a home, but to a person."

It would take 10 years for the prototype to be released as a commercial product, due in large part to the lack of existing towers and infrastructure that had to be built. Called the *DynaTAC 8000x*, it took 10 hours to charge for 30 minutes of talk time. This is the same phone Michael Douglas famously used in the movie *Wall Street* to talk to his desk-bound protégé while he watched the sunrise from the beach. It cost $3,995, which, adjusted for inflation, would be around $9,000 today.

SEE ALSO *Star Trek* Premieres (1966), iPhone (2007)

Martin Cooper with the first portable handset. Cooper made the world's first mobile phone call on April 3, 1973, to his rival Joel Engel at Bell Labs.

Xerox Alto

Butler Lampson (b. 1943), **Charles P. Thacker** (1943–2017)

Designed and built at the Xerox Palo Alto Research Center (PARC), the Alto was the world's first personal computer to be controlled by a graphical user interface—the forerunner of Apple's Macintosh OS, Microsoft's Windows®, and many other machines.

The Alto was based on the vision of interactive computing embodied in the oN-Line System (NLS) and presented by Douglas Engelbart at the December 9, 1968, demo to the Joint Computer Conference in San Francisco—the "Mother of All Demos." But whereas NLS required a massive mainframe to run multiple users and was difficult to learn, the Alto was created to be an easy-to-use personal computer.

The Alto was the forerunner of what would come. The system had a page-sized screen that displayed the computer's graphical user interface. The user interacted with the computer using a keyboard and mouse. He or she could store files on a personal hard drive, communicate with other systems over an Ethernet network, create documents with fonts and graphics using Bravo (the world's first What You See Is What You Get—WYSIWYG—word processor), and print them over a network to the first-ever laser printer.

The Alto also ushered in a revolution in software. Although the Alto's processor was initially programmed in machine language and a system programming language called *BCPL* (the forerunner to C programming language), PARC researchers used the Alto to develop Smalltalk, a sophisticated object-oriented language.

PARC made roughly 2,000 of the machines, donated 500 to various university laboratories, and used the rest inside Xerox to support research. In 1979, Jef Raskin (1943–2005), who started the Macintosh project at Apple, arranged for Steve Jobs and other key members of the Apple Lisa and Macintosh teams to visit the PARC lab so they could see a machine with bitmapped graphics and a mouse in action; they returned to Apple inspired, having seen the future. In February 1981, Bravo's lead developer, Charles Simonyi (b. 1948), quit Xerox and joined Microsoft to write a word processor that became Microsoft Word®. For his work designing the Alto's software, Butler Lampson was awarded the 1992 A.M. Turing Award.

SEE ALSO Object-Oriented Programming (1967), Mother of All Demos (1968), Laser Printer (1971), Microsoft and the Clones (1982), Macintosh (1984)

A woman edits an electronic document with a Xerox Alto computer, the first personal computer to be controlled by a graphical user interface.

Data Encryption Standard

Horst Feistel (1915–1990)

In the late 1960s, Lloyds Bank in the United Kingdom asked IBM to create an unattended cash-dispensing machine—what we now call an *automatic teller machine*, or ATM. IBM realized that it needed to encrypt the data sent between the bank and the machine, lest thieves splice the telephone wires and convince the machine to dispense all of its cash. So IBM charged its newly created cryptography research group headed by Horst Feistel with the task. The group created an algorithm called *Lucifer*.

Lucifer encrypted blocks of data using 128-bit long keys. The algorithm was unbreakable, as far as anyone knew, meaning that there was no way to find the secret key used to encrypt a message other than by trying all possible keys—an impossible task.

In May 1973 and again in August 1974, the US National Bureau of Standards (NBS) invited cryptographers to submit their algorithms in a competition to create a national encryption standard. Lucifer was the best submission that NBS received. But when NBS finally adopted the algorithm as the Data Encryption Standard (DES), two important changes were made at the request of the US National Security Agency (NSA): the key size was cut from 128 bits to 56, and the way that the algorithm used its keys became significantly more complicated. Some academics criticized the move, claiming that the NSA had intentionally weakened the algorithm.

It turns out that the agency had actually strengthened the algorithm. The NSA had discovered an attack on Lucifer using a classified cryptanalytic technique called *differential cryptanalysis*. But the NSA couldn't explain this in 1974. Academics independently discovered differential cryptanalysis two decades later.

DES remained in use well into the 1990s, when the nonprofit Electronic Frontier Foundation constructed a special-purpose DES-cracking machine for roughly $250,000. From that point on, it was clear that a single application of DES was not sufficient to keep secrets safe. In 1999, many users started using Triple DES, in which the algorithm is used three times, with three different keys, for an effective key length of 168 bits.

Today Triple DES has largely been replaced by the Advanced Encryption Standard (AES).

SEE ALSO Advanced Encryption Standard (2001)

Engraving of Lucifer by Gustave Doré, for John Milton's Paradise Lost. *The US National Bureau of Standards adopted the Lucifer algorithm as the Data Encryption Standard in 1974.*

First Personal Computer

Henry Edward "Ed" Roberts (1941–2010)

For all the technology contributions and influential people who helped launch the personal computer revolution, the Altair 8800 is generally considered the machine that sparked it. It was designed by American engineer Ed Roberts and his team at Micro Instrumentation and Telemetry Systems (MITS).

In 1974, the only way an individual could own a personal computer was to build one. Microprocessors had been on the market for three years, but the only way a hobbyist could get a personal computer was to draw a circuit diagram, fabricate a case, and purchase the needed parts from a dozen or more different companies. The Altair changed that.

Based in Albuquerque, New Mexico, MITS was founded in 1969 to make electronics kits for hobbyists. In 1974, the company introduced the world's first microcomputer kit. Roberts wanted a microprocessor more powerful than the Intel 4004 and 8008 already on the market and negotiated a low price for the up-and-coming Intel 8080 chip. Though they were normally $300 apiece, he was able to get them in bulk for $75 a chip. The Altair kit included a metal case, power supply, boards, and assembly instructions. It did not have a keyboard or monitor. Instead, programs and data were toggled into the computer using the front panel switches; results were displayed with lights. The user could only create programs to make the lights blink. (An RS-232 interface card was available as a separately purchased accessory.)

Unbeknownst to Roberts, there was a sizable customer base eager for that experience. Roberts projected he would initially sell around 200 computers but within seven months had shipped more than 5,000, due in part to the machine's appearance on the cover of *Popular Mechanics* magazine in January 1975 with the caption "World's First Minicomputer Kit to Rival Commercial Models." That edition caught the attention of Bill Gates and Paul Allen, who shortly thereafter approached MITS and offered to write the first programming language for the Altair. This would become Microsoft's initial product: Altair BASIC.

SEE ALSO First Microprocessor (1971), IBM PC (1981), Microsoft and the Clones (1982)

Cover of the January 1975 issue of Popular Electronics, *featuring the MITS Altair 8800.*

HOW TO "READ" FM TUNER SPECIFICATIONS

Popular Electronics

WORLD'S LARGEST-SELLING ELECTRONICS MAGAZINE   JANUARY 1975/75¢

PROJECT BREAKTHROUGH!

World's First Minicomputer Kit to Rival Commercial Models...
"ALTAIR 8800"   SAVE OVER $1000

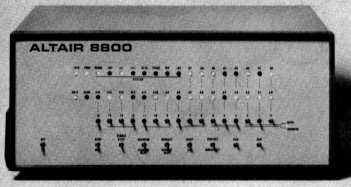

ALTAIR 8800

ALSO IN THIS ISSUE:

● **An Under-$90 Scientific Calculator Project**
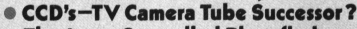
● **CCD's—TV Camera Tube Successor?**

● **Thyristor-Controlled Photoflashers**

TEST REPORTS:

Technics 200 Speaker System
Pioneer RT-1011 Open-Reel Recorder
Tram Diamond-40 CB AM Transceiver
Edmund Scientific "Kirlian" Photo Kit
Hewlett-Packard 5381 Frequency Counter

18101

Adventure

William Crowther (b. 1936), Don Woods (b. 1954)

Adventure, later renamed *Colossal Cave Adventure*, was an interactive, text-based simulation of exploring Kentucky's Mammoth Cave created by William Crowther, a programmer who helped develop ARPANET. *Adventure* wasn't just any simulation—it was a wildly popular game in which the player navigated through the cave on an expedition looking for treasure. Players typed short instructions into the command line using plain English—no need to read a manual—and then received responses back in plain English.

A caving enthusiast, Crowther had already mapped the inside of Mammoth Cave when he decided to use it as the subject for the game. He used the natural features and artifacts in the cave as guideposts for the player as he or she made decisions about where to go next. *Adventure* launched what would come to be known as "interactive fiction" games. These were games that weaved narrative, logic, and puzzles into a larger story that could branch in different directions, depending upon what the player wanted to do. *Adventure* was also the inspiration for other games, including *Rogue*, a rather addictive dungeon-exploration game that shipped with Berkeley UNIX and inspired its own subgenre ("roguelike") of dungeon-exploration games.

Originally developed for the PDP-10, *Adventure* consisted of 700 lines of FORTRAN code and 700 lines of data that described 78 map locations, 66 rooms, and 12 navigation messages. The first expansion of the game was done by Don Woods, a grad student at Stanford in 1976. With Crowther's permission, Woods amplified the fantasy elements in the game, reflecting his affinity for the writings of J. R. R. Tolkien. Different versions of *Adventure* were created over the years, and the game led directly to *Zork* and other popular games by Infocom, an MIT spinoff founded in 1979.

Adventure's historic influence is notable in the hacker community, which continues to perpetuate unique phrases and words from the game in other contexts. Two favorites include the magic word *xyzzy* and "YOU ARE IN A LITTLE MAZE OF TWISTY PASSAGES, ALL DIFFERENT."

SEE ALSO *Spacewar!* (1962), ELIZA (1965)

The Colossal Cave Adventure *video game—one of the first interactive fiction games—on a VT100 terminal.*

YOU ARE STANDING AT THE END OF A ROAD BEFORE A SMALL BRICK BUILDING.
AROUND YOU IS A FOREST. A SMALL STREAM FLOWS OUT OF THE BUILDING AND
DOWN A GULLY.
>ENTER BUILDING
YOU ARE INSIDE A BUILDING, A WELL HOUSE FOR A LARGE SPRING.
THERE ARE SOME KEYS ON THE GROUND HERE.
THERE IS A SHINY BRASS LAMP NEARBY.
THERE IS FOOD HERE.
THERE IS A BOTTLE OF WATER HERE.
>GET LAMP
OK
>INVENTORY
YOU ARE CURRENTLY HOLDING THE FOLLOWING:
BRASS LANTERN
>QUIT
DO YOU REALLY WANT TO QUIT NOW?
>Y
OK

YOU SCORED 27 OUT OF A POSSIBLE 350, USING   4 TURNS.
YOU ARE OBVIOUSLY A RANK AMATEUR. BETTER LUCK NEXT TIME.
TO ACHIEVE THE NEXT HIGHER RATING, YOU NEED 9 MORE POINTS.

.■

The Shockwave Rider

John Brunner (1934–1995)

For all of humanity's technical achievements, equally important are the voices that reflect upon the social changes and new norms that emerging technologies might bring. These tales can be particularly insightful when they envision an entire society that has yet to exist. One of the most famous of these is British author John Brunner's 1975 novel *The Shockwave Rider*. Influenced heavily by Alvin Toffler's 1970 nonfiction bestseller *Future Shock*, which concerns the negative impact of accelerated change and information overload on people, *The Shockwave Rider* describes in salient details a world in which data privacy and information management are abused by those in power and computer technology dominates individuals' everyday lives.

The story revolves around Nick Haflinger, a gifted computer hacker who uses his phone-hacking skills to escape from a secret government program that trains highly intelligent people in a dystopian 21st-century America. The government and elitist organizations maintain control of society through a hyperconnected data and information net that keeps the general population ignorant of the world around them. Prominent themes in the book include using technology to change identities, moral decisions associated with data privacy and surveillance, and the mobility of self when the value of personal space and individuality is deemphasized.

The Shockwave Rider is also notable for coining the phrase *worm* as a computer program that replicates itself and propagates through computer systems. In the book, Haflinger employs different types of "tapeworms" and "counterworms" to alter, corrupt, and liberate data in the net to his advantage.

The Shockwave Rider is generally credited with being an early influence on the emergence of the 1980s sci-fi cyberpunk genre, in which plots focus on unanticipated near-future dystopias, societal conflict, and warped applications of technology. Well ahead of its time, it shows how computer technology is not just a tool to extend human cognition and improve productivity, but also an instrument that can enable the worst extremes of human nature.

SEE ALSO "As We May Think" (1945), *Star Trek* Premieres (1966), Mother of All Demos (1968)

Cover of the 1976 Ballantine Books edition of The Shockwave Rider, *by John Brunner.*

24853/$1.50

JOHN BRUNNER

BALLANTINE BOOKS SF

**The Most Electrifying Science Fiction
Novel of the Year**

THE SHOCKWAVE RIDER

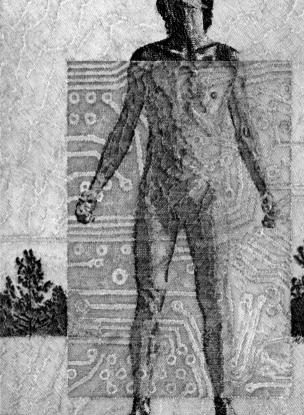

AI Medical Diagnosis

Edward Shortliffe (b. 1947)

MYCIN was the first expert knowledge system to prove that a computer program could outperform physicians and medical students in diagnosing a specific medical problem. MYCIN specialized in recommendations for antimicrobial therapy aimed at individual patients with severe blood infections, such as meningitis.

Research for MYCIN began in 1972 as physician and computer scientist Edward Shortliffe's doctoral thesis at Stanford. The program used early AI techniques based on rule-based knowledge representations of what a human expert knows. To create the rules, information engineers discussed patient case histories with antimicrobial medical experts. The engineers captured the data as a series of IF-THEN statements, which were then compiled into the MYCIN system. MYCIN provided judgments and recommendations to physicians by modeling the question-and-answer style typically seen in a doctor-patient exchange. The physician would sit at a computer terminal and answer questions about his or her patient in response to the questions generated by the system. Soon the machine would give a diagnosis, or at least a recommendation.

MYCIN had three interrelated parts. A consultation system provided therapeutic advice based on domain-specific knowledge. An explanation system provided the rationale, reasoning, and motivation behind the conclusion reached, the recommended therapeutic advice, and the line of questioning used for the diagnosis. And, finally, a knowledge acquisition system allowed an expert or physician to easily update the static knowledge base.

For the test that proved the system's accuracy, MYCIN's recommendations for antimicrobial treatment of 10 patients were compared with the choices made by nine microbial experts. Eight independent evaluators with expertise in the treatment of meningitis assessed the results. MYCIN received a score of 65 percent compared to the human experts, who received scores ranging from 42 percent to 62 percent.

Despite its success, MYCIN was never implemented in a working environment. The computational resources required for end users such as hospitals were infeasible at the time, and questions surrounding the ethics and legality of acting for or against therapeutic conclusions reached by MYCIN still had to be resolved.

SEE ALSO "As We May Think" (1945), DENDRAL (1965), Mother of All Demos (1968)

Magnetic resonance imaging (MRI) of the brain, performed to rule out eosinophilic meningitis in a 13-year-old boy. MYCIN specialized in recommendations for antimicrobial therapy for patients with infections such as meningitis.

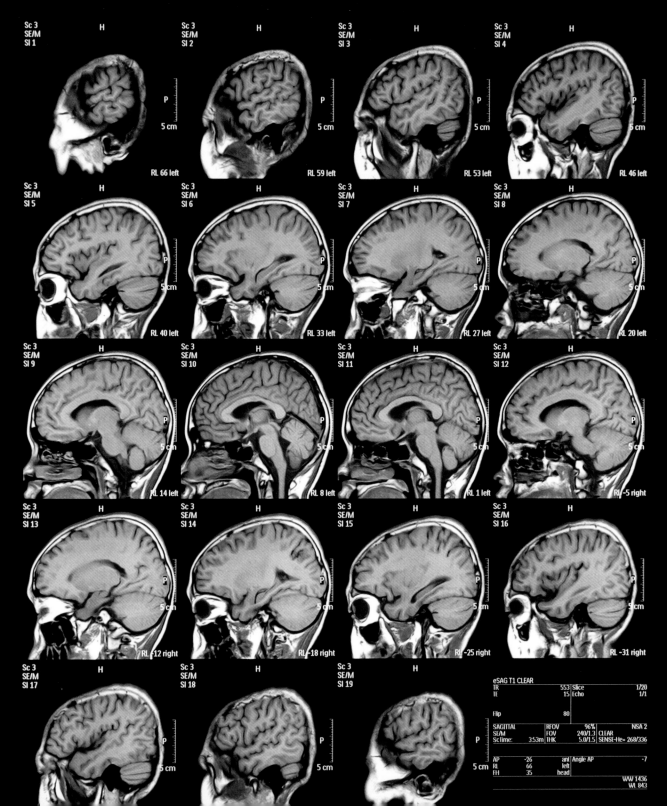

BYTE Magazine

Wayne Green (1922–2013), **Virginia Londner Green** (dates unavailable)

Wayne Green and his ex-wife, Virginia Londner Green, created *BYTE* magazine in 1975 based on the tremendous popularity of the computer-related articles that appeared in their amateur radio magazine, 73. The two hired Carl Helmers, the self-publisher of a monthly newsletter called the *Experimenter's Computer System* (ECS), to be the editor. *BYTE* took over the newsletter's mailing and, with the backing of Green Publishing, expanded significantly. *BYTE* is generally considered the first personal computing magazine, representing a passionate community of interest that would turn into a computer revolution.

The first issue of *BYTE* came out around the same time as the initial home-computer kits began to hit the market and had several firsts, including Microsoft's first advertisement. Early issues of *BYTE* covered topics such as: "Which Microprocessor for You?," "Assembling Your Assembler," and "Build a Graphics Display." The magazine also considered the more philosophical aspects of early personal computing, such as "What Is This Process—Designing a Program?" and the practical reasons one would want to build or have a PC in the home—for example, checkbook balancing, recipe converting, game playing, and even hosting a computerized remote security system.

BYTE was also famous for its illustrated covers by artist Robert Tinney (b. 1947), who would use his nontechnical eye to craft visual metaphors based on each issue's theme while accounting for the bigger concept of how computers fit into popular culture. *BYTE* had a listing of "computer clubs" in California, Colorado, Connecticut, North Carolina, New York, and Texas. Among them was Silicon Valley's "Homebrew Computer Club," which played a role in Steve Wozniak's invention of the Apple computer. And *BYTE* was home to science fiction author Jerry Pournelle (1933–2017), who wrote a popular column, The View From Chaos Manor. *BYTE* remained dominant into the 1990s, when declining readership and advertising revenue resulted in the sale of the magazine to CMP Media in May 1998 and suspended publication two months later. CMP made several attempts to run *BYTE* as a web-only publication until 2009, when it entered the halls of computer history.

SEE ALSO First Personal Computer (1974), Homebrew Computer Club (1975)

Cover of the July 1982 issue of BYTE, *the first personal computing magazine.*

BYTE
the small systems journal

JULY 1982 Vol. 7, No. 7
$2.95 in USA
$3.50 in Canada/£1.85 in U.K.
A McGraw-Hill Publication

COMPUTERS IN THE ARTS AND SCIENCES

Homebrew Computer Club

Fred Moore (1941–1997), **Gordon French** (dates unavailable)

Like the early American explorer Davy Crockett, "King of the Wild Frontier," the members of the legendary Homebrew Computer Club were trailblazers who helped initiate and lead the PC revolution. The club, whose theme was "Give to help others," was a place for like-minded enthusiasts to find each other and swap knowledge, demonstrate work, trade information about software and hardware design, share schematics and experiences, learn about computing publications, access parts through bulk equipment purchase, make build-it-yourself computing more accessible, and—most of all—push the envelope of what a homebrew computer could do.

The momentum behind the group's existence was fueled, in part, by the Altair 8800 kit, which hit the market around the same time the club was formed. As the first successful PC kit, the Altair was the missing link that opened a world of experimentation and innovation to individuals on a scale that had not previously existed.

The club's first meeting occurred in Menlo Park, California, in March 1975 in the garage of cofounders Gordon French and Fred Moore. As the club grew, meetings were moved to the Stanford Linear Accelerator Center in Menlo Park. Members included a parade of notable technologists, including Steve Wozniak and Steve Jobs, who gave away schematics of the Apple I computer at the club, as well as infamous phone phreaker John Draper (a.k.a. "Captain Crunch"), who possessed the useful skill of making free long-distance telephone calls.

Many of Homebrew's members were hardware junkies who had not given much consideration to the software side of what their inventions could do. "Software is a lot more difficult to build than hardware," noted Gordon French in the second issue of the Homebrew Computer Club's newsletter.

The Homebrew Computer Club brought together a community of enthusiastic talent whose pure curiosity, excitement, and sense of exploration about technology for technology's sake profoundly influenced what would grow to be ground zero of the PC industry.

SEE ALSO First Personal Computer (1974), *BYTE* Magazine (1975), Apple II (1977), Microsoft and the Clones (1982)

The cover of the Homebrew Computer Club's April 12, 1975, newsletter.

AMATEUR COMPUTER USERS GROUP   NEWSLETTER   HOMEBREW COMPUTER CLUB
Issue number two   Fred Moore, editor, 558 Santa Cruz Ave., Menlo Park, Ca. 94025   April 12, 1975

The Mythical Man-Month

Frederick Brooks (b. 1931)

In 1963, IBM was sinking huge amounts of resources into finishing its new OS/360 operating system in time for the launch of System/360. The new operating system—the most complex ever created—was designed to be a single, unified system for IBM's new family of computers. But while IBM's hardware was on target for the 1964 launch, the software wasn't.

Faced with delays and overruns, the project's manager did what any manager might do in that situation: he hired more programmers. Much to Fred Brooks's surprise, OS/360 fell even further behind schedule. So was born Brooks's Law: "Adding manpower to a late software project makes it later." The law appears in his classic collection, *The Mythical Man-Month: Essays on Software Engineering*, first published in 1975 and required reading for generations of computer science graduates.

The book, which has become the bible of software engineering, describes phenomena such as the "second-system effect," which is the tendency of designers to put into the second version of a program all the features that were dropped from the first version—resulting in 2.0 versions that are bloated and buggy. Brooks also details techniques that he developed to manage the hundreds of people who were working on the OS/360 project—many of which are still used today.

Brooks left IBM in 1964 to take a faculty position at the University of North Carolina at Chapel Hill, where he established the school's computer science department. IBM announced System/360 the same year. Brooks issued a 20th-anniversary edition of *The Mythical Man-Month* in 1995 with four new chapters. Foremost among them was his paper "No Silver Bullet," in which Brooks argues that there is no single technique ever invented in computing that has improved productivity, reliability, or simplicity by a factor of 10. Instead, he writes, the key to great design is to identify talented designers early in their careers, mentor them, and give them opportunities to design systems and interact with and stimulate other exceptional designers.

SEE ALSO IBM System/360 (1964)

Computer engineering professor Frederick Brooks speaks at the Turing Centennial Conference in Manchester, 2012.

Public Key Cryptography

Ralph Merkle (b. 1952), **Whitfield Diffie** (b. 1944),
Martin Edward Hellman (b. 1945)

For more than 2,000 years, cryptography's Achilles' heel was the need for two parties to privately, in person, agree upon a secret key (a string of letters and numbers) before they could send coded messages back and forth. This limitation didn't pose a problem for diplomats and generals, who received keys in person before heading off on their assignments. But with the birth of email in 1971, computer scientists realized that the only way for email messages to be secure would be for them to be encrypted.

In 1974, Ralph Merkle, then a senior at University of California, Berkeley, came up with an inelegant, yet groundbreaking solution to this problem as part of a class project. Merkle's approach let two people agree upon a cryptographic key by first exchanging millions of cryptographic puzzles over the internet. Merkle's professor didn't understand the significance of the solution, so Merkle dropped the course. Merkle wrote up his idea and submitted it to *Communications of the* ACM, then the premiere journal of computer science; the paper was rejected with the comment, "Experience shows that it is extremely dangerous to transmit key information in the clear."

The following year, at Stanford in an embryonic cryptography research group headed by professor Martin Hellman, student Whitfield Diffie pursued an idea similar to Merkle's but with a more efficient solution based on number theory. Now called the *Diffie-Hellman key exchange*, the system allows two parties to exchange specially crafted numbers from which each can derive the same encryption key. An outside observer— perhaps a wiretapper—cannot derive the key. The work was presented at the National Computer Conference in June 1976.

As for Merkle, his paper was finally published by *Communications* in 1978— complete with an apologetic note from the journal's editor.

Diffie and Hellman were awarded the A.M. Turing Award in 2015 for their key-exchange algorithm. Merkle went on to invent cryptographic hashing and hash trees— the basis of cryptocurrencies like Bitcoin. He won the IEEE Hamming Medal in 2010.

SEE ALSO RSA Encryption (1977), Bitcoin (2008)

Public key cryptography allows two parties to exchange specially crafted numbers from which each can derive the same encryption key.

Tandem NonStop

James Treybig (b. 1940)

The big idea behind James Treybig's Tandem Computers was to create computers that were so reliable, they never crashed. Delivering such computers—and doing it at a price that was economically competitive—was the technical magic that made Tandem, founded in 1974, one of the fastest-growing companies in America in the early 1980s.

Tandem took a fundamentally different approach to building a reliable computer than its peers. Most companies achieved reliability by having a master computer and a "hot-spare": if the master crashes, the spare takes over. This approach was wasteful, if nothing else; the spare was an entire computer that wasn't doing anything. It also didn't result in systems that were dramatically more reliable. If the master crashed because of a software problem, frequently the same problem would affect the spare.

Tandem avoided single points of failure by using a radically different design. The NonStop design called for the single mainframe to be built from multiple nodes, each of which had multiple processors interconnected with redundant communications paths. Each node, in turn, had between two and 16 individual processors, each of which had its own power supply, memory, backup battery, and input/output channel. Like the nodes, the processors were connected with redundant communications channels as well. Instead of sharing memory, the processors sent messages to each other.

In a typical operation, the different parts of the system watched over each other. If an error was detected, the system determined the module in which the error took place and isolated that module from the rest of the system. If a piece of hardware failed, the system would then run a diagnostic on the hardware and determine if it was a permanent failure requiring a replacement: failed boards could simply be unplugged from the system and replaced, without the need to shut down the entire system. Software was similarly designed to be highly compartmentalized, with data being constantly backed up, so that a software error could also be isolated and corrected.

Tandem completed its first NonStop system in 1975 and delivered it to Citibank® in May 1976. Customers soon discovered that the NonStop design, besides being inherently reliable, was also inherently scalable: by doubling the number of nodes, a customer could have a NonStop system that ran twice as fast.

SEE ALSO Connection Machine (1985)

Advertisement for the Tandem 16 NonStop computer, which achieved reliability by using multiple processors interconnected via redundant communications paths.

Dr. Dobb's Journal

Bob Albrecht (dates unavailable), **Dennis Allison** (dates unavailable),
Jim C. Warren Jr. (b. 1936)

Mention *Dr. Dobb's Journal* to programmers of a certain age and you may get a nostalgic sigh and smile. Founded by Bob Albrecht, a member of the legendary Homebrew Computer Club, *Dr. Dobb's* was a technical journal that focused on sharing knowledge about software development and building best practices among those who wrote code. According to the first volume, the publication served as a "communication medium concerning the design, development, and distribution of free and low-cost software for the home computer." Unusual in that it focused almost exclusively on programming rather than hardware, *Dr. Dobb's* first tackled one of the fundamental challenges of early personal computer limitations: memory. The first personal microcomputers, such as the Altair, only had 4 kilobytes of memory. Developing programs small and functional enough to work in that environment was a challenge worth solving to expand the functionality of the early machines.

Dr. Dobb's addressed the limited memory of early computers by publishing listings of programs that people could type directly into a computer and run. The first few editions focused on Tiny BASIC, a simple version of the BASIC programming language that needed only 3 kilobytes of memory. Its developer, Bob Allison, a lecturer at Stanford, wrote a series of articles for *Dr. Dobb's* that included the complete code for Tiny BASIC. The exposure and accessibility of the code enabled others to develop interpreters for processors other than the Intel 8080 processor used by the Altair 8800. After the series of Tiny BASIC articles, the name of the journal was changed to *Dr. Dobb's Journal of Computer Calisthenics & Orthodontia.*

An early editor of *Dr. Dobb's*, Jim Warren Jr., later published compendia of the journal. In it he captured his observations about changes and new features that emerged from the microcomputing field over the publication's first year, adding his vision of where the technology was heading and the special category of reader who chose to follow and contribute to it. *Dr. Dobb's* finally shuttered its doors in 2014 after 38 years.

SEE ALSO First Personal Computer (1974), Homebrew Computer Club (1975)

The cover of volume 1 of Dr. Dobb's Journal of Computer Calisthenics & Orthodontia.

Dr. Dobb's Journal of

COMPUTER

Calisthenics & Orthodontia

Running Light Without Overbyte

Volume One   People's Computer Company, Box E, Menlo Park, CA 94025   1976

A Reference Journal

for Users of

Home Computers

 Volume One

People's Computer Company

RSA Encryption

Ronald L. Rivest (b. 1947), **Leonard Adleman** (b. 1945),
Adi Shamir (b. 1952), **Clifford Cocks** (b. 1950)

The inventors of public key cryptography envisioned a future in which you could send somebody an encrypted email by looking up the person's public key in a directory, using the key to encrypt the email, and then sending it. When the person received the email, he or she would use his or her corresponding private key to decrypt it. There was just one problem: the Diffie-Hellman key exchange algorithm was interactive—it didn't have persistent keys that could be published. More work was needed.

Taking up the challenge, three friends—all professors at MIT—tried their hand. Over the following months, Ron Rivest and Adi Shamir developed many mathematical approaches for creating public and private keys. Each time, Len Adleman was able to crack the system, which would allow an attacker to decrypt a message from the public key, without knowing the corresponding private key.

In April 1977, the three were having Passover dinner and, naturally, discussing their project. After several cups of wine, they hit upon the solution: the private key would be made by choosing two large prime numbers and multiplying them together. Using basic number theory, they could show that cracking the public key would require the ability to factor the resulting number. As long as an attacker couldn't factor the number, messages encrypted with the public key would be uncrackable.

In the years that followed, the so-called RSA system was built into Netscape® Navigator®, leading to the commercialization of the internet, and into smart cards, protecting credit card transactions from fraud. The RSA trio shared the 2002 A.M. Turing award for their work.

But there is a threat on the horizon. Because quantum computers—computers that perform calculations using quantum mechanics—can factor numbers almost instantly, the US National Institute of Standards and Technology is working on a so-called "post-quantum cryptography" project that involves, among other things, finding a replacement for RSA.

British cryptographer Clifford Cocks independently discovered the same mathematical technique in 1973, which the Brits kept classified until 1997.

SEE ALSO Public Key Cryptography (1976), RSA-129 Cracked (1994), Quantum Computer Factors "15" (2001)

RSA encryption relies on public and private keys. The private key is created by multiplying together two large prime numbers; to crack the public key, an attacker would have to factor the resulting number.

Apple II

Steve Jobs (1955–2011), **Steve Wozniak** (b. 1950),
Randy Wigginton (dates unavailable)

If the Altair 8800 was the machine that put computers in the hands of individual hobbyists, then the Apple II was the machine that put computers in the hands of everyday people. Steve Wozniak, the lead designer, and Randy Wigginton, the programmer, demonstrated the first prototype at the legendary Homebrew Computer Club in December 1976 and, along with Steve Jobs, the team's financial wizard and chief promoter, introduced it to the public in April 1977 at the West Coast Computer Faire. The Apple II was the first successful mass-produced personal computer.

The Apple II was based on the Apple I, which Wozniak designed and built himself. The Apple I was sold as a single-board computer: purchasers needed to supply their own keyboard, monitor—or a television and a radio frequency (RF) modulator—and a case. The Apple II, in contrast, came with keyboard and case, although it still needed an RF modulator to display on a TV.

The Apple II was widely popular with techies, schools, and the general consumer. It offered BASIC in ROM, so users could start to write and run programs as soon the machine powered on. It came with a reliable audiocassette interface, making it easy to save and load programs on a low-cost, consumer-grade cassette deck. It even had color text, a first for the industry.

In 1978, Apple introduced a low-cost 5¼-inch external floppy drive, which used software and innovative circuit design to eliminate electronic components. Faster and more reliable than the cassette, and capable of random access, the disk turned the Apple II from a curiosity into a serious tool for education and business. Then in 1979, VisiCalc®, the first personal spreadsheet program, was introduced. Designed specifically to run on the Apple II, VisiCalc helped drive new sales of the computer.

The Apple II was a runaway success, with Apple's revenues growing from $775,000 to $118 million from September 1977 through September 1980. Apple ultimately released seven major versions of the Apple II. Between 5 million and 6 million computers would ultimately be sold.

SEE ALSO First Personal Computer (1974), *BYTE* Magazine (1975), Homebrew Computer Club (1975), VisiCalc (1979)

Apple II advertisement from the December 1977 issue of BYTE *magazine.*

Introducing Apple II. ™

First Internet Spam Message

Gary Thuerk (dates unavailable), **Laurence Canter** (b. 1953),
Martha Siegel (1948–2000)

On May 3, 1978, at 12:33 EDT, the world's first mass-mailed electronic marketing message—what we now call *spam*—blasted out to more than a hundred email accounts on the US Department of Defense's ARPANET. Sent by DEC employee Gary Thuerk, the email advertised an open house for DEC's new line of computers, the DECSYSTEM-2020.

Unaware that the early mail program had the ability to read email addresses from an address file, Thuerk entered all the addresses into the email's *To:* header; 120 fit in the header, and the remaining 273 overflowed into the message body, making the message both inappropriate *and* unattractive.

The reaction was overwhelmingly negative. The chief of the ARPANET Management Branch at the Defense Communications Agency responded with an email stating that the message "was a flagrant violation of the use of the ARPANET" and that the network was "to be used for official US Government business only."

But Richard Stallman at the MIT Artificial Intelligence Laboratory, a free-speech advocate, wrote in response to the complaints: "I get tons of uninteresting mail, and system announcements about babies being born, etc. At least a demo MIGHT have been interesting." Stallman did object, however, to the way the message was sent, writing, "Nobody should be allowed to send a message with a header that long, no matter what it is about."

The reference to unsolicited email as *spam* did not come into vogue until the 1980s. It originates from a 1970 *Monty Python* sketch about a group of Vikings in a cafeteria who drown out the other conversations in the room by repetitively singing "SPAM, SPAM, SPAM," referring to the canned-meat product manufactured by Hormel® Foods. The explosion in unsolicited bulk information has since spread to every digital medium. Although some of the messages remain "spam"—the same message being sent everywhere, to all recipients—increasingly messages are precisely customized to the individual recipients by algorithms that reference vast warehouses of personal data.

SEE ALSO First Electromagnetic Spam Message (1864), @Mail (1971)

The world's first unsolicited email, or "spam," was sent to ARPANET accounts on May 3, 1978. Today many email services are able to detect and isolate such messages into separate folders.

Sent Mail

Spam (372)

Trash

Minitel

Online games, shopping, chat, theater reservations, banking, and education—amazingly, this was everyday life for many French in the 1980s, well ahead of the internet and World Wide Web. In 1978, state-owned France Telecom came out with the Minitel—a small, mushroom-colored computer terminal that had a screen, keyboard, and traditional connection to landline telephone wires (but no microprocessor). The French government gave a free terminal to every French Telecom subscriber, and in 1982 Minitel went national. At its peak in the mid-1990s, there were approximately 25 million people participating in any of the 26,000 online services the system offered. And, yes, that included online pornography services, known informally as "Minitel Rose."

The technology behind Minitel was something called a *videotext*. It was not unique to France. Efforts at implementing versions of videotext were occurring in other places too, all with varying success, including Britain, Germany, Spain, Sweden, Japan, Singapore, Brazil, Australia, New Zealand, South Africa, Canada, and the United States. What was unique about Minitel was the government's enormous support—France Telecom claimed deploying Minitel would be cheaper than distributing printed phone books. It was a phenomenon well ahead of its time. There was even a service to order groceries online for same-day delivery. Minitel enabled the lines between ephemeral consumer need and physical fulfillment of that need to blur, well before industry giants such as Amazon came on the scene.

Minitel was finally shut down in 2012, its usage cannibalized by the internet, the World Wide Web, and mobile technology. At the end, Minitel's die-hard users were cattle farmers exchanging information about their herds and doctors communicating patient information to the national health service.

Minitel was a source of national pride in France. One speculation as to why Minitel never became the global alpha dog of the online revolution is that it was not an open platform. Still, Minitel was a communal phenomenon that gave the world a glimpse of how a core technology could achieve huge economies of scale if it hit the sweet spot with the end user.

SEE ALSO E-Commerce (1995)

Woman in a French village using Minitel during the 1987 Paris–Dakar Rally.

Secret Sharing

Adi Shamir (b. 1952), George Robert Blakley Jr. (b. 1932)

Let's say you put a copy of your will in a high-security safe with a combination lock. You want your lawyer to be able to open the safe only if you are dead. So you give the combination to your lawyer and hope that he or she honors your wishes. Alternatively, you can split the combination into two pieces and give the halves to two different lawyers. Only by working together can they open the lock.

But what if one of the lawyers is scatterbrained and loses his or her half? It might make more sense to hire three lawyers, create three sets of splits, and give each lawyer one split from two sets. This lets any two of the lawyers reassemble the secret. This approach to secret splitting becomes unworkable and rather ridiculous as the number of splits increases: if you have 11 lawyers and want any six to be able to open the lock, the key must be split into 462 pieces, with each lawyer receiving 252 pieces.

MIT professor Adi Shamir and Texas A&M professor George Blakley came up with similar solutions to this problem in 1979—solutions so elegant, efficient, and secure that they are still used today.

Shamir solved the problem with basic geometry. For example, let's say the "secret" to be protected is a single number. Draw a line on a graph that crosses the y-axis—a point called the *y-intercept*. Tell each lawyer a point on the line. Acting alone, no one can determine the y-intercept. But if any two get together, the y-intercept is trivial to compute—they just put their points on a graph and draw the line. Blakley's system is similar, but is based on defining a point in n-dimensional space.

What if you want three shares to reconstruct? In that case, instead of drawing a line on the graph, draw a parabola: whereas a line requires just two points to describe it, a parabola requires three. Now give each lawyer a point on the parabola: any three of them can reconstruct the curve and determine the secret y-intercept. Shamir's actual system is just a little more complicated than the description above. Instead of drawing nice curves on a two-dimensional number plane, it draws the curves in a finite number field, similar to the way the Diffie-Hellman key exchange and the RSA encryption algorithm are used in practice. And the numbers that are shared aren't the combinations to physical safes; they are encryption keys for encryption algorithms.

SEE ALSO Public Key Cryptography (1976), RSA Encryption (1977)

Secret sharing involves splitting confidential information into parts that are distributed among participants, who get some of or all the parts necessary to reconstruct the information.

VisiCalc

Dan Bricklin (b. 1951), Bob Frankston (b. 1949)

After graduating with an undergraduate degree in computer science from MIT in 1973, Dan Bricklin worked at DEC for a bit, then briefly at the FasFax Corporation, which manufactured cash registers. When he decided to go back and earn a master's degree in business administration from Harvard, he found that doing business analyses on paper was sheer drudgery. So Bricklin wrote a program on an Apple II in BASIC to automate the calculations and let him concentrate on the modeling.

Bricklin partnered with his friend Bob Frankston, who rewrote the program into the Apple II's machine code using a cross-assembler that he also wrote, which ran on a mainframe computer at MIT. In 1979, the two incorporated as Software Arts to further develop the program. A few months later, they licensed it to a company called Personal Software (soon renamed VisiCorp), which brought the program to the market. Over the following five years, more than a million copies of VisiCalc were sold.

There had been previous numeric-modeling tools on IBM mainframe computers, but VisiCalc was the first program to combine interactivity, automatic recalculation, in-place editing, and the look of a traditional spreadsheet. These features made VisiCalc incredibly easy to learn, and its advanced features for creating and editing formulas made it possible for nonprogrammers to build complex financial models almost immediately. Companies started buying $2,000 Apple II computers to run a $100 program.

But the Apple II was the wrong computer to launch a successful business application. The computer had a display of just 25 lines of blocky 40-column text and a puny 8-bit microprocessor, and it maxed out at just 48 kibibytes of RAM. In 1981, IBM introduced the IBM PC, with a 25-by-80 display and a 16-bit microprocessor that could address a maximum of 640 kibibytes of RAM—more than enough room to hold a complex financial model. Two years later, the Lotus Development Corporation released 1-2-3®, a spreadsheet that could take advantage of what the PC had to offer. Sales of VisiCalc plummeted from $12 million in 1983 to $3 million in 1984. In April 1985, Lotus bought Software Arts and immediately pulled the VisiCalc from the market.

SEE ALSO Apple II (1977)

The user's guide for VisiCalc, the first spreadsheet for personal computers.

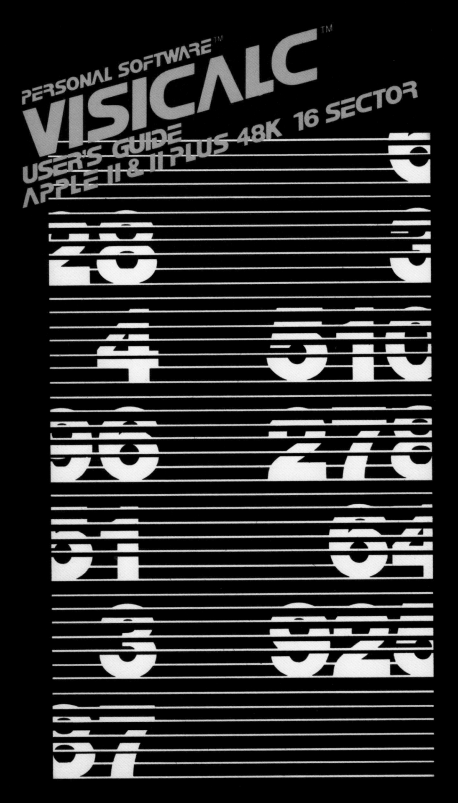

PERSONAL SOFTWARE™

VISICALC™

USER'S GUIDE

APPLE II & II PLUS 48K 16 SECTOR

Sinclair ZX80

Clive Sinclair (b. 1940), **Jim Westwood** (dates unavailable)

Costing just £79.99 for a kit and £99.99 for an assembled, fully functional computer, the Sinclair ZX80 was the United Kingdom's first mass-market PC—and the first computer to sell anywhere in the world for less than £100 (about US$200 at the time). Based on the Zilog Z80 microprocessor, the ZX80 came with 1 kibibyte of RAM and a 4-kibibyte ROM that contained a BASIC interpreter. The ZX80 was packaged in a white plastic case with a small membrane keyboard that was easy for children to type on, although a bit more difficult for adults. It included a built-in cassette interface for saving and loading programs, and a radio frequency (RF) modulator, so it could display directly on a consumer television set.

To save money, most of the RF modulation was done in software, so the screen annoyingly blanked out every time a key was pressed or a program was running. The display was 24 lines of 32 black-and-white characters. Each character could also display simple 2 × 2 graphics.

The ZX80 sold 50,000 units in its first year. The following year the computer's maker, Sinclair Research, launched the ZX81, a more stylish computer that fixed the screen-blanking problem by offering a so-called slow mode in which the screen never blanked out (but which stole computing cycles from other operations, hence the degradation in speed). The ZX81 also had an 8-kibibyte ROM that provided floating-point math for the BASIC interpreter, allowing the computer to be used for serious engineering computations. ZX81 sold 1.5 million units and cost £49.95 for the kit, £69.95 assembled. Both systems were designed by Jim Westwood and manufactured by Timex. In 1982, Sinclair released a color version of the machine, called the ZX *Spectrum*.

Many ZX machines had problems with overheating; some hobbyists responded by removing the computer from the case and putting it in a larger enclosure, adding an external keyboard in the process. Other expansions included 16 kibibytes of additional RAM and even a printer that sold for £49.95 and printed on aluminum-coated paper. Ultimately, more than a hundred companies sold aftermarket products for the ZX80/81.

Clive Sinclair was knighted for his success with the ZX80/81.

SEE ALSO BASIC Computer Language (1964)

The Sinclair ZX80, the United Kingdom's first mass-market personal computer.

Flash Memory

Fujio Masuoka (b. 1943)

Flash memory was invented in 1980 by Fujio Masuoka, then an employee at Toshiba® Corporation in Japan. Masuoka and his small team were trying to create memory that retained data in the absence of a power source, functioned as a solid-state device rather than a mechanical disc, and could store a lot of data at an affordable price (unlike magnetic cores). It would be four years before he was able to present his work to the broader semiconductor industry at an IEEE meeting in San Francisco.

It would take another two decades before flash would dominate the market for portable storage. In the meantime, there were all kinds of experiments in portable digital media, including Digital Audio Tape (DAT), MiniDiscs, and even tiny hard drives the size of matchbooks. All of these systems were mechanical, which made them subject to shock and vibration, impacted their reliability, and put constraints on their size and capacity.

Flash, in contrast, relied on the same basic semiconductor fabrication techniques that were used to create microprocessors and other integrated circuits. Each year the flash chips got smaller, cheaper, and faster, and increased in capacity. The invention of flash memory has made possible a whole range of new mobile devices, including smartphones, MP3 players, digital cameras, e-readers, and tablets. These uses had to wait until the price of flash fabrication dropped dramatically in the late 1990s.

Flash has also proven to be amazingly flexible. Individual chips can be soldered directly onto printed circuit boards like any other chip during manufacturing. Removable flash can be packaged first into CompactFlash® (CF) cards, and then into Secure Digital (SD) cards, which saw their capacity increase from 1 megabyte to 2 terabytes (a factor of 2,000) between 1999 and 2017.

Flash memory has also been integrated into simple, handheld devices with another technology—the Universal Serial Bus (USB)—creating portable storage devices that can be used with laptops, desktops, and other devices that do not have flash memory readers.

SEE ALSO USB Universal Serial Bus (USB) (1996), USB Flash Drive (2000)

A microphotograph of an MX25U4035Z 4-mebibit serial flash unit; the flash storage is the area on the left, while the control logic is on the right.

RISC

David Patterson (b. 1947), **John L. Hennessy** (b. 1952)

Professor David Patterson at the University of California, Berkeley, studied systemic inefficiencies inside the CPUs of his day and came up with an approach to optimize how computers managed machine instructions: the Reduced Instruction Set Computer (RISC). Many machine instructions were designed in the 1960s to make it easier for humans to program computers in assembly language. But by the 1980s, computers were mostly being programmed in high-level language. Patterson reasoned that the CPU could be radically simplified by removing the instructions that could be useful to a human programmer but not to a machine compiler.

With funding from the US military, Patterson started the RISC project in 1980. The following year a similar project started at Stanford under computer scientist John L. Hennessy. Both projects drew in part from ideas developed in the 1960s for the CDC 6600 by Seymour Cray.

Within a few years, both projects produced their first microprocessors. They were dramatically faster than the so-called Complicated Instruction Set Computers (CISC) on the market, most notably the x86 microprocessors manufactured by Intel. Still, the RISC chips remained a tiny part of the world's microprocessor market, largely because they couldn't run most of the software that had been written for Microsoft's DOS and Windows operating systems.

The Stanford team went on to create MIPS Computer Systems in 1984, a company that developed high-performance microprocessors that were adopted by Silicon Graphics® and (for a time) DEC. But MIPS had trouble competing against Intel, which had far more money to spend on research as a result of its larger customer base.

The big turning point for the RISC idea came in the 1990s, when chipmakers figured out how to put a RISC computer *inside* a CISC computer. That is, the CPU fetched x86 CISC instructions but translated them into RISC instructions inside the chip. This produced the best of both worlds: a computer that could run the legacy CISC code but had the speed and power advantages of RISC.

Today nearly all computers are RISC.

SEE ALSO Microprogramming (1951), Cray Research (1972), Microsoft and the Clones (1982)

Thirty-seven RISC microprocessors manufactured on a single silicon wafer at the IBM factory of Corbeil-Essonnes, France, February 1, 1992.

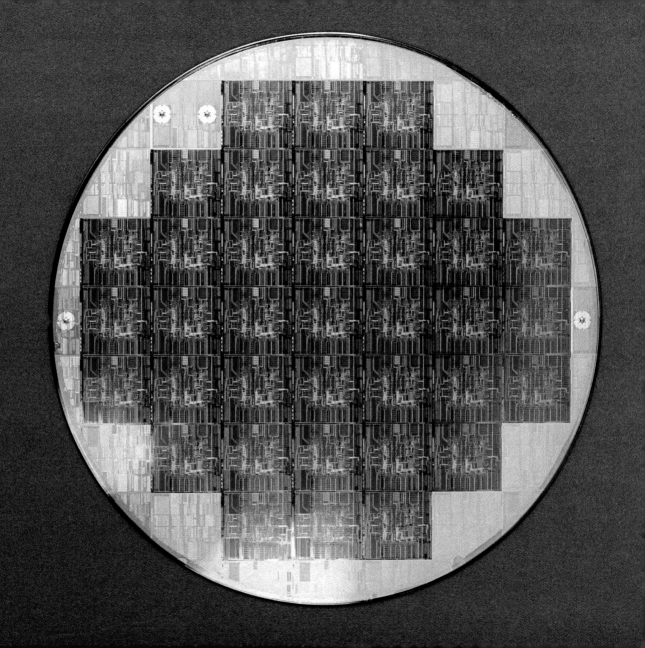

Commercially Available Ethernet

Robert Metcalfe (b. 1946), **David Boggs** (b. 1950),
Chuck Thacker (1943–2017), **Butler Lampson** (b. 1943)

While working on his PhD thesis, Robert Metcalfe ran across a paper on the ALOHANET wireless computer network that had been developed at the University of Hawaii. Eager to find out more, he flew to Hawaii to learn from ALOHANET's inventors.

Whereas ALOHANET sent data packets through the air, Metcalfe's design sent radio frequency energy through coaxial cables that had multiple taps, one for each computer, and a "terminator" at each end. Called *Ethernet*, after the nonexistent "ether" that many 19th-century scientists thought was the medium through which light propagated, the network proved to be a simple, cheap, and fast way to connect computers within a room or a building—a local area network (LAN).

Although Metcalfe is sometimes called the inventor of the Ethernet, the actual patent was filed by Xerox, where Metcalfe worked during and after graduate school, with David Boggs, Chuck Thacker, and Butler Lampson listed as co-inventors. Metcalfe left Xerox in 1979 and formed 3Com®, which worked with DEC, Intel, and Xerox to make Ethernet a computing-industry standard. The Institute for Electrical and Electronic Engineers (IEEE) adopted Ethernet as the IEEE 802.3 standard in June 1983.

The Ethernet standard specifies both the physical connection between computers and the logical structure of the packets that the network carries. But Ethernet doesn't specify the higher-level network protocols. As a result, DEC and Xerox invented their own network layers—DECNet and XNS (Xerox Network Systems). All of these proprietary networking technologies eventually lost out to Internet Protocol (IP) running over Ethernet.

By the end of the 1980s, companies were introducing versions of Ethernet that could run over twisted-pair wiring instead of coaxial cable. Eventually standardized as 10Base-T, twisted pair dramatically reduced wiring costs and increased reliability, because each computer had its own wire that led back to an Ethernet "hub." The original Ethernet ran at 10 megabits per second, but in 1995 so-called "fast Ethernet" running over twisted pair at speeds of 100 megabits per second was introduced, followed by a gigabit in 1999, and 10 gigabits per second (running over optical cables) in 2002.

SEE ALSO First Wireless Network (1971)

Ethernet cables and network switches.

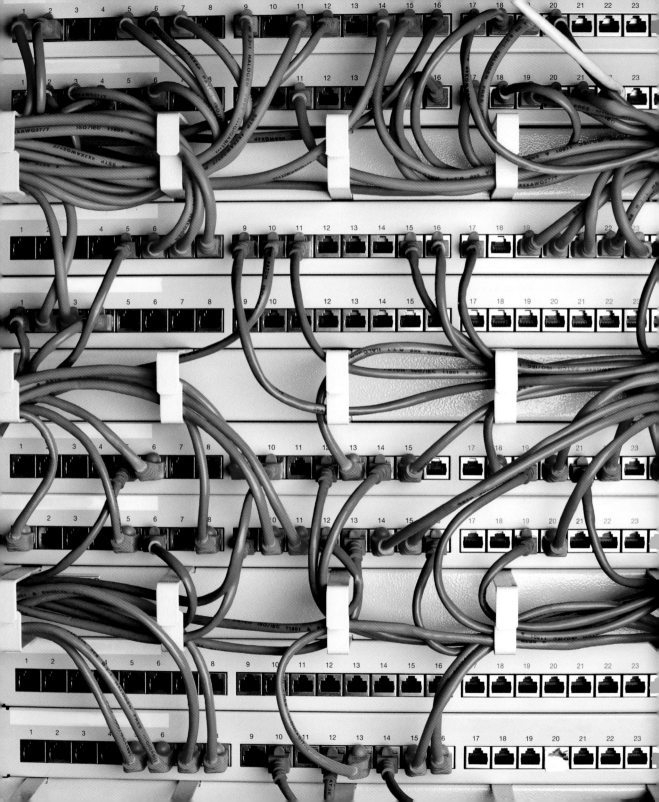

Usenet

Tom Truscott (b. 1953), **Jim Ellis** (1956–2001)

Designed by Tom Truscott and Jim Ellis at Duke University in 1979 and implemented the following year, Usenet is a distributed peer-to-peer electronic bulletin board system with no central control and few rules, other than the community standards of the users.

The network initially consisted of two computers connected by UUCP, the UNIX-to-UNIX Copy Protocol. But it was easy to add more computers to the network, and people did. Unlike the ARPANET, which was based on permanent connections, UUCP was based on computers making telephone calls to one another and transmitting messages that had been queued for delivery, so there was little cost to add new machines. Usenet layered a distributed message board on top of the modem-based network. Any message written and posted on one computer would be queued and eventually copied to every other computer on the network with a simple message-flooding protocol.

The Usenet network divided the message board into multiple *newsgroups*. Any user on any connected computer could write and "post" an article to a local message board. Readers could instantly read the article on the same computer, and it was soon copied to every other. Suddenly anyone on the network could be a publisher, and people who were geographically separated could meet others who shared their interests and carry out computer-moderated discussions.

Usenet popularized the development of formally curated Frequently Asked Questions (FAQ), and some of the groups were prone to the occasional personal attacks. Lacking central control, Usenet evolved community norms, volunteer moderators, and occasional intervention by the system administrators of the more-powerful backbone sites, the critical sites through which most of the articles had to flow, which gave the administrators the ability to censor articles (or authors) as they wished.

Usenet's eclipse was not caused by the World Wide Web but by the success of Usenet itself: throughout the 1990s and the 2000s, Usenet's *alt.binaries* newsgroups grew in size, with some of their content containing illegal pornography. The backbone providers responded by first banning the *alt.** hierarchies and eventually all of Usenet.

SEE ALSO ARPANET/Internet (1969), Telebit Modems Break 9600 bps (1984)

Usenet was an electronic bulletin board system, distributing queued messages to computers that called into the network.

IBM PC

William C. Lowe (1941–2013)

IBM nearly sat out the microcomputer revolution. In 1980, the field was dominated by companies such as Apple, Atari, and Commodore selling 8-bit computers for the home and office. One day William Lowe, the director of IBM's Boca Raton Labs, was approached by Atari about the possibility of having IBM rebrand and sell one of Atari's computers as its own—presumably a win-win deal for the two companies. Lowe took the idea to his management, who promptly turned it down and gave him 12 months to put together IBM's own microcomputer.

Calling it *Project Chess*, Lowe went on to break every rule in IBM's book. Instead of running the project with hundreds of engineers, Lowe launched with a team of 12. Instead of relying on IBM-developed hardware and software, the computer would be built with off-the-shelf components and already-existing software. Rather than selling it through IBM's extensive dealer network and servicing it with IBM employees, the computer would be sold through retail stores, whose employees would be trained to fix the machines when they broke. But Lowe decided to make repairs a rarity: every component of every machine was tested to the extreme, with the result of delivering to customers a reliable machine with zero defects.

The IBM Personal Computer 5150, as it was eventually branded, featured a 16-bit processor with 16 kibibytes of user memory, expandable to 256 kibibytes. It had a full-size, professional keyboard and could be ordered with a beautiful, high-quality monochrome display for text processing or a color screen capable of graphics. There were five user-serviceable expansion slots, opening the door for a rich aftermarket of both hardware and software. A fully working prototype was completed in April 1981 and taken to Seattle for Microsoft to continue working on the software.

IBM announced the computer on August 12, 1981, priced at $1,565 for a basic system. IBM received orders for 40,000 units the first day and sold more than 750,000 by the end of the first year.

SEE ALSO Microsoft and the Clones (1982)

The IBM Personal Computer, on display at the Musée Bolo in Lausanne, Switzerland.

Simple Mail Transfer Protocol

Jonathan B. Postel (1943–1998), **Eric Allman** (b. 1955)

1861

Email was the internet's first "killer app." From its very early days, the ability to exchange email with other scientists, professionals, and funding agencies was one of the key reasons that universities and businesses vied for a network connection. In the early years, there were stories of people turning down jobs or graduate school appointments to avoid losing access to network email.

Because email preceded the internet, network compatibility problems for sending messages emerged when it finally did come along. Different systems used different character sets (ASCII versus EBCDIC, for example) and even different word lengths (8 versus 12 bits). Some systems allowed special characters in usernames that other systems couldn't even display. Getting these systems to communicate and exchange email in a reliable, repeatable manner was a major undertaking, with many protocols developed, fielded, and then replaced.

As its name implies, the Simple Mail Transfer Protocol (SMTP) was a simplified version of the more complex Mail Transfer Protocol. Invented by Jonathan Postel, one of the architects of the internet, the simplification involved breaking the act of sending a message from one computer to another into the explicit steps of establishing a connection, specifying where the message was coming from, individually specifying each of the message recipients, and finally sending the message. The protocol's simplicity made programs easier to develop and debug.

Soon after SMTP was published, Eric Allman at the University of California, Berkeley, added SMTP support to a program he had written for delivering email called *sendmail*. Berkeley included sendmail along with support for the TCP/IP networking protocol in its UNIX operating system, making it possible for universities and businesses with internet connections to send and receive internet email using the SMTP standard.

SMTP has proven to be a remarkably durable standard. Although it has been updated several times, mail can still be sent and received using the basic protocol that was developed nearly four decades ago. This durability has come at a price, however. SMTP has no protections against receiving unwanted email, which led directly to the rise of spam.

SEE ALSO @Mail (1971), First Internet Spam Message (1978)

Computers on the internet exchange mail using the Simple Mail Transfer Protocol (SMTP).

Japan's Fifth Generation Computer Systems

Japan's Ministry of International Trade and Industry (MITI)'s Fifth Generation Computer Systems (FGCS) project was a targeted 10-year research investment in logic programming, parallel computing, and artificial intelligence. The $450 million targeted investment of government funds was designed to push Japanese computer manufacturers and researchers to the forefront of computing by subsidizing the development of technologies that were not commercially feasible at the start of the project but were thought to be game-changing after development.

Key to MITI's plan was the development of data-flow and logic programming techniques that were not popular in the West but that many academics thought would be the next big thing. Such technologies were widely believed to be better suited to parallel processing, an approach for speeding computers by performing multiple operations simultaneously. Wrapped up in this approach was a huge investment in AI. MITI's leaders wished to create computers that could speak and understand natural human language, represent knowledge, prove mathematical theorems, and reason as people do. MITI picked the technologies and the companies for the project. The software would be written in Prolog, an arcane computer language that had elegant mathematical properties but ran quite slowly.

Many US companies were spooked. After all, in the early 1980s, strategic investments by the Japanese government helped Japan to capture 70 percent of the world's market for 64-kibibyte (KiB) dynamic RAM chips. Was Japan poised to repeat the experience, this time with AI?

Ten years later, MITI shuttered FGCS and published to the internet the software it had created. It turned out that the millions that Japan had invested to make Prolog run fast was no match for the billions that the US market had available to make Intel's x86 architecture run even faster. Meanwhile, the very premise of using Prolog as the basis of AI was called into question. Summing up Prolog's failure to deliver, AI researcher Carl Hewitt wrote: "The laws of thought are inconsistent."

SEE ALSO *Artificial Intelligence* Coined (1955), Artificial General Intelligence (AGI) (~2050)

This parallel inference engine could run a single Prolog program simultaneously on 512 parallel processors.

AutoCAD

Michael Riddle (dates unavailable), **John Walker** (b. 1950)

Some of the most spectacular, physics-defying buildings in the world—such as the Sydney Opera House—owe their existence to computer-aided design (CAD) programs such as AutoCAD. AutoCAD was one of the earliest and most influential software programs for design and blueprinting of buildings and structures. It dramatically increased productivity, unlocked the practical limitations of architectural and engineering design, and enabled the real-world fabrication of structures that otherwise would have been impossible to translate from paper.

Released in 1982, one of the most revolutionary aspects of AutoCAD was how and where it could be used. Rather than running on a large mainframe with a separate graphics controller (standard for CAD software at the time), AutoCAD was released as an application that ran on a personal computer. Architects and others who depended on CAD now had a tool that not only was more convenient to access, but also offered features that would let them easily make modifications and updates as projects evolved or requirements changed. For those who previously worked from paper with drafting tables and other "analog" design tools, AutoCAD was transformational. Later versions would include many more features and capabilities, such as collaboration functions for dispersed teams and individuals as well as technical feedback that informed decisions about the real-world structural integrity of a proposed design.

AutoCAD built upon decades of research in computer graphics, much of it funded by the US government. The program itself was based on a 1979 program called *Interact CAD* designed by Mike Riddle. Along with programmer John Walker and 13 other cofounders, they created a company called *Autodesk*®. AutoCAD was one of five different desktop automation tools that the founders planned to pursue as products. AutoCAD was the first one completed and launched to market. It was introduced at COMDEX (Computer Dealers' Exhibition) in Las Vegas and described as the first CAD program to run on a PC. The product would quickly grow to become the most popular CAD software in the world.

SEE ALSO Sketchpad (1963), IBM PC (1981)

Computer-aided design programs allow architects and others to design and blueprint buildings and structures, such as the front elevation of the building shown here.

+22,95

+22,65

+20,05

+19,92

+17,85

+14,95

+14,25

+12,05

+11,25

+9,15

+8,30

+6,25

+5,50

+3,35

+2,75

+2,75

±0,00=1,38

±0,00

First Commercial UNIX Workstation

Andy Bechtolsheim (b. 1955), **Scott McNealy** (b. 1954), **Vinod Khosla** (b. 1955), **Bill Joy** (b. 1954)

In 1980, Andy Bechtolsheim was a graduate student at Stanford, where he designed a computer system for the university network modeled somewhat on the Xerox Alto. The idea was to build single-user, high-performance, networked, graphical workstations using largely off-the-shelf parts, such as the new high-performance, 32-bit microprocessors from Motorola and the UNIX operating system. The computer worked, and together with Scott McNealy and Vinod Khosla, two graduates of the Stanford Graduate School of Business, Bechtolsheim founded Sun Microsystems. A few months later, they were joined by Bill Joy, one of the key UNIX developers from the University of California, Berkeley.

At a time when high-capacity hard drives were still fantastically expensive, Sun pioneered the idea of the *diskless workstation*—a computer that downloaded its operating system, programs, and all user files over its high-speed local area network. But rather than keeping the technology as its proprietary advantage, Sun was true to its academic roots: the company presented highly technical papers explaining how its Network File System (NFS) worked and then gave away the source code. Sales took off.

Sun was soon up against the limits of what the microprocessors from Motorola and Intel could deliver, so the company decided to develop its own RISC-based computer: the Scalable Processor Architecture (SPARC). For a while in the 1990s, Sun's workstations were the fastest single-user computers, its database servers were faster than IBM's fastest mainframes, and the fastest existing supercomputer was based on Sun's SPARC technology. But over time, Intel's larger customer base let the company spend more money on research and development, and the relative performance of SPARC faltered. Sun ported its UNIX operating system to Intel's processors but couldn't compete against much cheaper Intel-based systems running Linux®. Sun was ultimately bought by the Oracle Corporation®, a database vendor.

SEE ALSO UNIX (1969), RISC (1980), Linux Kernel (1991)

Sun-1 workstation, featuring Ethernet networking, custom memory management, high-resolution graphics, and a 16-bit, 10-megahertz Motorola 68000 CPU.

PostScript

Chuck Geschke (b. 1939), **John Warnock** (b. 1940)

For more than a hundred years, printers were relatively simple devices: they received a character in Baudot or ASCII over a wire, and then printed it. Printing a page of text involved receiving (and printing) a few thousand characters. The laser printer changed all that.

At 300 dots per inch, a standard printed page has roughly 8.5 million pixels. But sending millions of 0s or 1s from a computer to the printer for every page is both slow and inefficient, because most pages are composed mostly of white space and text; it makes much more sense to send a font to the printer, and then send characters.

PostScript® is a language that specifies how to send fonts and characters to the printer—but it is also much more. It is a specialized computer language for describing printed documents. Instead of receiving a sequence of characters or a bitmap from the computer, the printer receives and runs a special-purpose computer program that has the side effect of producing the desired page. Because it is based on executing code, PostScript makes it possible to describe complex pages and effects very succinctly.

PostScript itself was based on a language called *Interpress* that Chuck Geschke and John Warnock developed at the Xerox Palo Alto Research Center (PARC). But Interpress was designed solely for the early Xerox printers—a very limited market. So in 1982, Geschke and Warnock left Xerox PARC and founded Adobe Systems®, where they created and then commercialized PostScript as a universal page-description language for everything from low-cost home and office printers to high-end phototypesetters.

The first printer to ship with PostScript was the Apple LaserWriter in March 1985. Together, the Macintosh and the LaserWriter let small businesses and professionals easily and rapidly create high-quality typeset documents, giving birth to desktop publishing. The PostScript language became an industry standard and helped make Adobe one of the world's most important software publishers.

Eleven years later, Adobe created the Portable Document Format (PDF) as a simplified, more modern version of PostScript.

SEE ALSO Laser Printer (1971), Xerox Alto (1973), Desktop Publishing (1985)

The PostScript language makes it easy to create complex illustrations with text, graphics, and color.

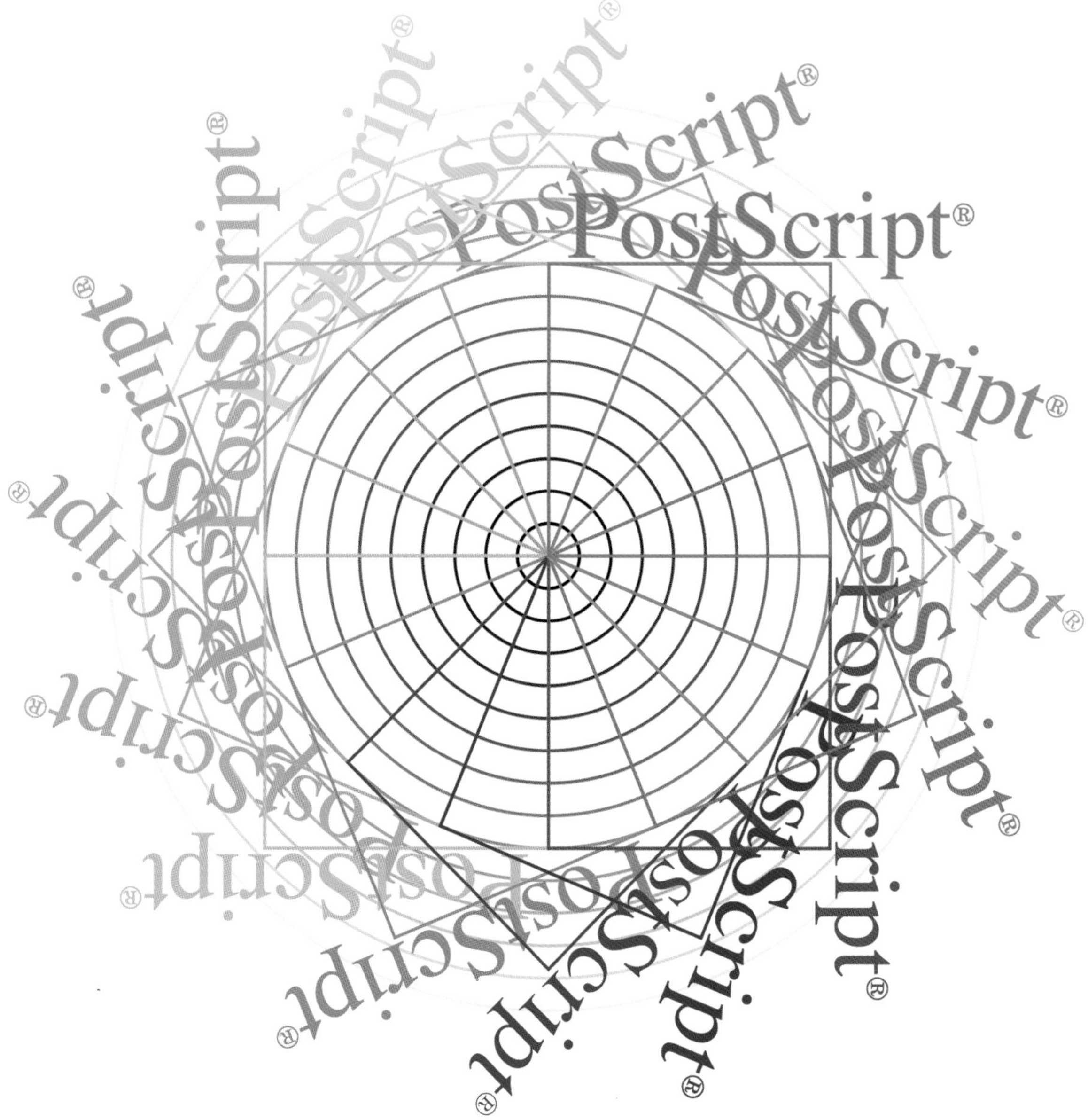
PostScript®

Microsoft and the Clones

Tim Paterson (b. 1956), Bill Gates (b. 1955)

When IBM released its PC in 1981, the machine came with BASIC in ROM and an optional floppy disk drive with the PC Disk Operating System (PC-DOS). But IBM didn't write BASIC or PC-DOS—both came from a scrappy little company in Redmond, Washington, called Microsoft, whose major product until that point had been versions of the BASIC computer language running on different kinds of microcomputers.

In the spring of 1980, IBM looked around and saw that the dominant operating system for microcomputers was an 8-bit operating system by Digital Research Inc. (DRI) called *CP/M* (Control Program/Monitor). IBM was building a 16-bit micro, so the company needed a 16-bit version of CP/M. Unable to ink a deal with DRI, IBM instead signed a contract with Microsoft in July 1980 to make a 16-bit operating system for the PC that would be functionally similar. The goal was to make it easy for software developers to port their programs from CP/M to the new IBM micro.

There was not enough time for Microsoft to write its own operating system, so Microsoft bought a license from Seattle Computer Products (SCP), one of Microsoft's BASIC customers. SCP had also tried and failed to license a 16-bit version of CP/M, which wasn't ready at the time, so Tim Paterson, one of SCP's programmers, had written his own operating system called *QDOS* (Quick and Dirty Operating System) for the company's computer. SCP later renamed QDOS as 86-DOS. Microsoft licensed 86-DOS for $25,000, and then bought all the rights for another $75,000 just before the PC's release.

Microsoft's deal with IBM allowed Microsoft to license DOS to others—and it did. Within a year, Microsoft had licensed DOS to 70 other companies under the name *MS-DOS*. Suddenly there were dozens of computers on the market that could run the exact same software as the IBM PC—and that cost a fraction of the price. Soon companies making and selling PC-compatible computers were popping up all over the planet.

Because IBM did not own the licensing rights to MS-DOS, it took only a few years before the countless PC clones using MS-DOS flooded the market, hurting IBM's position as the dominant player in the very PC market it had created. In December 2004, IBM finally announced that it was exiting the PC market—and its stock rose by 1.6 percent after the news.

SEE ALSO IBM PC (1981), Microsoft Word (1983)

Bill Gates, cofounder of Microsoft.

First CGI Sequence in Feature Film

Alvy Ray Smith (b. 1943)

For all the incredible computer-generated imagery (CGI) milestones in film, one of the most spectacular and technologically significant was the Genesis sequence in *Star Trek II: The Wrath of Khan*. The sequence is pure movie magic—that intangible thing that transports audiences to a place of suspended disbelief—due in part to the amazing visual effects of computer-animated cinematography.

The Genesis sequence begins with a CGI retinal identification scan of Captain Kirk's eyeball that triggers the start of a video describing the Genesis Project. The Genesis device, in the form of a missile, shoots through dark space into a barren planet. An explosion occurs on impact, sending a shockwave across the surface, spreading a rippling fire growing around the entirety of the planet. The point of view (POV) of the audience sweeps along for the ride, and as it pulls out, the planet is now covered with life where none existed. Trees, water, forests, animals—all come from nothing. There are additional POV shots showing flybys around the now-Earthlike planet, covered with streaks of blue and white.

At the time, this sequence was groundbreaking for a movie. While more technically complex than other CGI efforts before it, what sent the Genesis sequence into the history books for movie fans and tech geeks was how well it paired with the story, dialogue, and accompanying music. It both was dramatic to watch and advanced the plot at a critical point in the movie. The Lucasfilm® computer division did the work in collaboration with George Lucas's Industrial Light & Magic (ILM®). The sequence was rendered on DEC VAX computers, direct descendants of the PDP-1, with some frames taking more than five hours to complete.

The Genesis sequence was designed and directed by Alvy Ray Smith, who, along with Ed Catmull (b. 1945), would go on to cofound Pixar Animation Studios® as a spinoff entity, building upon the work done at ILM.

SEE ALSO PDP-1 (1959), *Star Trek* Premieres (1966), Pixar (1986)

The "Genesis" sequence in Star Trek II: The Wrath of Khan *was a landmark achievement in the use of CGI in film.*

STAR TREK II
THE WRATH OF KHAN

National Geographic Moves the Pyramids

Weird things can happen when representations of reality are composed of little more than easily modified 1s and 0s. The cover of *National Geographic*'s February 1982 edition showed a stunning shadowscape of the Egyptian Pyramids at Giza with mysterious adventure seekers riding atop a camel train. It was a beautiful, haunting photo of not just one, but two of the world's wonders overlapping on top of each other, in a sort of Indiana Jones–like presentation.

And it was fake. According to *National Geographic*, a photo editor had taken a horizontal photograph of the Pyramids submitted by one of their journalists and cropped it for a vertical cover. The result was pyramids pushed closer together. The altered picture was visually stunning. *National Geographic* originally explained the crop as "retroactive repositioning of the photographer." Since then, according to its website, the magazine has "made it part of our mission to ensure our photos are real."

With the transition of images from film to digital, the scope and potential extent of photo-editing capabilities becomes vast. Like all other industries that have had to find new norms for their craft when executed through digital means, new responsibilities and standards of appropriateness have evolved along with the greater flexibility at every stage of the creative and production processes that digitization provides. The National Press Photographers Association (NPPA) weighed in, citing other examples where digital alteration occurred, including sports, news, fashion, and other areas that have a lot to gain (or lose) from how the public perceives and automatically accepts as truth the visual stories they promote about their domain.

National Geographic did its editing with a Scitex digital photography system costing between $200,000 and $1 million. In less than a decade, editing tools such as Adobe Photoshop® would be available for desktop computers. It would take another decade for the democratization of online publishing, allowing anyone with a connected computer to distribute distorted photographs instantaneously on a global scale.

SEE ALSO *Blog* Is Coined (1999)

The original landscape photograph by Gordon Gahan of a camel train in front of the Pyramids at Giza.

Secure Multi-Party Computation

Andrew Chi-Chih Yao (b. 1946)

In 1982, computer scientist and theorist Andrew Yao posed a curious mathematical problem: suppose two millionaires meet for the first time. Can the millionaires determine who has more money without revealing each one's net worth to the other or to anyone else? The answer, surprisingly, is yes.

In the language of mathematics, Yao had figured out a way for the inequality $a<b$ to be evaluated by a protocol executed between two parties, where one party knew a and the second knew b, without either revealing their secret. Called *the millionaire's problem*, this thought experiment marked the beginning of what we now call *Secure Multi-Party Computation* (SMC).

Four years later, Yao produced an even more remarkable result: *any mathematical function of two inputs* can be securely evaluated with the same privacy guarantee. And because functions of more than two inputs can be expressed as a combination of two-input functions, and because any computer program can be expressed as a combination of mathematical functions, Yao's work generalized to all possible programs run by any number of participants. For this work, he won the A.M. Turing Award in 2000.

This is significant because Yao's solution may one day facilitate all kinds of e-commerce and digitally enabled activities that today rely on trusted intermediaries. For example, today eBay® knows the bids of all users involved in an auction, and eBay's users must trust that eBay won't reveal those bids inappropriately. With an auction based on Secure Multi-Party Computation, such trust would be unnecessary.

In the years that have followed, Yao and others have worked to create efficient implementations of SMC, as well as protocols that have other security properties—for example, the ability to detect when one party is cheating. Today, SMC is just beginning to make its way from the laboratory into realistic applications. For example, in 2016, the Boston Women's Workforce Council used a secure multiparty system to allow employers in Boston to measure the size of the wage gap in Boston between men and women by computing the average wage paid to men and women in various job categories.

SEE ALSO Secret Sharing (1979), Zero-Knowledge Proofs (1985)

Professor Andrew Yao teaches students at Tsinghua University in Beijing.

TRON

Steven Lisberger (b. 1951), Bonnie MacBird (b. 1951), Alan Kay (b. 1940)

Like a modern Alice in Wonderland, game designer Kevin Flynn is kidnapped and beamed into computer company ENCOM's mainframe, where he is forced to fight for his life as a player in the games he created.

The video-arcade craze was in full swing when Walt Disney Productions released *TRON* in the summer of 1982. The movie resonated with an audience enamored with the world of technology, just as "high-tech" became accessible to the masses. While the look and feel of the movie worked brilliantly for some, the story was simply too ahead of its time for others. Critical reviews were mixed. With its glowing space-age combat suits and high-concept narrative about a regular guy doing battle with anthropomorphized software, the movie was celebrated by technophiles and panned by many critics as a stunning visual display without a credible plot.

TRON broke ground in computer animation. It was the first movie to include whole scenes that were synthetically generated, totaling about 20 minutes of the movie, as well as scenes that seamlessly mapped live-action actors into a computer-generated world. Several animators refused to work on *TRON*, fearful that digital animation would soon put conventional, hand-drawn animation out of business. At the time, the technique was considered so radical that the Academy of Motion Picture Arts and Sciences disqualified the movie as a contender for a special effects award nomination because it felt that computer-aided visuals were a cheat.

Computer pioneer Alan Kay from Xerox PARC was a consultant on the film and helped to edit the movie script on the "Alto"—a prototype personal computer. Written by Bonnie MacBird and directed by Steven Lisberger, the film did not do well at the box office. The arcade games based upon the movie were incredibly popular, however, and out-grossed the movie—a harbinger of things to come. The film went on to develop a cult following among computer game geeks and spawned a franchise including the sequel *TRON: Legacy*, released in 2010.

SEE ALSO *Rossum's Universal Robots* (1920), *Star Trek* Premieres (1966), Xerox Alto (1973)

Poster from the movie TRON, written by Bonnie MacBird and directed by Steven Lisberger.

A world inside
the computer
where man
has never been.

Never before now.

TRON   A LISBERGER-KUSHNER PRODUCTION
STARRING JEFF BRIDGES BRUCE BOXLEITNER DAVID WARNER CINDY MORGAN AND BARNARD HUGHES
EXECUTIVE PRODUCER RON MILLER MUSIC BY WENDY CARLOS STORY BY STEVEN LISBERGER AND BONNIE MACBIRD
SCREENPLAY BY STEVEN LISBERGER PRODUCED BY DONALD KUSHNER DIRECTED BY STEVEN LISBERGER
SONGS BY JOURNEY FROM WALT DISNEY PRODUCTIONS TECHNICOLOR®
FILMED IN SUPER PANAVISION® 70
DOLBY STEREO Read the Ballantine Book Original motion picture sound track available: CBS Records and Tapes
IN SELECTED THEATRES Released by BUENA VISTA DISTRIBUTION CO., INC. © 1982 Walt Disney Productions

PG PARENTAL GUIDANCE SUGGESTED
SOME MATERIAL MAY NOT BE SUITABLE FOR CHILDREN

PRINTED IN U.S.A.

820081

Home Computer Named Machine of the Year

Otto Friedrich (1929–1995)

TIME® magazine's "Man of the Year" franchise took a digital detour in 1982, naming the personal computer the "Machine of the Year." Subtitled "The Computer Moves In," it was the first time the iconic magazine had named a nonhuman as the year's most influential force. As the editors stated, "There are some occasions, though, when the most significant force in a year's news is not a single individual but a process, and a widespread recognition by a whole society that this process is changing the course of all other processes."

The momentum behind the decision is understood best through journalist Otto Friedrich's 11-page cover story, an homage to personal computers. Friedrich's article encapsulated the general public's wonder and excitement about the opportunities to be had with a personal computer and made explicit through colorful examples how doctors, lawyers, housewives, and even a former NFL player improved the quality of their work and created new business opportunities around knowledge and computer power.

While the article's description of what the PC could do was impressive, even more profound was its message about what a person *could become* by using a PC to extend his or her cognitive and physical capabilities. At its most fundamental level, 1982 was the year the public understood that their own human limitations might no longer be limitations at all.

The anxiety and exhilaration that accompany breakthroughs and weighty transitions were also visible throughout the issue, which touched on the function (and value) of face-to-face community, the potential neurological effects on the brain if humans didn't have to spend time thinking about routine work, notions about how computers might change the business of crime, rumination about how the young use computers differently than those over 40, and the long-term impact of computers on employment.

For many, the idea that a machine might replace the "Man of the Year" crystallized the sense that an uneasy shift was underway in America. For others, it crystallized the notion that the need to have a "competitive edge," whether professional or personal, was about to accelerate to warp speed, and with it the pace of everyday life.

SEE ALSO Apple II (1977), IBM PC (1981), Nintendo Entertainment System (1983)

A father and son use an early personal computer. The widespread use of the home computer in the early 1980s inspired TIME *magazine to announce the personal computer as "Machine of the Year" on its January 3, 1983, cover.*

The Qubit

Stephen Wiesner (b. 1942), **Benjamin Schumacher** (dates unavailable)

Modern computers are made from devices that store, transport, and compute with bits. A bit can be a 0 or a 1; a register that holds 8 bits can therefore have 256 (2^8) possible values, but only one at a time. Things are different at the quantum level, where the state of a quantum particle such as an electron, photon, or ion is better described by something called a *probabilistic wave equation*. For example, electrons have a value called *spin*, and similar to a child's top that can spin clockwise or counterclockwise, an electron can have its spin aligned "up" or "down."

A quantum computer can use the spin of an electron to represent a bit of information. But because the spin of an electron is really represented by a probability wave, the hypothetical bit can have the values of 0 and 1 *at the same time* — at least, until its value is measured. This phenomenon, called *superposition*, is not captured by the concept of a "bit," which can be either a 0 or a 1. Instead, it is called a *qubit*, for "quantum bit." A computer made from 8 quantum bits could simultaneously represent 256 possible numbers while seeking a solution.

Besides using subatomic particles to represent information, quantum computers rely on quantum entanglement, another aspect of quantum mechanics. Derided by Albert Einstein in a 1935 paper (and possibly earlier) as *spukhafte Fernwirkung* (spooky action at a distance), entanglement makes it possible to connect a collection of qubits in such a way that they can be used to solve a mathematical computation. But doing so requires that the qubits be isolated from outside forces — for example, by keeping the qubits very cold and protected from electrical noise. In practice, the more qubits in a quantum computer, the harder it is to build.

Stephen Wiesner invented the concept of quantum information in the 1970s: his mental experiment involved a hypothetical bank note containing an unforgeable quantum serial number that was written with entangled qubits. The term *qubit* was coined by Benjamin Schumacher in his 1995 article "Quantum Coding." Previously, qubits were called *two-level quantum systems*. The first quantum computer was demonstrated in 1995.

SEE ALSO The Bit (1948), Quantum Computer Factors "15" (2001)

A quantum machine, with the mechanical resonator situated in the bottom left of the chip, and the coupling capacitor — the smaller white rectangle — between the mechanical resonator and the qubit.

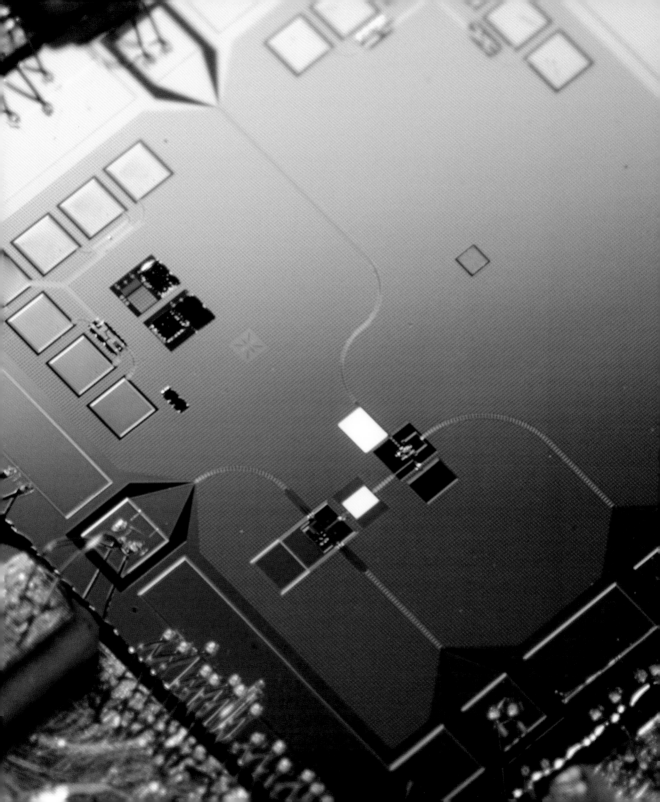

WarGames

Lawrence Lasker (b. 1949), **Walter F. Parkes** (b. 1951),
John Badham (b. 1939)

WarGames was *the* movie that transformed the computer nerd into a hero. A blockbuster starring Matthew Broderick and Ally Sheedy, the movie follows high school hacking whiz David Lightman, who almost starts World War III between the United States and the Soviet Union when he breaks into a military supercomputer and challenges it to a game of *Global Thermonuclear War*.

During an online troll for free video games, David unwittingly breaks into a North American Air Defense (NORAD) system that controls the entire US nuclear arsenal (using a back door left by the original programmer, of course). Believing he has found a way to sneak through the back door of a toy company's network, he challenges the computer to play a game that can be "won" only by not playing at all. Only after targeting Las Vegas and Seattle with Soviet missiles does David realize the game is real and the computer he is playing against has taken matters into its own hands as it escalates the crisis in an attempt to "win" the game.

The movie left such an impression on President Ronald Reagan that he asked the chairman of the Joint Chiefs of Staff, General John W. Vessey Jr., if it really was possible to break into sensitive US government computers. After investigating the plausibility of such a scenario, the general came back with his answer: "Mr. President, the problem is much worse than you think." Not long after that, US lawmakers published the *Computers at Risk* report, which established the beginning of the US defensive computer security program.

Written by Lawrence Lasker and Walter Parkes, and directed by John Badham, *WarGames* left an impact on generations of future coders, many of whom went on to work in Silicon Valley—where, in 2008, Google held a 25th-anniversary screening. The movie's legacy has lived on with computer hackers as well—for example, the name of the famous Las Vegas hackers' convention DEFCON is an homage to the film.

SEE ALSO SAGE Computer Operational (1958)

Poster from the movie WarGames, *written by Lawrence Lasker and Walter Parkes, and directed by John Badham.*

MISSILE WARNING

SUB-LAUNCH DETECTION

Is it a game, or is it real?

WarGames

A Leonard Goldberg Production · A John Badham Film · "WARGAMES" · MATTHEW BRODERICK · DABNEY COLEMAN · JOHN WOOD · ALLY SHEEDY

Written by LAWRENCE LASKER & WALTER F. PARKES · Director of Photography WILLIAM A. FRAKER, A.S.C. · Music by ARTHUR B. RUBINSTEIN · Executive Producer LEONARD GOLDBERG · Produced by HAROLD SCHNEIDER

Directed by JOHN BADHAM · Produced in association with Sherwood Productions · Panavision · Metrocolor · Read the Dell Book · DOLBY STEREO

3-D Printing

Charles (Chuck) Hull (b. 1939)

Charles "Chuck" Hull invented 3-D printing, also known as *additive manufacturing*, while working for a company that made hard coatings for tables. Working with polymers that were cured and hardened by exposure to ultraviolet light, he realized he could create complex shapes one layer at a time by shining a UV laser into a vat of polymer and using a computer to control the light's position. He filed US Patent 4,575,330 on August 8, 1984.

With additive manufacturing, material is added together to create an object rather than subtracted from a block or mass of material through cutting, milling, or drilling. Many different materials can be used, including resins cured by laser, plastics melted at an extrusion nozzle, powdered metal, and even food. All 3-D printing techniques, however, have two common starting points: digital files—the "blueprints" that are used by a computer system to specify where the additive material will be deposited—and a mechanism that can deliver the material layer by layer to create the three-dimensional "thing."

The smallest items that have been 3-D printed are nanosized sculptures that fit on the forehead of an ant. Large printers using carbon fiber are now being used to create structures that require precise curves over an extended distance, such as a trim tool designed for building the airplane wing of a Boeing 777 or wind turbine blades for power generation. NASA demonstrated zero-gravity 3-D printing on the International Space Station, when it printed a ratchet wrench in 2014. This means people in isolated locations—such as space—can fabricate things they need that would otherwise be out of their reach.

Eventually 3-D printing may change the entire manufacturing process. Rather than a manufacturer turning raw materials into objects and then transporting them to a store and finally the end user, items may be fabricated in regional or local centers, or even in the home. Chefs are getting in on the act, too. Today some supermarkets are using 3-D printers to make cake decorations; in the future, artificial organs may be 3-D printed as well, perhaps built with cells grown from the patient, to avoid the chance of rejection.

SEE ALSO AutoCAD (1982)

3-D printing, also known as additive manufacturing, *can produce objects in an enormous variety of shapes and colors.*

1983

Computerization of the Local Telephone Network

Caller ID made its debut in the 1980s, along with a whole bunch of other services: call return, caller ID blocking, and even call waiting. These services became possible because of the digitization of the local telephone network, in particular the design and implementation of a switch designed by Western Electric called the *Number 5 Electronic Switching System* (5ESS®).

The 5ESS was revolutionary for the telecom industry, essentially turning the business model upside down and helping to transform the telephone network into a sprawling digital ecosystem where voice, data, video, image, and any other format of communication represented by 1s and 0s could be transmitted simultaneously. The technology behind the 5ESS could also expand or contract the number of lines, depending upon demand—and do it without an army of telephone company engineers having to install or rip out cables.

The 5ESS took 20 years, 5,000 workers, and 100 million lines of system source code to develop. The skills needed to pull this off represented a spectrum of technical expertise. AT&T split into separate companies in the middle of the project, which meant that it was not enough suddenly to design and build the new switch—it had to be packaged and marketed to a whole new kind of customer, the Regional Bell Operating Companies (RBOCs). Whole swaths of the 5ESS effort involved training salespeople, technicians, and executives on how this new technology worked and how to talk about it, depending upon the level of detail their job required.

For the consumers, the 5ESS meant a more complicated phone system, with more features that could be purchased and the sudden need to think about the impact of those features on their privacy. Caller ID let people know who was calling before they picked up the phone, but it also revealed the caller's phone number. Telephone calls would never be the same. The economic backdrop that was occurring at the time of 5ESS's debut was far from tranquil. After the breakup of the Bell System, 5ESS was moved to the AT&T Network Systems division, which was then divested to Lucent Technologies and then finally acquired by Nokia®.

SEE ALSO Strowger Step-by-Step Switch (1891), Digital Long Distance (1962)

Services such as caller ID were made possible by the digitization of the local telephone network.

First Laptop

It came to market more than three decades ago, but the specifications of RadioShack's®
TRS-80 Model 100 computer are still pretty sweet. The world's first laptop computer
weighed just 3 pounds and had a battery that lasted 20 hours. It came preloaded with
a rudimentary word processor and a basic spreadsheet program. The system's storage
was 100 percent solid state (no spinning disk) and expandable. The computer booted
instantly. There was even a serial interface and a telephone modem, allowing the
computer to connect and exchange information with another computer in the same
room or on the other side of the world. There was also a cassette interface for storing
data on audiotape.

The Model 100 was designed for computing on the go. It quickly became a popular
tool of journalists and others who needed to pack light, travel to a remote location, and
write; more than 6 million units sold, with prices starting at $1,099.

The built-in word processor allowed the user to type, edit, and dump the current
document to the phone line, but not much else. There was a Microsoft BASIC
interpreter and a telecommunications package for connecting to remote systems, an
address book, and a to-do list organizer. The computer booted quickly because its
applications were burned into read-only memory. Its solid-state storage was nothing
more than a few static random access memory (SRAM) chips with a battery to prevent
them from forgetting their data: the Model 100 could be purchased with 8 kibibytes of
RAM, expandable to 24 kibibytes. The computer's display was an 8-line, 40-character
liquid-crystal display that showed black-and-white text, with limited graphics support.

Inside the Model 100 was an 8-bit Intel 80C85 microprocessor. The computer's
operating system and applications all fit within 32 kibibytes of read-only memory
(ROM); much of the code in the ROM was personally written by Bill Gates.

Meanwhile, a healthy aftermarket of peripherals, software, and books evolved for
the Model 100, which sold until 1986, when it was replaced by the slimmer Model 102.

SEE ALSO BASIC Computer Language (1964), Microsoft and the Clones (1982)

The RadioShack TRS-80 Model 100 had a flat-top design and sold more than 6 million units.

MIDI Computer Music Interface

Dave Smith (dates unavailable), **Ikutaro Kakehashi** (1930–2017)

Before the invention of the Musical Instrument Digital Interface—better known as *MIDI*—there were technical and practical challenges to producing a piece of music comprising different instruments and sounds. Musicians could use an orchestra of people, or they could combine sounds from different synthesizers with a multitrack recorder, a cumbersome, inefficient process. Of course, there was also the one-man band. And then came MIDI.

As the micro revolution in the 1970s developed into a full-fledged industry, advances in computers started disrupting traditional market sectors, such as the music industry. As the popularity of synthesized sound in music composition and the number of manufacturers that designed hardware and software to produce it grew, so did the complications associated with making all of these technologies talk to one another in order to create the end product—music. Dave Smith and Ikutaro Kakehashi (both musicians and engineers), along with others in the industry, recognized this challenge and pushed for development and acceptance of a single digital communication protocol to solve it.

The protocol was designed to enable many synthesizers or other electronic instruments to be interoperable with one another through a single interface or "master controller" that managed how the original sound sources were mixed, manipulated, and edited. Essentially, the protocol has a set of commands telling different instruments to play different notes. To create MIDI messages, a composer could simply play a sequence on a keyboard: the messages would be sent over the interface. Those MIDI messages could then be saved in a MIDI file, which, when played back over the interface, would recreate the music. MIDI made it easy for traditional sectors of the music industry to computerize, while it revolutionized the creative possibilities for music composition. It also helped popularize the home recording studio and put music creation into the hands of more users, because now it was possible to compose and record great music without having to be a virtuoso performer: the computer did the playing.

SEE ALSO Diamond Rio MP3 Player (1998)

Using MIDI, the computer can follow the melody that the musician plays and generate a synthetic accompaniment.

Microsoft Word

Charles Simonyi (b. 1948), **Richard Brodie** (b. 1959)

It would probably be a struggle to find a computer user who has not at some point encountered Microsoft Word. It is far from the only word processor in existence, and it wasn't the first, but it nonetheless has triumphed as the world's dominant word processing program for decades.

Word was developed by two former employees of the Xerox Palo Alto Research Center—Charles Simonyi and Richard Brodie—both of whom created what *was* the world's first What You See Is What You Get (WYSIWYG) word processor, Bravo. WYSIWYG means that the word processor runs on a graphical screen, showing proportionally spaced fonts, boldface, italics, and more.

Word 1.0 (originally named Multi-Tool Word) was released in October 1983 for MS-DOS and Xenix operating systems. But it wasn't the first: competing programs included WordPerfect®, a text-mode word processor developed for the Data General minicomputer, and WordStar, a word processor created for 8-bit microcomputers, both developed in the late 1970s and ported to MS-DOS in 1982. All three programs ran in text mode; Word didn't run in graphical mode on the PC until 1989, when Microsoft released Word for Windows.

Microsoft did all it could to push the adoption of Word, including bundling free copies of the program in the November 1983 issue of *PCWorld* magazine, the first time a disk was ever distributed by magazine.

The first WYSIWYG version of Microsoft Word was Word for the Apple Macintosh, which Microsoft shipped in 1985. The program was so popular—and so much better than Word on MS-DOS—that it helped fuel sales of the Macintosh.

Word benefitted as Microsoft's operating systems grew in popularity—first MS-DOS, then Windows—and as tools in the Microsoft Office Suite expanded. Over time, the more people who used Word, the more difficult it became for their collaborators to use any other word processor—a prime example of what economists now call a *network effect*.

SEE ALSO "As We May Think" (1945), Mother of All Demos (1968), Xerox Alto (1973), IBM PC (1981), Microsoft and the Clones (1982), Desktop Publishing (1985)

Microsoft Word, part of the Microsoft Office suite, has become one of the world's most iconic and popular word-processing programs.

Nintendo Entertainment System

Fusajiro Yamauchi (1859–1940), **Hiroshi Yamauchi** (1927–2013)

Founded by Fusajiro Yamauchi in 1889 as Nintendo Koppai, a maker of Japanese Hanafuda playing cards, this company went from a small, paper-based game publisher to a behemoth consumer electronics institution known for its iconic game titles, including *Pokémon*®, *Super Mario Bros.*®, *Tetris*®, and *Donkey Kong*®.

Nintendo® has had a profound impact on both the gaming industry and how people interact with computer gaming technologies. It practically created mobile electronic gaming and helped to introduce the fusion of entertainment and mobile, as well as the concept of multiplayer engagement across computer devices. Few other companies were following this strategy at the time Nintendo was pushing such concepts.

First sold in 1983 in Japan as the Famicom or Family Computer, the Nintendo Entertainment System hit the US market in 1985 and the European market in 1986. In South Korea, it was called the *Hyundai Cowboy*. That cowboy would go on to reinvigorate the US gaming industry, which had been in a decline since 1983. Nintendo president Hiroshi Yamauchi is credited with making the Nintendo Entertainment System a success, in part by insisting that the company sell a low-cost, cartridge-based game console, rather than a more expensive home computer complete with keyboard and floppy disk drive.

The Nintendo Game Boy®, released in 1989, was one of the first portable consoles and helped to introduce the idea of having a mobile entertainment experience rather than a TV-tethered one. Revolutionary at the time, despite its monochrome screen, the Game Boy helped lay the foundation for the gaming experiences on smartphones today. Nintendo also produced the first device to have a satellite modem add-on for a game console. It was a 1995 feature offered on the Super Famicom console in Japan. Called *Satellaview*, it broadcast games between 4 p.m. and 7 p.m. and offered a variety of titles, including a few exclusives that were available only during a specific time slot.

Nintendo software has also proved useful for real-world applications: in 2009, the Japanese police used the "Mii" feature on the Wii® that allows people to create their own avatars to create a picture of a hit-and-run suspect for a wanted poster.

SEE ALSO Augmented Reality Goes Mainstream (2016)

Graffiti by an unidentified artist of Nintendo's Mario and Luigi, from Super Mario Bros., *on a wall in the city center in Bristol, England.*

Domain Name System

Paul Mockapetris (b. 1948), **Jon Postel** (1943–1998), **Craig Partridge** (b. 1961), **Joyce Reynolds** (1952–2015)

When a web browser starts up and searches for the Google home page (*www.google.com*) a *domain name system* (DNS) query packet is sent to a computer called a *name server*, asking for the address of the name server that handles the *com* domain. The name server then sends a response containing the address of the *com* domain server. This process repeats, as your computer asks the *com* server for the address of the *google.com* name server, and finally asks the *google.com* name server for the address of *www.google.com*.

Although this lookup process may seem unwieldy, it has a singular advantage: it provides for a single, distributed, redundant database that can hold addresses for the billions of hosts that make up the modern internet.

Before DNS, the mapping between hostnames and the computer's internet address was kept in a file called *HOSTS.TXT*, the master version of which was kept on a single computer at the Stanford Research Institute Network Information Center (SRI-NIC). System administrators at different labs manually updated the file as necessary by sending partial updates to the SRI-NIC administrator. Computers all over the internet would then download copies of the file for local use. The whole process was awkward and error prone, and the HOSTS.TXT file at SRI-NIC was perennially out of date. Worse, as the internet started to grow exponentially, so too did the length of the HOSTS.TXT file, which made downloads of the file increasingly slower. Meanwhile the updates started coming more and more frequently.

Paul Mockapetris, a researcher at the Information Sciences Institute (ISI) at the University of Southern California, came up with the idea of the DNS system in the spring of 1983; the first successful test of the software was June 23, 1983. Mockapetris, Jon Postel (also at ISI), and Craig Partridge, a researcher at BBN, then refined the system. The basic protocols are still in use today, more than three decades later.

In October 1984, Postel and Joyce Reynolds, another ISI researcher, created the initial set of generic top-level domains: *.arpa* for the legacy hosts, *.edu* for education, *.com* for commercial, *.gov* for civilian government hosts, *.mil* for the US military, and *.org* for nonprofits.

SEE ALSO ARPANET/Internet (1969)

DNS provides for a single database that can hold addresses for the billions of hosts that make up the internet.

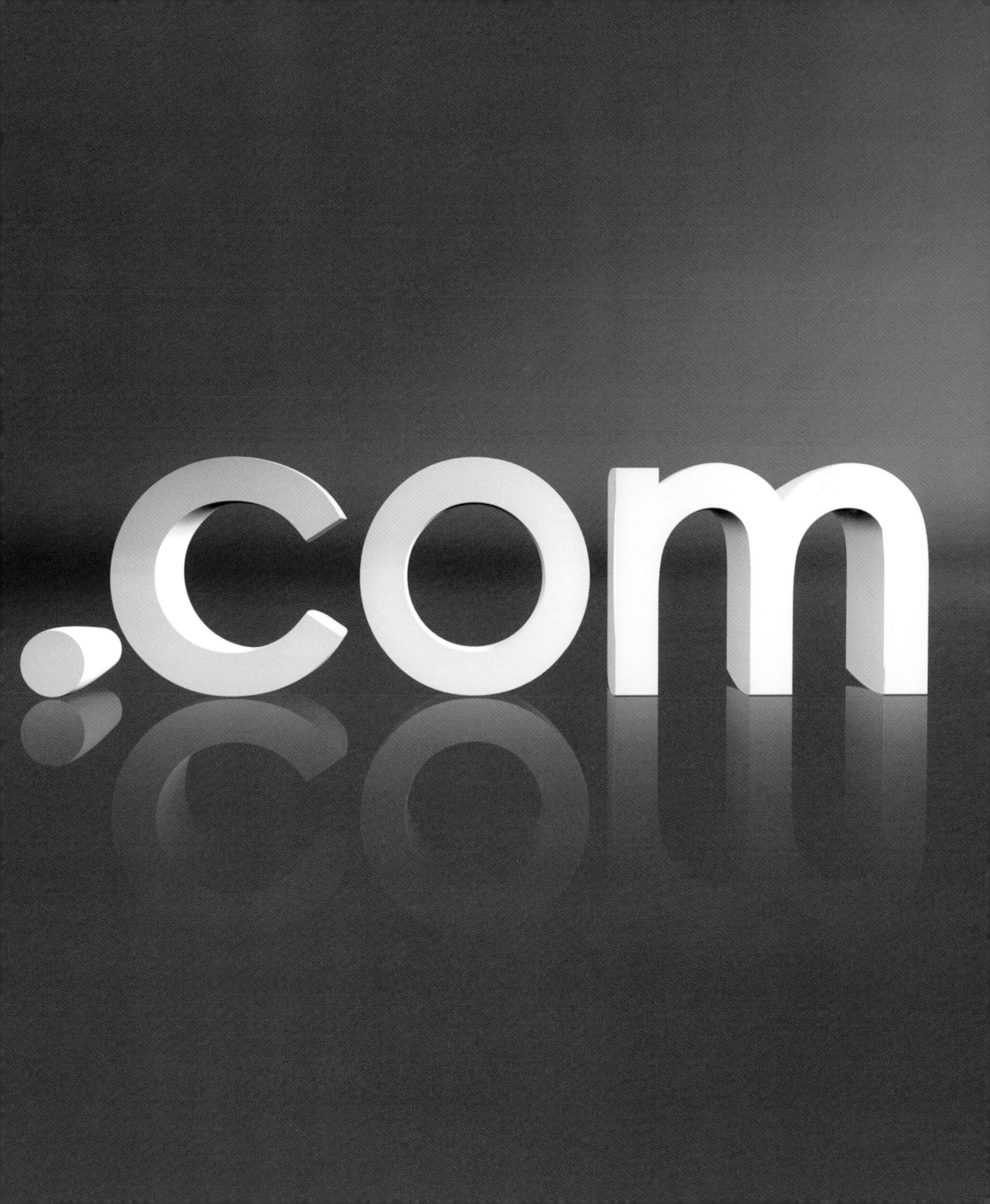
.com

IPv4 Flag Day

Jon Postel (1943–1998)

The original ARPANET relied on Interface Message Processors (IMPs) to forward packets and a combination of complex protocols. These protocols included the ARPANET Host-to-Host Protocol (AHHP), the Network Control Protocol (NCP), and the Initial Connection Protocol (ICP). All communications between hosts had to go through IMPs, and IMPs were expensive.

The first ARPANET IMP was installed in September 1969, and the network was in regular operation by 1971. Just two years later, in 1973, work started on a complete network redesign. At the core of the new design was the Internet Protocol (IP), a connectionless protocol that moved packets between hosts. IP was designed to be fast, predictable, and extensible, but it didn't guarantee that packets would ever be delivered. Instead, IP provided a "best-effort network" that followed the end-to-end principle—intelligent endpoints are responsible for managing their own communications. Also called the *stupid network*, this network had just one job: to move packets from one host to another.

Above the IP layer was the Transmission Control Protocol (TCP), which allowed two computers on the ends of the network to communicate over a reliable stream of 8-bit characters. Programs using TCP simply created a connection to a "socket" on a remote computer and then sent and received characters as necessary. The computer's operating system then packaged the characters into packets, sent the packets, and retransmitted the packets if they got lost. TCP made it easy for programmers to develop services such as remote terminal, file transfer, and email.

For many years, ARPANET supported both NCP and TCP/IP. In 1981, Jon Postel warned the ARPANET's users that ISI, the network's operator, could disable NCP on the entire ARPANET simply by telling the IMPs to drop NCP traffic. In the middle of 1982, he did just that, briefly—and the sky didn't fall. A few months later he did it again, this time for two days. Finally, Postel made the change permanent on January 1, 1983, which became known as "flag day."

SEE ALSO Interface Message Processor (IMP) (1968), ARPANET/Internet (1969), NSFNET (1985), ISP Provides Internet Access to the Public (1989), World IPv6 Day (2011)

January 1, 1983, became known as "flag day," when ARPANET permanently transitioned from NCP to TCP/IP.

Text-to-Speech

Dennis H. Klatt (1938–1988)

Text-to-speech (TTS) systems are computers that read typewritten text and then speak it out loud. The first English text-to-speech system originated in Japan in 1968. But it was DECtalk, a standalone appliance for turning text to speech, that commoditized this technology. The invention helped many people, including those unable to talk due to medical reasons or disabilities. While a lot of basic research fails to transition into practical applications, DECtalk was a success story.

Much of DECtalk's core capability was built on the text-to-speech algorithms developed by Dennis Klatt, who had joined MIT as an assistant professor in 1965. Packaged into a hardware appliance by the Digital Equipment Corporation, DECtalk had asynchronous serial ports that could connect to virtually any computer with an RS-232 interface. DECtalk was kind of like a printer, but for voice! Two telephone jacks let users hook DECtalk up to a telephone line, allowing DECtalk to make and receive calls, speak to the person at the other end of the phone line, and decode the touch tones of their responses.

TTS systems such as DECtalk work by first converting text to phonemic symbols and then converting the phonemic symbols to analog waveforms that can be heard by humans as sound.

When it launched, DECtalk's price tag was approximately $4,000, and it came equipped with a variety of different speaking voices. Over time, the number of voices expanded, with names such as Perfect Paul, Beautiful Betty, Huge Harry, Frail Frank, Kit the Kid, Rough Rita, Uppity Ursula, Doctor Dennis, and Whispering Windy. An early user of the DECtalk algorithm is the world-famous British physicist Stephen Hawking, who lived with amyotrophic lateral sclerosis, or ALS. Unable to talk, his voice is best recognized as "Perfect Paul." The National Weather Service also used DECtalk for its NOAA Weather Radio broadcasts.

SEE ALSO Electronic Speech Synthesis (1928)

After he was unable to speak because of the progress of a degenerative nerve disease, physicist Stephen Hawking used a text-to-speech device as his voice.

Macintosh

Jef Raskin (1943–2005), Steve Jobs (1955–2011)

With a bitmapped display, an integrated floppy drive, 128 kibibytes of RAM, and a mouse, the Apple Macintosh was the first mass-market personal computer to feature a graphical user interface. Despite its relatively small 9-inch, 512 × 342-pixel monochrome display, the bundled MacWrite word processor and MacPaint drawing program ignited the desktop publishing revolution. More importantly, the Mac's uncompromising emphasis on ease of use and visual design set standards for the computing industry that remain to this day. It was a computer that could be used without reading a manual or receiving formal instruction.

Jef Raskin started the Macintosh project at Apple in 1979. Raskin, a former computer science professor, was hired in 1978 to manage Apple's technical publications. But Raskin was a master engineer, and he soon assembled a team to create an industry-shattering, low-cost ($1,000), highly usable computer, naming it after his favorite apple. Meanwhile, the rest of Apple was hard at work on the Apple III, the planned successor to the Apple II, and the Apple Lisa.

Raskin arranged for Apple engineers to make two visits to the Xerox Palo Alto Research Center (PARC) to see the Alto workstation. The Alto was well known in California's Silicon Valley, of course, but Raskin saw to it that Apple's engineers were given detailed explanations of the Alto's internals. Seeing how the system operated at such an intricate level spurred them to dramatically change their thinking.

Upon return from Xerox PARC, the engineers gave the Lisa a graphical user interface (the Mac had a GUI from the beginning). As the Lisa prototype firmed up, Steve Jobs was thrown off the team—so he turned his attention to the Mac. Personality clashes followed, and Raskin left Apple in 1982.

Released to the world on January 24, 1984, Apple positioned the Mac as an easy-to-use PC that stood in opposition to IBM's systems. Despite the steep price of $2,495, and despite the relative paucity of software, Apple nevertheless managed to sell 250,000 Macs in 1984. The millionth Mac rolled off the production line in March 1987 and was given by Apple to Raskin in recognition of his contribution.

SEE ALSO Laser Printer (1971), Xerox Alto (1973), IBM PC (1981)

The original 1984 Apple Macintosh 128-kibibyte computer.

VPL Research, Inc.

Jaron Lanier (b. 1960)

Founded by polymath Jaron Lanier and several friends, VPL (Virtual Programming Language) Research was the first virtual reality (VR) startup. Based in Silicon Valley, its seminal hardware and software were the first commercial products of their kind and helped bring the technology into mainstream consciousness. VPL is credited with creating the first real-time surgical simulator (a virtual knee and gallbladder) and creating the first motion picture capture suit, among its many varied contributions. VPL helped the city of Berlin plan reconstruction; it helped the oil and car industries simulate products and scenarios; it worked with NASA to experiment with flight simulation; it worked with Jim Henson to prototype a new Muppet; and it even worked with the International Olympic Committee to try to design a new sport that would take place in VR (that one did not come to pass).

The company's core products included the "EyePhone," which was a set of large goggles that covered the eyes to facilitate the virtual experience, and the DataGlove, whose sensors synced with the EyePhone so the user could move virtual objects. There was also the DataSuit, which enabled full-body movement. The employees of VPL—known as *Veeple*—had a vision to enable people to share virtual experiences simultaneously and help people realize a newfound appreciation for the beauty of physical reality once they left the virtual environment. As described by Lanier, before VR, there was never really anything to compare reality to.

Influenced and inspired by giants in the field such as Ivan Sutherland, Lanier was responsible for coining the term *virtual reality*, in part to distinguish VPL's work from the concept of virtual worlds, which was understood as an individual alone in a virtual space, as opposed to a communal experience. VPL in turn inspired others: examples in popular culture include the movies *The Lawnmower Man* (the main character is based in part on Lanier) and *Minority Report* (which used representations of VPL's DataGlove and EyePhone as integral elements in the futuristic storyline). By the mid-1990s, VPL's heyday had passed, but its vision of an immersive world inside computers is inescapable, in movies, television shows, computer games, and more. In 1999, Sun Microsystems acquired the worldwide rights to VPL's patent portfolio and technical assets.

SEE ALSO Head-Mounted Display (1967)

VPL's virtual reality gear, including data glove, goggles, and body suit, exemplified "gloves and goggles" virtual reality.

Quantum Cryptography

Charles H. Bennett (b. 1943), **Gilles Brassard** (b. 1955)

In a 1984 paper, "Quantum Cryptography: Public Key Distribution and Coin Tossing," physicists Charles Bennett and Gilles Brassard came up with a new approach for distributing cryptographic keys that relied on the fundamental laws of quantum physics, rather than on couriers carrying locked suitcases or the difficulty of factoring large numbers. Thus, unlike public key cryptography, quantum key distribution (QKD) is unaffected by advances in number theory, factoring, or quantum computers.

Bennett and Brassard described a protocol, now called *BB84*, for sending photons between two people, typically referred to as Alice and Bob. Alice, the sender, modulates the quantum state of each photon with a randomly generated bit using a randomly chosen measurement basis. Bob receives the photons and measures each, choosing his own random measurement basis for each. Bob and Alice now share all of their measurement bases: for the photons where Alice and Bob randomly chose the same bases—approximately 50 percent of the time—the encoded bits that Bob measured should match the bits that Alice sent. To find out, Bob tells Alice half of the values that he measured. If these match, then Bob and Alice use the second half of the bits as their cryptographic key.

What if there is a person in the middle, eavesdropping on the communications? If the attacker merely measures the photons, then under Heisenberg's uncertainty principle, there's a significant chance that each photon's state may be changed, and the intrusion will be detected when Bob tells Alice his measurements.

Building a working system requires precise control over single photons. That wasn't possible in 1999, when MagiQ Technologies, Inc.®, a Somerville, Massachusetts, startup, incorporated to build a working system. But it was possible four years later, when MagiQ announced the commercial availability of its equipment in 2003.

Since then, QKD and other approaches of using quantum mechanisms to secure data are increasingly catching the interest of financial institutions, research firms, and governments. In 2017, the Chinese government demonstrated quantum cryptography using an alternative protocol based on quantum entanglement.

SEE ALSO Vernam Cipher (1917), Quantum Computer Factors "15" (2001)

The eye of an observer reflected in a mirror in quantum cryptography apparatus. Photographed at the lab of Anton Zeilinger at Vienna University, Austria.

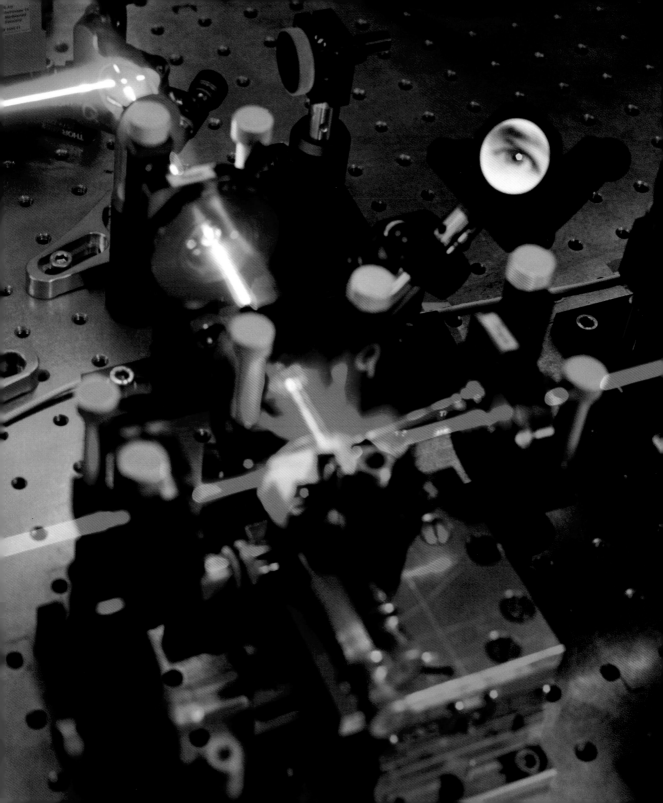

Telebit Modems Break 9600 bps

The TrailBlazer modem, from the US-based company Telebit, was a computer-networking breakthrough. In 1984, at a time when most dial-up computer users were just beginning to make the transition from modems that could send 1200 bits per second (bps) to those that could send 2400 bps, Telebit introduced a modem that could transfer data over an ordinary phone line at speeds between 14,400 and 19,200 bits per second.

The secret to the TrailBlazer's speed was its proprietary channel-measuring protocol. At the time, phone calls traveled over many analog wires to reach from one end to another, creating something communications engineers called a *channel*, and every channel was slightly different. Telebit's Packetized Ensemble Protocol (PEP) divided that channel into 512 different analog slots. When one TrailBlazer sensed it was communicating with another, the two modems would measure the channel and determine which of those slots could be used for high-speed data transfer. In any given instance, the modems would allot the majority of slots to the modem transferring the most data. The TrailBlazer also had direct support for the UNIX-to-UNIX-Copy protocol (UUCP), making it a hit with Usenet sites.

In 1985, each TrailBlazer cost $2,395. The modems frequently paid for themselves in the first year, however, through savings on long-distance charges.

The TrailBlazer triggered what came to be known as the "modem wars." They started when Telebit's primary competitor, US Robotics Corporation®, introduced its own 9600 bits per second modem for $995 in 1986. The two modems were not compatible. Telebit responded by slashing the price of the TrailBlazer to $1,345 in 1987.

The industry knew the path to riches would come only from a larger, multivendor market—and that required standardization. The first high-speed standard was the V.32 9600 bits per second in 1987; prices for external models dropped to $400. A succession of faster and lower-priced models followed until, finally, the International Telecommunication Union (ITU) released the draft V.90 standard in February 1998, which supported 56-kilobit-per-second download speeds to consumers from specially equipped internet service providers (ISPs). This was as fast as was theoretically possible over an analog phone line without the use of compression.

SEE ALSO The Bell 101 Modem (1958), Usenet (1980), PalmPilot (1997)

This spectrograph shows the tones made by a pair of modems during the first 22 seconds of a particular high-speed connection.

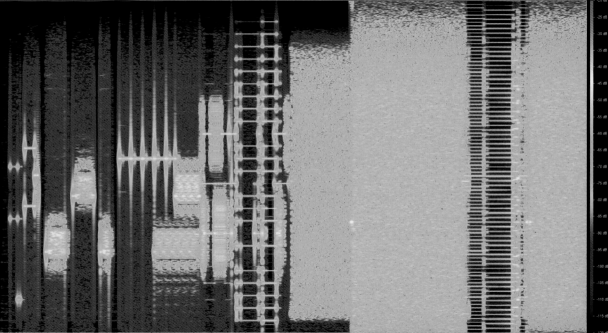

Verilog

Most computer languages describe how to assemble bits in a computer's memory—for example, the arrangement of instructions in a computer's memory in the case of C, or elements of text on a page in the case of HTML.

Hardware description languages (HDLs), in contrast, describe the arrangement of wires, resistors, and transistors that make up electronic circuits. A hardware designer writes the HDL "program" with a text editor. The program is then translated into a circuit diagram and eventually into a layout, which is used to create a chip mask and eventually an integrated circuit.

In the early 1980s, the development of very-large-scale integration (VLSI)— microelectronics with staggering numbers of tiny transistors—necessitated the creation of a new kind of HDL that could represent not just wiring diagrams, but more complicated structures such as clocks, registers, state machines, and complex behaviors. In addition to a layout and semiconductor mask, tools were needed that could simulate the HDL design before burning it into silicon. These tools, called *simulators*, could run an HDL design on a conventional computer. Although simulators ran designs much slower than the final silicon and were not always accurate (especially for large circuits), it was much faster—and cheaper—to test designs in simulation, rather than creating potentially buggy silicon.

Verilog was one of the first successful languages that allowed designers to create, simulate, test, and eventually produce running silicon for complex circuit designs. Gateway Design Automation created the Verilog language and its first simulator; the company then licensed the language to other companies that wanted to make Verilog tools.

Verilog's primary competitor is VHDL (VHSIC Hardware Description Language), developed in 1987 by the US Department of Defense. VHDL is more restrictive and pedantic than Verilog, resulting in circuit design programs that are typically larger and harder to write, but that have a higher chance of being correct—that is, the simulated circuit matches the behavior of the actual circuit. Available for use without a license, VHDL was widely adopted. In 1989, Cadence Design Systems acquired Verilog, and then responded to the threat of VHDL by releasing Verilog into the public domain— making the language dramatically more popular and allowing it to be standardized as IEEE Standard 1364.

SEE ALSO Field-Programmable Gate Array (1985)

A 37-line Verilog program that describes a simple electronic circuit.

+ Verilog Code

```verilog
1  module mux_from_gates ();
2  reg c0,c1,c2,c3,A,B;
3  wire Y;
4  //Invert the sel signals
5  not (a_inv, A);
6  not (b_inv, B);
7  // 3-input AND gate
8  and (y0,c0,a_inv,b_inv);
9  and (y1,c1,a_inv,B);
10 and (y2,c2,A,b_inv);
11 and (y3,c3,A,B);
12 // 4-input OR gate
13 or (Y, y0,y1,y2,y3);
14
15 // Testbench Code goes here
16 initial begin
17    $monitor (
18      "c0 = %b c1 = %b c2 = %b c3 = %b A = %b B = %b Y = %b",
19      c0, c1, c2, c3, A, B, Y);
20    c0 = 0;
21    c1 = 0;
22    c2 = 0;
23    c3 = 0;
24    A = 0;
25    B = 0;
26    #1 A = 1;
27    #2 B = 1;
28    #4 A = 0;
29    #8 $finish;
30 end
31
32 always #1 c0 = ~c0;
33 always #2 c1 = ~c1;
34 always #3 c2 = ~c2;
35 always #4 c3 = ~c3;
36
37 endmodule
```

Connection Machine

Danny Hillis (b. 1956)

Processing more data faster is one of the fundamental challenges in computing. From the 1940s until the 1990s, most improvements came from faster components, clock cycles, and storage systems. But another approach to processing more data faster is to break up the problem into many smaller problems and process in parallel, with a fleet of machines.

Parallelization is possible because many computing problems are similar to knitting sweaters. If you needed four sweaters in a month, you could hire the world's fastest knitter to churn them out, knitting one each week. Or you could hire 12 fast knitters to work simultaneously on the sleeves and the bodies and then join them all together on Friday. And if you needed 10,000 sweaters? In that case, you could hire 50,000 knitters of average skill: they don't have to be fast, and it doesn't really matter if some of them fail, provided there is a clever approach for properly organizing the effort.

This is the kind of parallelism behind Thinking Machines. Based on a PhD thesis by its cofounder Danny Hillis at MIT, the company's first supercomputer combined 65,536 puny 1-bit microprocessors, each "connected" through a massively parallel network.

Called the *CM-1*, it proved difficult to program, because few programmers could visualize algorithms that run efficiently on massively connected 1-bit processors. In 1991, the company released the CM-5, which used 1,024 standard 32-bit microprocessors. Programming was easier, and with so many processors, the CM-5 was one of the fastest computers in existence.

Although the idea of combining thousands of computers together in a single machine worked well in the 1990s, a decade later the dominant approach for solving big problems was grid computing—connecting thousands (or millions) of conventional computer systems over Ethernet with specialized software. Using conventional systems that were not very fast and subject to failure was the electronic equivalent of hiring "average knitters" in the sweater example. Instead of designing complicated parallelizing hardware, like the CM series, the challenge shifted to designing clever software.

SEE ALSO Hadoop Makes Big Data Possible (2006)

Removing a panel of the CM-2, exposing one of the "subcubes" containing 16 printed circuit boards, each containing 32 central processing chips, each chip containing 16 processors, for a total of 8,192 processors in the sub-cube, or 65,536 processors in total.

First Computer-Generated TV Host

Annabel Jankel (b. 1955), Rocky Morton (b. 1955)

This is the world of Max Headroom, a digitized manifestation of fictional star reporter Edison Carter, who, in the film and sci-fi series, was seriously injured by his employer in a botched cover-up. When Network 23's technical boy-genius, Bryce Lynch, scans Carter's brain into the company's computer, the result is a hyperactive, eccentric, somewhat sentient avatar that pontificates about social order and topics of little consequence, such as the IQ of a moth.

British directors Annabel Jankel and Rocky Morton conceived of Max when they were hired to find ways of linking music videos together for England's Channel 4. They started with a movie, *Max Headroom: 20 Minutes into the Future*, and expanded it to a Channel 4 show that ran for four seasons. Max quickly became one of the most popular television personalities in England. Max also showed up in the United States in a third series that was broadcast in 1987 and 1988.

Max Headroom's title character became a cultural phenomenon and 1980s pop icon, appearing in Coca-Cola® advertisements and an episode of *Sesame Street* in 1987. Played by Canadian actor Matt Frewer, the real Max was definitely *not* computer generated: Frewer required four hours of makeup and prosthetic application each time he played Max on TV and in endorsement deals, guest appearances on *Late Night with David Letterman* and other shows, and simultaneous covers on *Newsweek* and *MAD* magazine in 1987.

In all of its various incarnations, *Max Headroom* wrestled with issues surrounding commercial advertising and news manipulation in the service of high ratings and revenue. The show used technology and the logic of an artificially intelligent personality to paint a dark picture of society, where standards of integrity and decency have given way to an insatiable appetite for salacious reporting and entertainment.

Although somewhat campy, *Max Headroom* was avant-garde in proposing that a person's brain could be scanned into a computer and produce a (somewhat) sentient being. Max Headroom's pedigree tapped into a deeper set of questions that resonated with the general public, including how far technology could be taken.

SEE ALSO *Metropolis* (1927)

Matt Frewer as Max Headroom, the computer-generated television host who became a pop-culture icon in the 1980s.

Zero-Knowledge Proofs

Shafi Goldwasser (b. 1958), **Silvio Micali** (b. 1954),
Charles Rackoff (b. 1948)

How do you prove that you know a secret without revealing the secret? Three computer scientists, Shafi Goldwasser, Silvio Micali, and Charles Rackoff, figured a way in 1985, establishing a new a branch of cryptography rich with applications that are only now beginning to be realized.

Zero-knowledge proofs are a mathematical technique for demonstrating facts about a proof without revealing the proof itself, provided that the demonstration involves interaction between two parties: the *prover* who wants to provide that some mathematical statement is true, and a *verifier* who checks the proof. The verifier asks the prover a question, and the prover sends back a string of bits, called a *witness*, that could be generated only if the statement is true.

For example, consider assigning colors to the states or countries on a map. In 1976, mathematicians proved that any two-dimensional map can be colored using just four colors such that no countries that touch are colored with the same color. But doing the same with just three colors is much harder and can't be done with all maps. Until the invention of zero-knowledge proofs, the only way for a person to show if a specific map could be colored with three colors was to do just that: produce the map colored with three colors.

Using zero-knowledge proofs, the witness demonstrates a specific map colored with just three colors has no instance of two touching countries that are the same color, and it does this without revealing the coloring of *any* country.

Building a practical system from zero-knowledge proofs requires the application of both cryptography and engineering—hard work, but some practical systems have emerged, including password-authentication systems that don't actually send the password, anonymous credential systems that allow a person to establish (for example) that the credential holder is over 18 without revealing his or her age or name, and digital money schemes that let people spend digital coins anonymously but still detect if an anonymous coin is spent twice (called *double spending*).

For their work in cryptography, Goldwasser and Micali won the A.M. Turing Award in 2012.

SEE ALSO Secure Multi-Party Computation (1982), Digital Money (1990)

Zero-knowledge proofs let a prover demonstrate possession of a fact without revealing that fact. For example, you can prove that there exists a way to color a map with just three colors, without revealing the completed map.

FCC Approves Unlicensed Spread Spectrum

Michael J. Marcus (b. 1946)

The wireless spectrum is divided into different bands, each with its own physical properties and regulatory rules. In the 1970s, transmitting with a radio in most bands required getting licensed by the government. The unfortunate exception was citizens band (CB) radio, which was virtually unusable because it was so crowded.

Then, in 1981, an engineer at the US Federal Communications Commission (FCC) started an official notice of inquiry to explore using a portion of the radio frequency (RF) spectrum for spread spectrum communications. The engineer, Michael J. Marcus, doggedly pursued the idea for four years at the FCC. His vision was to legalize the use of spread spectrum technology, an approach developed during World War II to hide wireless communications from enemy monitoring and jamming, to allow for research and the development of civilian spread spectrum systems.

But making unlicensed spectrum available proved controversial with incumbent manufacturers and wireless users. In the end, Marcus won his rulemaking on May 9, 1985, but the use of spread spectrum was largely limited to so-called industrial, scientific, and medical (ISM) bands at 900 megahertz, 2.4 gigahertz, and 5.7 gigahertz. Because that band was also used by commercial equipment such as microwave ovens and radar systems, there was too much interference to make it usable for other applications.

Still, the genie was out of the bottle. In March 1991, the NCR Corporation® started selling a wireless network product called *WaveLAN* that provided 2 megabits per second over the unlicensed 900-megahertz band. The technology proved popular, leading to the development of the 802.11 Wireless LAN Working Committee. The industry created the Wi-Fi Alliance in 1999, licensing the Wi-Fi® trademark and logo to products that passed an interoperability test. While the original Wi-Fi systems could stream data at a maximum of 2 megabits per second, equipment implementing the standard approved in December 2013 can go as fast as 866.7 megabits per second.

SEE ALSO First Wireless Network (1971)

The wireless spectrum, which once required a license from the government for use, began opening up to the public in 1985.

NSFNET

ARPANET was a restrictive club. The precursor to today's internet, ARPANET was open only open to organizations directly supporting the US Department of Defense (DOD), which was picking up the bill. Expanding the club meant expanding the funding base. The US National Science Foundation (NSF) did just that in 1980, when it gave a consortium of universities $5 million to create the Computer Science Network (CSNET), and again in 1985, when it commissioned the National Science Foundation Network (NSFNET). The networks interconnected at universities that had both DOD and NSF contracts.

CSNET featured a mix of technologies with the short-term goal of providing computer science departments with email and limited-access remote terminal capability. The number of hosts jumped from three in 1981 to 24 in 1982 and 84 in 1984. But the network's fundamental goal was to get its academic members access to the ARPANET so they could exchange email and use the other resources.

NSFNET had the much broader goal of creating a nationwide network—and, in particular, providing researchers around the US with access to the NSF's five supercomputing centers. The original NSFNET went live in 1986 with seven 56-kilobit-per-second links—and they were almost immediately saturated. In 1988, the network was expanded to 13 nodes interconnected by 1.5-megabit-per-second T1 links. Within three years, those T1 links were upgraded to 45-megabit-per-second T3 links. By that point, NSFNET linked 3,500 networks at 16 sites. Few users noticed when ARPANET was shut down on February 28, 1990—the non-Defense users had all migrated.

While NSFNET could purchase faster network links, there was no hardware or software that could run at such speeds. "No one had ever built a T3 network before," said Allan Weis, president of Advanced Network and Services, Inc., the nonprofit created to manage the network, as reported in the NSFNET final report. It took years of work before the T3s were running at full speed.

But while NSFNET greatly improved access to the internet, it also had an "acceptable use policy" that limited the use of the network to the support of research and education—and prohibited commercial ventures.

SEE ALSO ISP Provides Internet Access to the Public (1989)

A visualization study of inbound traffic measured in billions of bytes on the NSFNET T1 backbone for the month of September 1991, ranging from purple (0 bytes) to white (100 billion bytes).

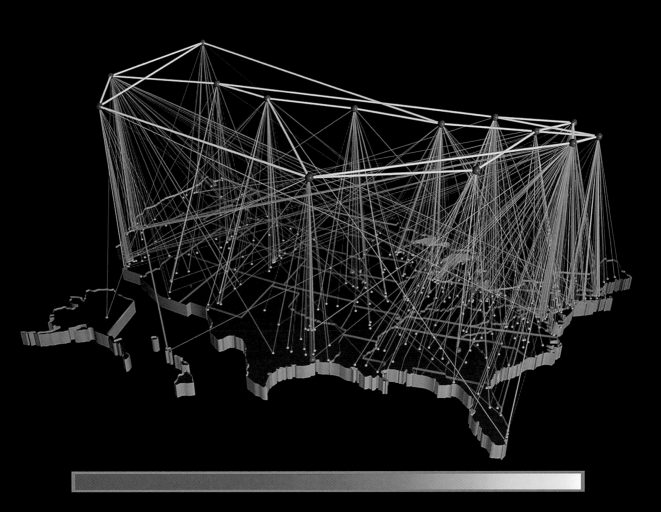

Desktop Publishing

Paul Brainerd (b. 1947)

Prior to desktop publishing, consumers and small businesses alike had two choices when it came to publishing: they could have something typed, or they could go to a copy shop to have it typeset on expensive, professional equipment. Desktop publishing allowed anybody to produce beautiful documents with fonts and graphics. It was a disruptive communication publishing technology that became the force behind low-budget magazines, newsletters, and pamphlets and helped train a generation of graphic artists who would be ready for the web and the coming age of social media.

The wave that started the desktop publishing era began well before the tsunami of products hit in 1985. Research at Xerox PARC in the 1970s developed foundational technologies while small-scale efforts by individuals and small newspapers developed a variety of computerized text layout methods and capabilities. In the late 1970s and early 1980s, small publishers were using "daisy wheel" printers with proportional fonts for basic typesetting, but text was still cut out with X-ACTO® knives and pasted down on white paperboard before it was taken to the printer.

In 1984, the Apple Mac debuted and Hewlett-Packard introduced the first desktop laser printer, the LaserJet®. Paul Brainerd, who had previously worked at a company that developed publishing software for newspapers, realized that the combination of microcomputers, laser printers, and the right software created the possibility for individuals to become their own publishers—something he called *desktop publishing*. That summer, Brainerd put together a team and created the Aldus Corporation, named after Aldus Manutius, a 15th-century Venetian scholar and printer.

The following year, Aldus put PageMaker® for the Mac on the market, considered the first desktop publishing application, while Adobe released PostScript, which would become the industry standard for page description language (PDL). Apple started to sell the LaserWriter. Suddenly individuals could do their own typesetting and even print small batches themselves.

SEE ALSO Laser Printer (1971), Xerox Alto (1973)

The first IBM PC version of Aldus PageMaker, seen here, included a free copy of Windows 1.0.

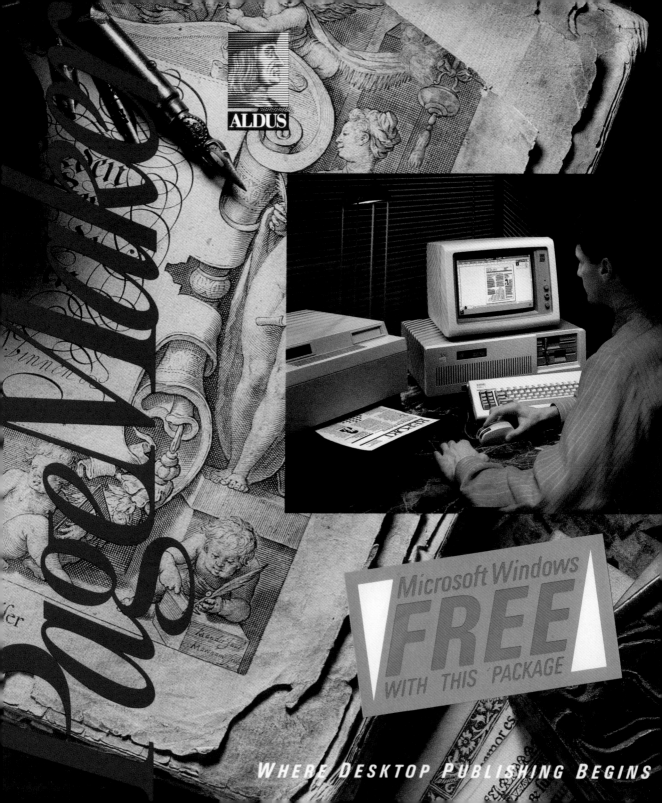

PageMaker

ALDUS

WHERE DESKTOP PUBLISHING BEGINS

Field-Programmable Gate Array

Ross Freeman (1948–1989), **Bernard Vonderschmitt** (1923–2004)

Many kinds of calculations can be implemented in either hardware or software. Typically, hardware is faster but more complicated, while software is slower but easier to create and debug. That's because hardware typically has many circuits and wires that perform the calculation in parallel. Software, in contrast, runs as a series of instructions within a computer's CPU: the same circuits and wires are repurposed over the course of the calculation for many different uses. Software remains more popular than hardware for solving problems, because it's typically easier to develop and change.

But what if hardware could be programmed just like software? Programmable hardware could implement special algorithms for video processing. It could run AI algorithms for image recognition at high speed, and presumably with less power. Programmable hardware could also be used to replace complicated circuit board designs with hundreds of individual components with a single, programmable chip.

That's the idea of a field-programmable gate array (FPGA). The chips contain programmable logic cells that can be connected as needed. Once they are wired up, the gates can work like the circuits inside any other silicon chip—with one big difference. If the wiring diagram isn't correct, or if it needs to be changed, the program can be erased and the FPGA reprogrammed with a new configuration.

FPGAs are typically more expensive to purchase and program than application-specific integrated circuits (ASICs). But because they can be programmed in the lab and then reprogrammed as necessary, they make innovation dramatically cheaper and faster—especially in applications where only a few integrated circuits are needed, such as a prototype. Otherwise, the cost of replacing a buggy circuit would be prohibitive, like on a spacecraft. That's why NASA used FPGAs on its Mars *Curiosity* rover.

Ross Freeman and Bernard Vonderschmitt cofounded Xilinx® in 1985 and created the first commercially viable FPGAs, winning Freeman a place in the National Inventors Hall of Fame.

SEE ALSO Silicon Transistor (1947), First Microprocessor (1971), Verilog (1984)

The FPGA at the center of this circuit board can be programmed to pulse lights, spin motors, and synthesize music.

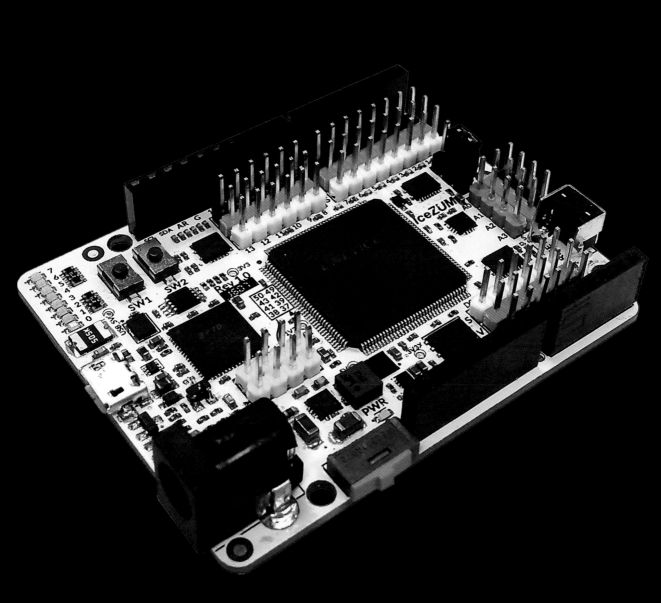

GNU Manifesto

Richard Stallman (b. 1953)

Richard Stallman, a staff programmer at the MIT AI laboratory, became a proponent of what we now call *open source software*—software distributed in a form that is easily modified by its users. His interest in open source started in 1980, when the lab's xerographic printer (XPG), an early laser printer, was replaced with a Xerox 9700. Unlike with the XPG, Xerox had not provided MIT with the new printer's source code, claiming it was proprietary intellectual property. This prevented Stallman from modifying the 9700 to tell people in the lab when the printer was finished with their print jobs, or if the printer had jammed.

The following year, Stallman found himself in a fight between two groups of programmers who had left the AI lab. One was making proprietary software in the Lisp computer language for the new AI workstations that the lab had purchased; the other was working on the older Lisp software that the lab had designed. Stallman felt that creating software that was not open to anyone who might want to use it damaged the lab's hacker culture; he responded by cloning the proprietary software and making it freely available.

In 1984, Stallman gave up on Lisp and the Lisp machine—they had both failed in the marketplace—and decided to create a new community devoted to creating and using software that could be freely shared, modified, improved, and learned from. He called this software *free software*—free as in *free speech* or *freedom*, although by design the software could be shared without cost. Stallman based his new work on the popular UNIX operating system, which would let it run on many different kinds of computer hardware. He called the project *GNU* (really G.N.U.), a recursive acronym that means "GNU's Not UNIX."

Stallman wrote the "GNU Manifesto" announcing his project and published it in the March 1985 issue of *Dr. Dobb's Journal*. He invited others to join him, either with their time or by donating money to the cause. The nonprofit Free Software Foundation (FSF) was incorporated seven months later, housed at the MIT AI lab. Several companies, including Hewlett-Packard, donated computers to the FSF to help fund the project.

SEE ALSO *Dr. Dobb's Journal* (1976), Linux Kernel (1991)

Photograph of the GNU homepage as seen through a magnifying glass. GNU (which stands for "GNU's Not UNIX") is a free operating system based on UNIX.

http://www.gnu.org/

GNU

...rating Syste...

...censes   Education   Softwa...

...+ GNU   Pr...   What is GNU?

GNU is a Unix-like operating system that is free software—
freedom. You can install versions of GNU (more precisely,
which are entirely free software. What we provide.

AFIS Stops a Serial Killer

Carl Voelker (dates unavailable), **Raymond Moore** (dates unavailable), **Joseph Wegstein** (b. 1922)

In the late 19th century, law enforcement realized they could use fingerprints to determine if a person convicted of a crime was a repeat offender or a first-time offender (first time caught, at least). So in 1924, the US Federal Bureau of Investigation (FBI) created its Identification Division, and by the early 1960s it had fingerprints from 15 million criminals. The Identification Division was drowning in its success: each day it received 30,000 new 10-print cards to search the database, a task that took a technician approximately 18 minutes for each card.

FBI Special Agent Carl Voelker went to the US National Bureau of Standards in 1963 to see if there was a way to create an Automated Fingerprint Identification System (AFIS) using information technology. There he met with Raymond Moore and Joseph Wegstein. Moore and Wegstein realized they would have to create a new scanner to read the fingerprint cards, develop software to extract the characteristic points of a fingerprint used for identification, and finally develop software to match the prints. The first two tasks were put out for bid and developed at the Cornell Aeronautical Laboratory and the Autonetics division of North American Aviation; the matching software was developed personally by Wegstein. Five years later, Rockwell Autonetics was tasked with creating five high-speed card readers that the FBI used to scan its 15 million criminal fingerprint cards, making them electronically searchable for the first time.

Similar systems were developed in the United Kingdom, France, and Japan, although these systems were designed primarily to match partial prints left at crime scenes against fingerprint cards, rather than for identity verification. NEC Corporation (formerly the Nippon Electric Company) designed a system for the Japanese National Police; similar systems were installed in San Francisco and Los Angeles, both of which saw burglary rates drop as prints left behind at break-ins could finally be used to identify a suspect. Then in 1985, the Los Angeles AFIS system identified the "Night Stalker" serial killer, Richard Ramirez, from a fingerprint he left on the mirror of a car he had stolen, ending his murderous crime spree.

SEE ALSO First Digital Image (1957), Algorithm Influences Prison Sentence (2013)

The Los Angeles Automated Fingerprint Identification System (AFIS) identified the "Night Stalker" serial killer from a fingerprint on the mirror of a stolen car.

Software Bug Fatalities

Atomic Energy of Canada Limited (AECL) manufactured a line of radiation therapy machines used for treating cancer. The Therac-25 was a lower-cost, smaller, and more modern version of the company's Therac-20. Both machines could generate a beam of low-power electrons, or they could crank up the power and slam the electrons into a piece of metal to produce therapeutic x-rays. But whereas the Therac-20 used a series of switches, sensors, and wires to implement its control logic and safety interlocks, the Therac-25 relied on software.

On April 11, 1986, a patient named Verdon Kidd was being treated for skin cancer on his ear with a Therac-25 when something went horribly wrong with his treatment. Instead of receiving a low-dose electron beam, he received a wallop of radiation. He saw a flash in his eyes, felt something hit him in the ear, and screamed in pain. The Therac-25, meanwhile, displayed the innocuous message "Malfunction 54." Kidd died 20 days later on May 1 from the massive radiation dose he had received, the first person known to have been killed by a software bug in a medical device.

The initial investigation by AECL revealed that if the Therac-25's operator hit the up arrow at a certain point in the program's operation, the machine would energize the electron beam in its high-intensity setting without moving the x-ray target into place. AECL's response was to remove the up-arrow key and cover the switch with electrical tape, an action the US Food and Drug Administration (FDA) later called inadequate.

In all, three patients died and two more received life-threatening doses of radiation—between 15,000 and 20,000 rads instead of the therapeutic dose of 200 rads—due to a subtle error in the Therac-25's programming called a *race condition*, when two parts of a program sometimes execute in the wrong order. Additional work by University of Washington professor Nancy Leveson and her student Clark Turner exposed the need to establish rigorous engineering procedures for software development, testing, and evaluation. Nevertheless, in an article written about the Therac-25 incident 30 years after the fact, Leveson concluded that US regulators still did not have standards in place for reliably preventing harm that might be caused by medical device software.

SEE ALSO Software Engineering (1968)

A radiation therapy mask showing laser lines for targeting cancer cells in the brain. Radiation doses must be carefully controlled to avoid harm to patients, such as the fatal incidents caused by software errors in the 1980s.

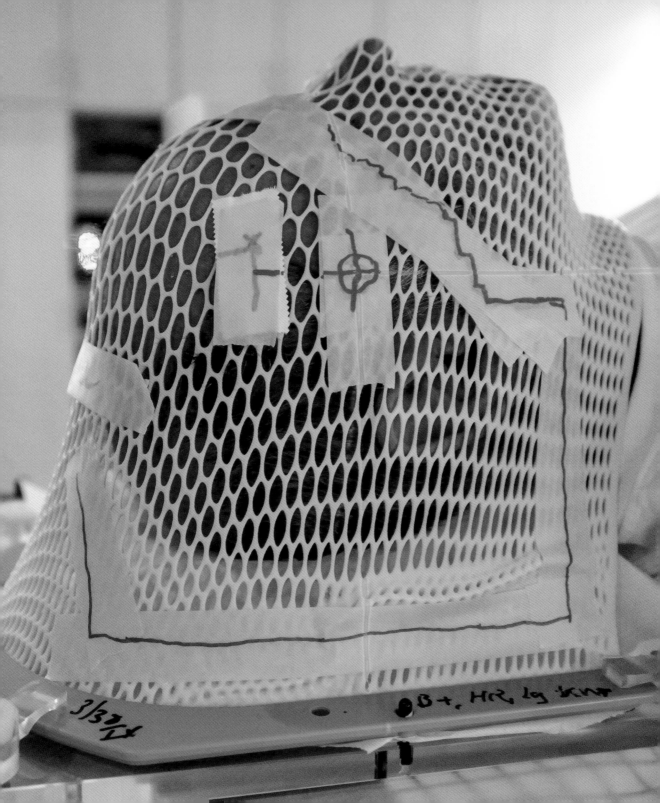

Pixar

Ed Catmull (b. 1945), Alvy Ray Smith (b. 1943), Steve Jobs (1955–2011)

Pixar may be best known for creating movies such as *Toy Story*, *Cars*, and *Inside Out*, among others, but it is also famous for the pioneering computer animation techniques and technologies it designed to bring its creative visions to life.

Now a subsidiary of the Walt Disney Company®, Pixar originated at the New York Institute of Technology's Computer Graphics Lab, where George Lucas found and hired away Ed Catmull and Alvy Ray Smith to run the computer division at Lucasfilm. Lucas sold the computer division to Steve Jobs in 1986, and it became an independent company called *Pixar*. Disney purchased Pixar on January 25, 2006, at a valuation of $7.4 billion.

Pixar's proprietary animation rendering technology—RenderMan®—is an industry standard that has received both scientific and technical awards for advances in realistic visual effects, including lighting, shading, and shadowing. The program also embodies techniques for processing the large amounts of 3-D data required for an animated movie. RenderMan's first trip to the red carpet was in 1989, when Pixar's short film *Tin Toy* became the first computer-animated film to receive an Oscar® when it was named Best Short Film (Animated). Since then, RenderMan has been used in numerous Academy Award®–winning films and was acknowledged with its own Oscar in 2001, when Ed Catmull, along with his colleagues Robert L. Cook and Loren Carpenter at Pixar, received an Academy Award of Merit "for significant advancements to the field of motion picture rendering as exemplified in Pixar's RenderMan."

In 2015, the traveling exhibit *The Science Behind Pixar* opened as a collaboration between Pixar and the Boston Museum of Science. The exhibit showcased the science, technology, engineering, art, and mathematics (STEAM) that Pixar uses to create its films. The exhibit is organized around Pixar's production pipeline and includes content on modeling, rigging, surfaces, sets and cameras, animation, simulation, lighting, and rendering. Related is Pixar's partnership with the NSF on a research project to help educate people about computational thinking. Leveraging six exhibit experiences from *The Science Behind Pixar*, the aim is to help students learn how to break down a challenge into pieces that can be understood and carried out by a computer.

SEE ALSO Sketchpad (1963)

A collection of popular films from Pixar Animation Studios.

Digital Video Editing

Similar to the music industry, the movie and television industries were hugely affected by advancements in computational power and techniques for digitally capturing and manipulating analog phenomena such as sound and images. Originally, movies were filmed on actual celluloid film. Editing was done by cutting and slicing the physical material that the entertainment content was captured on.

With the advent of videotape and electronic editors such as the Ampex® in the early 1960s, tape and film could be edited without physically lacerating and rejoining it for new sequences. In 1971, CBS and Memorex® teamed up to create the hugely expensive video editing system called the *RAVE* (Random Access Video Editor) that allowed an operator to create a rough edit with a computer and then have the system splice together a high-quality cut. But it wasn't until the creation of digital video editing software in the late 1980s that the modern era of movie and TV editing really took off.

In 1987, two Massachusetts companies in particular led the charge—EMC and Avid®—with the Avid/1 machine quickly becoming the dominant change agent and leading platform in the industry. In addition to boosting productivity and saving storage space, digital editing gave artists and technicians greater creativity and control in putting together compelling stories. When an editor is working with all digital files—audio, image, and video—there is enormous flexibility in sequencing, modifying, altering, and leveraging special effects to bring everything together in a cohesive narrative. In addition, digital video editing does not use original source content, which preserves the original material from degradation or harm.

Over the years, the importance of digital editing technologies to the movie and television industry have been acknowledged through a variety of engineering and creative awards. In 1993, Avid received a technical Emmy® award from the National Academy of Television Arts & Sciences for its Media Composer® editing system. And in 1999, Avid received an Academy Award for its motion picture editor, Film Composer.

SEE ALSO MIDI Computer Music Interface (1983)

The editing board in Studio 1 of the SAE Institute in Amsterdam gives the operator full control over sound and video during the editing process.

GIF

Steve Wilhite (dates unavailable)

File formats are specifications that describe the content and ordering of information inside a computer file. Programmers use these specifications to write software that reads and processes files, and that writes data back to disk so it can be read by other software.

In 1987, Steve Wilhite led the team that created the GIF (Graphics Interchange Format) file format to let users of the CompuServe® information system download and display color images on their home computers. Because it was designed for a dial-up service, he used an advanced compression algorithm called *LZW* that had been published a few years before. This made GIF images tiny compared to other formats. Two years later, Wilhite's team developed GIF89, extending the format with transparency and animations.

In 1992, a team led by Marc Andreessen at the University of Illinois at Urbana-Champaign's National Center for Supercomputing Applications (NCSA) started building Mosaic, the first browser that could display images on web pages. GIF was one of two file formats that the Mosaic web browser supported, and the only one that could display color. When the Mosaic team left to create Netscape Communications Corporation and the first mass-marketed web browser, they took with them their fondness for GIF, but they made a minor change to the format: a "loop" flag that caused the animated GIF to repeat endlessly. GIF's popularity rose with the popularity of the internet.

But GIF was not without problems. GIF had been designed for displaying bold graphics, charts, and logos—not photographs. The format supports only a limited number of colors—typically 256—and the compression is designed to shrink solid areas and patterns, not the gradients typical in photographs. Other formats, like JPEG, were needed to accommodate the rise of digital photography.

GIF's second problem was legal: Unisys®, the company that invented LZW, patented the algorithm and started demanding royalties from major users as early as 1995. The Unisys patent expired on June 20, 2004, making GIF free forever.

As for how the word *GIF* should be pronounced, Wilhite says it's with a soft g, as in "choosy programmers choose GIF," a reference to a peanut butter commercial that was popular at the time.

SEE ALSO JPEG (1992), First Mass-Market Web Browser (1992)

A GIF (Graphics Interchange Format) file typically supports 256 colors.

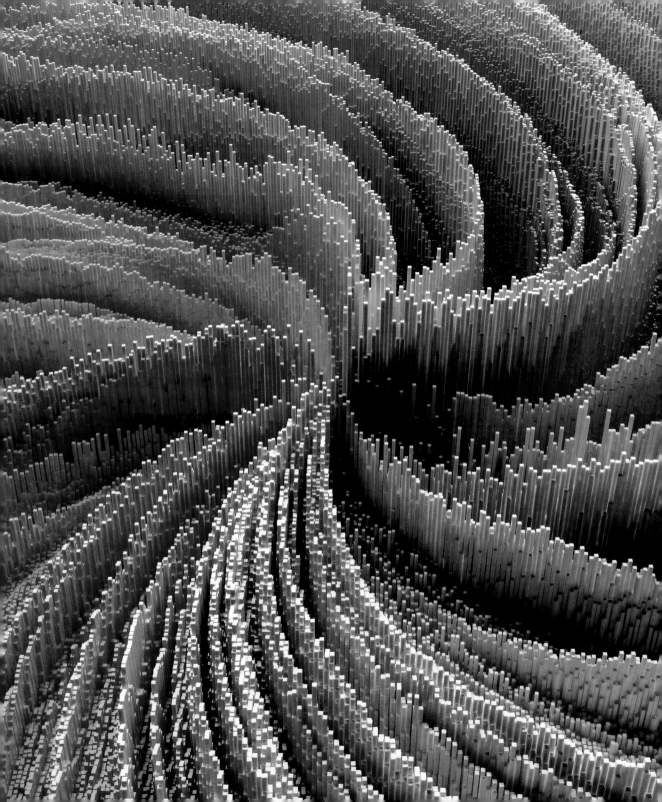

MPEG

Leonardo Chiariglione (b. 1943)

MPEG is a collection of associated standards for coding and compressing audio-video data. As more people used video to communicate and the applications that leverage those communications became more commercially important, so too did the need for an international standard that provided users with a uniform way to encode and transmit the content. Without it, interoperability, compatibility, and market growth in this sector would have been stymied or its evolution limited. Examples of innovation that were outgrowths of these standards include MP3 players, CDs, DVDs, Blu-ray Discs®, tablets, phones, cable boxes . . . and the list goes on.

The term MPEG stands for the Moving Picture Experts Group, the committee established to develop and standardize the technologies that underpin the protocols (which themselves have names like MPEG-1, MPEG-2, MPEG-3, and MPEG-4). Established in 1988 as an international group led by Dr. Leonardo Chiariglione, the MPEG working group fell under the JTC1—Joint Technical Committee—composed of the International Electrotechnical Commission (IEC), responsible for overseeing and managing the standards for electronic or electric technologies, and the broader International Organization for Standardization (ISO).

The original MPEG-1 standard was set in 1992 for establishing compression of lossy images and sound, in which unnecessary information is discarded with little perceived degradation in quality at low bit rates—specifically 1.5 megabits per second. The protocol was primarily associated with making video CDs and transmitting digital cable and satellite. MPEG-1 is also popularly associated with the MP3 music standard. The MP3 (MPEG-1 Audio Layer III) audio compression format is a patented audio codec developed by German engineer Karlheinz Brandenburg (b. 1954) and others for compressing digital audio that, uncompressed, takes up a lot of space. The ability to squeeze the file size down with minimal loss of quality had a variety of practical applications, perhaps most notably the ease of transmitting music files in an era of limited bandwidth capacity. It was standards such as these, in combination with other evolving network technologies, that would eventually help usher in peer-to-peer file sharing and a host of other collaborative innovations.

SEE ALSO Diamond Rio MP3 Player (1998), Napster (1999)

MPEG standards for coding and compressing audio-video data led to innovations such as CDs, DVDs, Blu-ray Discs, and more.

CD-ROM

James Russell (b. 1931)

The optical disc was *the* breakthrough storage technology for three decades in a row. Originally developed for storing digitized audio in the 1970s, optical discs were adopted to store computer data in the 1980s and then digital video in the 1990s. During this time, the physical format remained standard: a polycarbonate disc, 120 millimeters in diameter, with a central hole and a spiral track—like a vinyl record, but high tech. Read with a laser, optical discs used error-correcting codes to recover from reading errors caused by dust and minor scratches. The industry even developed writable media after a time.

The key inventions that made the compact audio disc possible date back to the 1960s, when James Russell, a classical music enthusiast working at Battelle Memorial Institute's Pacific Northwest National Laboratory, demonstrated a system that could digitize music, store it on optical media, and play it back. The consumer market that drove the adoption and commoditization of the technology was the result of a partnership between Philips® Electronics and Sony®, both eager to avoid a repeat of the "format wars" they had fought over the VHS and Beta tape recording formats in the 1970s.

Sony introduced the first audio CD player on October 1, 1982. The machine cost around $900, and the discs retailed for $30 each at a time when records were typically priced under $10. But CDs sounded so much better than the average vinyl record that the technology was a success.

For computers, the big breakthrough came in 1988, with the release of the CD-ROM standard. With greater error correction than audio CDs, a CD-ROM stored 682 megabytes of data—more than 450 3.25-inch floppy disks. Libraries started purchasing database applications delivered on CD-ROM. By the mid-1990s, CD-ROMs became the dominant media for distributing software, and writable CDs (CD-Rs) and rewritable CDs (CD-RWs) became a popular format for both backups and exchanging information.

Optical discs were just entering their fourth generation of technology—with storage in the 50-gigabyte range—when the growing availability of high-speed residential broadband created a better way for consumers to get their music, video, and software.

SEE ALSO Error-Correcting Codes (1950), DVD (1995), USB Flash Drive (2000)

The compact disc could be used both to access data from CD-ROMs and to play music from conventional CDs.

Morris Worm

Robert Tappan Morris (b. 1965)

On the morning of Thursday, November 3, 1988, researchers and system administrators all over the internet arrived at work to find their computers mysteriously sluggish, even nonresponsive. People couldn't log in. Systems would be rebooted, be functional for a few minutes, and then slow down again to a crawl. Technologists soon realized the problem: their systems were under attack, over the internet, by a piece of software that probed for vulnerabilities and, finding them, copied itself to the system and set about attacking others. The program was called a *worm*, taking its name from the tapeworm of John Brunner's novel *The Shockwave Rider*.

The worm was the creation of a 23-year-old Cornell University graduate student, Robert Tappan Morris—a computer whiz kid whose father just happened to be chief scientist of the National Security Agency's National Computer Security Center.

Although Brunner had hypothesized about network worms in the 1970s, and computer viruses had existed in the wild since 1982, nobody had ever seen something like this in real life. The worm had four different ways of breaking into computers, and once it was in, it tried to crack passwords and find other vulnerable machines to attack. The worm even had code that could detect if it was already running on a machine so that it wouldn't reinfect the system. This code was flawed, however, and the worm reinfected the infected systems many times over. The result was that many copies of the worm ended up running on the vulnerable systems, causing them to slow to a drag.

The attack was front-page news, and it was the first time that many people in the United States had even heard of the internet. A study by the US General Accountability Office concluded that 6,000 of the 60,000 computers on the internet at the time had been infected; it took most sites two days to completely eradicate the running program.

It's widely believed that Morris released the worm as an experiment—and to show internet system administrators that security needed to be taken more seriously. Because of the worm, many institutions (including the US government) created computer security emergency response teams (CERTs). As for Morris, after serving 400 hours of community service and paying a fine, he eventually received his PhD from Harvard and became a professor at MIT, where he was awarded tenure in 2006.

SEE ALSO *The Shockwave Rider* (1975), NSFNET (1985)

A computer worm is a malware program that replicates itself and spreads to other computers, often hiding in parts of the computer's operating system that are not visible to the user.

World Wide Web

Tim Berners-Lee (b. 1955)

The World Wide Web transformed the internet from an academic curiosity into a dominating technology touching the lives of virtually every person on the planet. Although variations of the web's key elements previously existed, the explosive growth of the web was due almost entirely to Tim Berners-Lee's vision of a worldwide information commons, combined with the web browser and web server that he created while working at CERN, the European Laboratory for Particle Physics.

The web combined the ideas of hypertext—text with links—and electronic publishing, with a critical twist: the information publisher and the reader didn't have to be on the same computer. Instead, individual web documents (as they were called at the time) were designed to be downloaded over the internet using the Hypertext Transport Protocol (HTTP) that Berners-Lee invented. The documents themselves were designed to be authored in a relatively simple subset of the Standard Generalized Markup Language (SGML) that Berners-Lee also invented, called the *Hypertext Markup Language* (HTML). Unlike other hypertext systems, HTML links were embedded directly in the text of a document. For inventing the World Wide Web, Berners-Lee was awarded the 2016 A.M. Turing Award.

The web was successful because, unlike other efforts at the time, it had few technical or legal encumbrances. Any computer connected to the internet could run a web server, which any internet user could reach by downloading and running a web browser. As a result, organizations and individuals could publish information to the global community without having to ask anyone's permission.

The web became the internet's second killer app (after email), and it soon far exceeded the first. Indeed, by the mid-1990s, people and businesses were connecting to the internet so that they could access the web, and companies were being formed for the singular purpose of creating and operating websites. Within 10 years, the web had become the greatest single engine of education, communication, and wealth creation that the world had ever seen. Nothing would ever be the same again.

SEE ALSO First Mass-Market Web Browser (1992)

Through the World Wide Web, people and data are constantly connected, revolutionizing how the world communicates and exchanges information.

SimCity

Will Wright (b. 1960)

Who knew that urban planning could be so much fun? Produced by Maxis Software, the iconic city-building game *SimCity*® was originally developed for the Commodore 64 in 1985 but not released until 1989, when it simultaneously appeared on the Atari ST, the Amiga, and DOS-based IBM PCs. Part of the Electronic Arts® brand, *SimCity* is one of the most popular and influential games of all time. It ushered in a new type of interactive entertainment and spawned a series of even more popular simulation games called *The Sims*®. It also played a role in increasing interest in thoughtful city planning and design and helped promote the "new urbanism" movement, which seeks to make cities friendlier for walking, biking, and outdoor recreation.

Created by Will Wright, *SimCity* relies on the player's creativity to build a successful, thriving city. Players choose and sequence the parts of their city that they wish to build, taking into consideration features that will enable the residents to live and interact with each other in harmony—for example, building a main road in proximity to a housing development, deciding the tradeoffs between choosing a clean power source or one that has some pollutant, or building a water tower so that homes and businesses can operate.

Wright's idea for *SimCity* was born out of a longtime fascination with complex adaptive systems and system dynamics. The premise of *SimCity* expanded the definition of what most people recognized as a video game, including the game mechanics used to motivate players and enable forward momentum as the simulation progressed. Part of the genius in Wright's product was the fact that people genuinely enjoy the mental process of creating things, and then seeing those things interact with and respond to a world of their own making. The victory condition for the player was as much an organic, personal goal as it was an external, explicit objective stated by the game.

SimCity also helped give rise to a new type of gamer, those people who were focused on strategy and creation. Today, *SimCity*'s legacy extends beyond the gaming and computer industries to a larger set of communal and societal concerns that in 1989 were just emerging in the public's consciousness.

SEE ALSO *Game of Life* (1972)

Will Wright, cofounder of the game development company Maxis and designer of SimCity, *photographed in his Emeryville, California, headquarters for* Computer Gaming World *magazine.*

1989

ISP Provides Internet Access to the Public

Barry Shein (b. 1953)

There were only a few options for accessing the internet in 1989. Students and staff at internet-connected universities had access, of course. So too did employees at select research labs and defense contractors. A few US government agencies had connections, as did a smattering of institutions scattered throughout Europe and Asia. What if you wanted to log in to a computer in another state without making a long-distance phone call? What if you wanted to develop a new internet protocol or application? In those cases, you were generally out of luck.

Barry Shein changed all of that in November 1989 when he launched the world's first commercial ISP, calling it—what else?—"The World." Shein was well connected and respected for his work getting Boston University on the internet. But no longer affiliated with a university, Shein was running a small consulting firm and dial-up bulletin board in Brookline, Massachusetts, just on the outskirts of Boston.

One day, the head of UUNET Technologies, a startup that provided high-speed internet access to businesses, asked Shein if UUNET could put some telecommunications equipment in Shein's machine room to serve business customers in the Boston area. Shein offered UUNET the space for free, provided that UUNET let Shein connect The World directly to the internet. Shein told UUNET that he wanted "all the bits I can eat"— flat-rate pricing for him and his dial-up customers.

The agreement was struck, and suddenly anyone who wanted access to the exclusive internet club could get it for $20/month.

Many of the internet's old guard were not happy with The World crashing their exclusive party. Then Shein got a call from someone at the National Science Foundation, the US government organization in charge of the internet at the time. Despite NSF's "acceptable use policy" that prohibited commercial use of the internet, the caller said that Shein could legitimize The World's connection by calling it "an experiment." And so The World officially became the world's first commercial dial-up internet service provider—and an early experiment in e-commerce.

SEE ALSO NSFNET (1985)

A pile of old modems, routers, and network equipment, displaying serial, phone, audio, and Ethernet connectors.

GPS Is Operational

Roger Easton (1921–2014)

With the launch of the first global positioning system (GPS) in 1978, the world was on its way to eliminating disorientation. Although the original goal was to provide radio location and navigation for US military planes and ships, today's GPS receivers are the size of a small coin and provide location information not just for government vehicles, but for civilian vehicles, pedestrians, and even inanimate objects such as buildings.

Each GPS satellite contains an atomic clock and electronics that beam the satellite's identifier and its exact time down to the planet 20,000 kilometers below. The signals travel at the speed of light, meaning they take roughly 0.06 seconds to reach the surface. Each receiver has an almanac that allows it to calculate each satellite's exact position based on the current time. Because the receiver also has an accurate clock, it can subtract the time that it gets from each satellite from the current time and determine the distance to each satellite. Knowing these distances, along with the satellites' actual positions, allows the receiver to calculate its own position. Although the first test satellite was launched in 1978, it wasn't until 1990 that sufficient production satellites were in orbit that the terrestrial GPS receivers could function reliably.

The idea of using radio waves for navigation dates back to World War II, when the Allies developed increasingly sophisticated systems to help bombers reach their targets. The satellite-based system was designed in the 1960s as a navigation and targeting system by Roger L. Easton, a scientist at the Naval Research Laboratory. It was only after the 1983 intentional downing of Korean Air Lines flight 007, which had unwittingly strayed into the Soviet Union's airspace, that President Reagan decided to make GPS freely available to the international community. Even then, GPS satellites were designed to transmit two signals: an unencrypted, less-accurate civilian signal designed for general use, and a more accurate encrypted signal intended for the US military. The two classes of service were called *selective availability*. Unexpectedly, radio-navigation use soon became dominated by civilians. In May 2000, President Bill Clinton ended the general use of selective availability, clearing the way for GPS's growing use as a consumer navigation system.

SEE ALSO First Wireless Network (1971)

Block II are the second generation of satellites that make up the Navstar Global Positioning System, known as GPS. Built by Rockwell International, they were the first fully operational GPS satellites.

Digital Money

David Chaum (b. 1955)

Today credit and debit cards are the primary tools people use to make purchases. But these cards do more than transfer value: for every transaction, they leave a lasting record containing the identity of the purchaser and the seller. That record can be a deterrent to many kinds of activities, both legal and illegal.

In the physical world, people can also use paper cash and coins to buy goods and services. Unlike the cards, cash and coins are anonymous: once the transaction is completed, there is nothing that ties the identity of the purchaser to the seller. Cash and coins are difficult to forge or copy, and they cannot be spent twice.

When he was a graduate student in the 1980s, David Chaum, a mathematician and cryptographer, was vexed by the challenge of replicating the qualities of cash and coins in some kind of digital currency. Chaum created DigiCash, the first practical implementation of digital cash to protect privacy and anonymity in the online world, and then founded DigiCash Inc. to promote the technology.

In Chaum's system, people create their own digital coins, each with a small denomination and a unique serial number. They then blind the serial number, have the bank digitally sign the coin—simultaneously removing the coin's denomination from the person's bank account—and finally unblind it. To spend the coin, the person gives it to the merchant, who deposits the coin into his or her own account (which has to be with the same bank). The bank can verify its signature, and it deposits the money into the merchant's account. The bank also records the coin's serial number, to prevent the merchant (or the original user) from depositing a second copy of the coin.

The idea of DigiCash never took off. A variety of factors contributed to DigiCash's end, including poor business decisions by the company and a market that was not yet ready to fully embrace the concept of electronic money. The company filed for bankruptcy in 1998, and in 2002 its assets were sold.

It would be years before electronic currencies such as Bitcoin and alternative payment systems such as PayPal® would enter the scene and gain a strong enough foothold in the market to realize the network effects required for them to take off.

SEE ALSO Bitcoin (2008)

With DigiCash, people created their own digital coins, each with a small denomination and unique serial number.

Pretty Good Privacy (PGP)

Phil Zimmermann (b. 1954)

Pretty Good Privacy (PGP) is an email encryption program developed by Phil Zimmermann, a peace activist and computer programmer who cared deeply about the privacy rights of global citizens.

In 1991, Zimmermann learned there was an anticrime bill being debated in the US Senate to require companies selling encryption products in the United States to include "trap doors" in their software so that government investigators could get copies of the unencrypted messages—the so-called plain text.

Zimmermann foresaw codebreaking warrants being used against people like him who were lawfully protesting the government's policies. So he decided to write a program to let people exchange encrypted email.

Zimmermann called his program *Pretty Good Privacy* and released version 1.0 on June 5, 1991. The program was buggy and had security vulnerabilities that were later discovered (and fixed), but there was just enough functionality to let people create public/private key pairs, distribute public keys over the internet, and then use those public keys to send each other encrypted mail. And as near as anyone could tell at the time, the messages sent by PGP were uncrackable by any government.

In 1993, RSA Security, Inc., the company created by the three MIT professors who had invented the RSA algorithm, complained to the US government that PGP violated patent 4,405,829, "cryptographic communications system and method," assigned to MIT and licensed to RSA Security. The government responded by launching an investigation of Zimmermann for illegally exporting cryptographic software in violation of laws restricting the export of munitions. That investigation lasted until January 11, 1996, when the government announced it was giving up on the prosecution. Four years later, the US Department of Commerce revised the export control regulations, making it legal to export encryption software in source code form.

Today, the PGP standard, implemented by both PGP and its compatible cousin, the GNU Privacy Guard, is one of the dominant systems for exchanging encrypted email.

SEE ALSO RSA Encryption (1977), GNU Manifesto (1985)

Pretty Good Privacy provided padlock-like security for everyday email messages.

Computers at Risk

David D. Clark (b. 1944)

At the behest of the Defense Advanced Research Projects Agency (DARPA), several organizations, including the National Research Council and the Commission on Physical Sciences, Mathematics, and Applications, authored a frightening 320-page report, *Computers at Risk: Safe Computing in the Information Age.*

"Making computer and communications systems secure is a technical problem," the report stated. But, it "is also a management and social problem." Simply put, the issue wasn't just that many computer systems were not secure—they were not *securable.*

That was a big problem, the report's authors wrote, because society was growing increasingly dependent upon networked computer systems: "As computer systems become more prevalent, sophisticated, embedded in physical processes, and interconnected, society becomes more vulnerable to poor system design, accidents that disable systems, and attacks on computer systems."

David D. Clark was the chair of the committee that authored the report, which was divided into chapters on topics such as technologies needed to achieve security, criteria to evaluate computer and network security, and why the security market had failed. Ideas suggested by the report's authors included establishing an Information Security Foundation, emergency response teams, and defining the tactics of high-grade threats. *Computers at Risk* also discussed how the US Department of Defense's computer system evaluation framework did not provide a concept for security that would be adequate for either industry or the private sector.

The overarching message was that computer system security requires a planned approach that considers the problem from a holistic perspective. Within this broad scope, it aimed to provide explanations that put the strategic computer security problem into context. *Computers at Risk* promoted measures to enhance a computer system's trustworthiness, mitigate vulnerabilities, and enable people to maintain vigilance as America's computer and information networks became increasingly connected and valuable.

Coming just two years after the Morris Worm, the report all but predicted the information system security challenges that the next 25 years would bring.

SEE ALSO *WarGames* (1983), Morris Worm (1988)

The 320-page report Computers at Risk *issued dire warnings regarding the security of computer and communications systems.*

Linux Kernel

Linus Torvalds (b. 1969)

On August 25, 1991, an undergraduate at the University of Helsinki in Finland sent a message to the comp.os.minix Usenet newsgroup saying that he was creating a free operating system for the Intel 80386 microprocessor. The message would go on to state that the system's kernel could already run two important programs that were part of the up-and-coming GNU operating system: the GNU "bash" shell (the program in which users type their commands) and the GNU C Compiler (the program that converts a programmer's code to machine code).

Intel's 80386 was the first mass-market 32-bit processor—the first microprocessor that could run advanced software without the memory limitations of previous micros. But the only operating systems available for the 80386 were proprietary. Microsoft's Windows was a limited operating system, and while various versions of UNIX were available for the 80386, they were expensive and did not come with source code.

In 1991, the hacker/hobbyist world was waiting for the Free Software Foundation (FSF) to finish Richard Stallman's long-promised GNU operating system. Launched in 1985 with the GNU Manifesto, the FSF's few hired programmers and thousands of volunteers were creating a clone of the hacker-friendly UNIX operating system. Alas, they had run into trouble creating that critical piece of the operating system called the *kernel*—the master control program that interfaces with the computer's hardware and arbitrates execution between all of the other programs running on the system. The *Hurd*, as the kernel was called, was a technically ambitious piece of computer science research, but making it all work was eluding Stallman's team.

With all of Hurd's delays, many people were interested in this new kernel that Linus Torvalds claimed to have written. Although his kernel worked, it still required support from MINIX, the operating system on which Torvalds had written the kernel. Those dependencies were removed by the middle of 1992, at which time it was possible to run a completely free operating system on Intel's hardware. Since then, interest in Linux has continued to grow, and it is now the most widely used operating system in the world, largely as a result of its adoption by Google for the Android operating system.

Meanwhile, the Hurd is still under development.

SEE ALSO UNIX (1969), GNU Manifesto (1985)

Linus Torvalds holds up a license plate with the Linux name on it on the exhibit floor of the Linux World Conference in San Jose, California, on March 2, 1999.

Boston Dynamics Founded

Marc Raibert (b. 1949)

Boston Dynamics is a vanguard robotics company, maker of marvels such as Atlas, the autonomous 6-foot (1.8-meter) backflipping biped humanoid, and BigDog, a quadruped designed to function as a pack mule, capable of traversing landscapes too rough for any wheeled vehicle while carrying 340 pounds (154 kilograms) and moving at 4 miles (6 kilometers) an hour.

Founded by Marc Raibert, a former professor at both Carnegie Mellon and MIT, Boston Dynamics started as a spinoff from MIT. Much of Boston Dynamics' early funding came from the US military, which wanted to find alternative solutions for missions that either were inherently dangerous for humans or required superhuman strength in an environment designed for bipedal navigation. The company also benefitted from decades of research into robotics and embodied intelligence at both academic and industrial research laboratories.

Notable robotic accomplishments include Cheetah, a running quadruped that broke the 2012 robot land-speed record by clocking in at 18 miles (29 kilometers) per hour. There is also Handle, a 6.5-foot (2-meter) robot with two legs on wheels that can jump 4 feet (1 meter) vertically, pick up and carry 100 pounds (45 kilograms), and maneuver with the balance and dexterity of both two-legged and four-legged creatures. PETMAN, the predecessor to Atlas, was designed to test the material effects of chemical weapons on protective safety suits. And then there is SandFlea, an 11-pound (5-kilogram), 6-inch (15-centimeter) -tall device on four wheels that can jump 30 feet (9 meters) in the air and land on a ledge. It can then jump back down, surviving falls from that proportionately great height.

In 2013, Google X (now part of Alphabet) purchased Boston Dynamics for an unknown sum. Four years later, in 2017, Alphabet sold the company to SoftBank®, the mammoth Japanese technology company responsible for the creation of Pepper, a robot that can interpret and respond to a range of human emotions. Boston Dynamics has yet to commercialize its robots.

SEE ALSO Isaac Asimov's Three Laws of Robotics (1942), Unimate: First Mass-Produced Robot (1961)

Front view of the 6-foot (2-meter) biped humanoid robot Atlas, created by Boston Dynamics and DARPA.

JPEG

The large file size of digital photographs was a persistent problem for decades. The pixelated picture of Steve Jobs displayed on the monochrome screen of the original Macintosh computer required only 21,888 bytes of memory to display. The same photo in full color would have required 525,312 bytes of memory—five times more than the early Mac's 131,072 bytes of random access memory.

France's Minitel network was especially vexed by this problem. That's because realistic digital images didn't just require a lot of memory to store—they also required a lot of time to transmit over the network. So in 1982, Minitel convened a "joint photographic experts group" (JPEG) to explore the problem of image compression and develop an approach for crunching a digital photograph into an image much smaller than might seem possible at first.

The experts from academia and industry set about creating something called a *lossy compression algorithm*. This meant that when a compressed image was decompressed, the resulting image wasn't a perfect match for the original image—some information was lost—but it just needed to look good enough for the human eye. The algorithm developed by JPEG traded off absolute faithfulness to the original colors in exchange for compressed images that are much smaller than they might otherwise be. The JPEG algorithm was tunable: the amount of compression could be specified, from very little compression (which produced big files filled with details) or very high amounts of compression (resulting in files that were small but, when decompressed, produced images that had many compression artifacts).

While other lossy compression algorithms were developed in the 1980s, what made JPEG a success was that it was open: the format was publicly documented, and companies could distribute software that implemented JPEG without paying royalties. That made JPEG a natural choice for digital cameras and the World Wide Web. As a result, today JPEG is pretty much everywhere.

SEE ALSO Minitel (1978), MPEG (1988)

Today's internet is filled with photographs of mischievous cats, shared by their owners using the JPEG lossy compression algorithm.

First Mass-Market Web Browser

Marc Andreessen (b. 1971), Eric Bina (b. 1964)

The World Wide Web was conceived as an information-sharing technology to make it easy for scientists to exchange knowledge and collaborate. One piece of this technology is a web browser, the software necessary to access, retrieve, and view the information wanted by the end user. Prior to Marc Andreessen and Eric Bina's Mosaic browser—developed at the National Center for Supercomputing Applications (NCSA) at the University of Illinois at Urbana-Champaign—browsers were difficult for nontechnical people to use, only served up text, and were largely limited to UNIX workstations. Mosaic changed all that, igniting the internet boom and popularizing the "web" among the general public.

Mosaic really made the web "World Wide." NCSA, and thus Mosaic, was funded by the High Performance Computing Act of 1991, also known as the *Gore Bill* because it was sponsored by then-senator Al Gore.

Existing browsers such as Midas, Viola, and Lynx were somewhat challenging to install and hard to use without technical expertise. And the web was mostly text: pictures in a web page were represented as links that had to be clicked to be opened, after which they were viewed in separate windows. Mosaic was simple. Getting it to run did not require any specialized skill set, and its interface was intuitive and easy for people to navigate. Links to different pages were highlighted and underlined in blue, and for the first time, graphics appeared alongside text. This addition of inline graphics is considered one of the key drivers that made the web take off and grow. Suddenly web pages were visually appealing and highly creative mediums for communication. What made this possible was Andreessen's creation of a new HTML tag called *IMG*.

Initially available only on UNIX, versions of Mosaic soon came out for Amiga, Apple Macintosh, and Microsoft Windows. In 1994, Marc Andreessen and others from the Mosaic team left Illinois to form what would become known as Netscape Communications, where they created the Netscape Navigator browser. Like its ancestral roots, Navigator broke new ground and was for a period of time the world's most popular browser until it was overtaken by Microsoft's Internet Explorer®. As enhanced commercial browsers had taken off, NSCA ceased development and support for Mosaic in 1997.

SEE ALSO World Wide Web (1989), AltaVista Web Search Engine (1995), Google (1998)

The Mosaic web browser. Links to different pages were highlighted and underlined in blue.

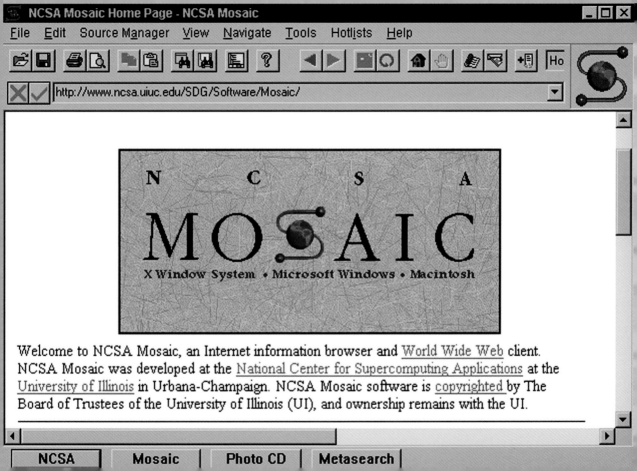

Unicode

Mark Davis (b. 1952), **Joe Becker** (dates unavailable),
Lee Collins (dates unavailable)

In 1985, Mark Davis and a team of engineers at Apple tried to create the first "Kanji Macintosh"—one that could display the kanji characters used to write modern Japanese. The challenge, they soon discovered, was not translating the English menus to Japanese: it was representing the characters *inside* the computer's memory.

The team discovered that different techniques were used for representing the tens of thousands of Japanese, Chinese, and Korean characters. Some characters were represented with a single byte, others with two, and there were "shift" codes to switch from one set to another.

It turned out that a group of engineers at Xerox were working on the same problem. They had started working on a database to map the identical characters in Japanese and Chinese to make it easier to create new fonts—something now called *Han unification*.

Davis met with Joe Becker and Lee Collins from Xerox in 1987. They agreed that the industry needed a single coding for all the world's alphabets. Becker coined the word *Unicode* for the project, for "unique, universal, and uniform character encoding."

Now with a name, the group started working on a set of technical principles. Unicode would be based on ASCII but as 16 bits instead of 7. Sadly, this meant that plain text files would double in size, but it also meant that plain text would be able to include *all* of the accented Latin characters used throughout Europe.

In August 1988, Becker presented the initial design, in a paper called "Unicode 88," to an international UNIX users' group association meeting in Dallas.

The idea for Unicode took off. A nonprofit Unicode Consortium was incorporated in 1990 to develop, maintain, and promote software internationalization standards, and in particular the Unicode standard. Today, Unicode is *the* worldwide standard for mapping codes to characters. Unicode has been expanded to cover dead languages such as Phoenician and fictional languages such as Klingon. It has thousands of symbols. Most recently, Unicode has been actively expanded to include emojis.

SEE ALSO Baudot Code (1874), ASCII (1963), Macintosh (1984)

Unicode has been expanded to include a new, popular form of communication: emojis.

Apple Newton

Michael Tchao (dates unavailable), **John Sculley** (b. 1939)

In 1993, electronic organizers had limited capabilities and functionality. They could hold names, addresses, and phone numbers. The Apple Newton was something much more ambitious: a complete reimagining of personal computing into a handheld, portable device that allowed the user to access and store information, write, be creative, and invent. Instead of storing data in files, the Newton used an object-oriented "soup," allowing different applications to seamlessly access each other's data in an intelligent, structured manner. In one demo, a person could receive an email message with Apple Mail, find the dates and times in the message, and use that information to schedule an appointment between the sender and recipient.

Newton was best known for the integrated stylus that it used for input and its ability to recognize English handwriting, both print and cursive. Because handwriting recognition was computationally intensive, Apple invested in a new, low-power microprocessor from a British company: the Acorn RISC Machine (ARM).

Apple's engineers had been working on various versions of a portable computer since 1987. The project had caught the attention of Apple's CEO John Sculley, who had created a concept video called "Knowledge Navigator" for the 1987 EDUCOM educational computing conference. Michael Tchao, a manager at Apple, pitched to Sculley, during a plane trip in 1991, the idea of creating an actual digital assistant.

Today the Newton is generally regarded as one of Apple's flops. Cartoonist Garry Trudeau mercilessly mocked the problems in the computer's handwriting capabilities in his popular *Doonesbury* comic strip. The Newton was never able to shake the reputation of having poor handwriting recognition, even when the acknowledged problems were largely addressed in version 2 of the machine's operating system.

Newton's other problem was its size: it was too large to fit in a pocket but too small to replace a desktop computer for serious computing. In many ways, it was just too different. Even though Apple sold 50,000 units the first three months, sales did not live up to expectations. Steve Jobs killed the product when he returned to Apple in 1997.

SEE ALSO Touchscreen (1965), PalmPilot (1997)

The Apple MessagePad 100, developed for the Apple Newton platform, displayed at the Musée Bolo in Lausanne, Switzerland.

First Banner Ad

Andrew Anker (dates unavailable), **Otto Timmons** (b.1959),
Craig Kanarick (b. 1967)

A clever little banner at the top of hotwired.com (then the online version of *Wired* magazine) on October 27, 1994, is generally regarded as the origin of this somewhat annoying species of web creature. It was a small black rectangle with rainbow-colored letters that read, "Have you ever clicked your mouse right HERE?" with an arrow pointing to a block of white letters that read "YOU WILL." And people did. Almost half who saw it, in fact.

Sponsored by AT&T, it was part of a larger campaign about where the future was going and the role that *Wired*'s readers would play in making it a reality.

As the World Wide Web gained in popularity, *Wired* magazine, one of technology's leading voices, simply had to have a web presence. The challenge was funding it. *Wired* was funded by a combination of revenue sources—newsstand sales, advertising, and subscriptions. *Wired* couldn't grow newsstand sales from the web content. This was a concern, because the real promise of the web was attracting a new class of readers.

Publishers believed that there was a lot of money to be made, if only they could figure out the business model. Andrew Anker, *Wired*'s chief technology officer at the time, decided the path to revenue should be through advertising. But what did that look like? In the early days of digital marketing, finding a way to become part of the online scene was half the battle.

The "YOU WILL" banner was designed by Craig Kanarick and Otto Timmons, who went on to found the Razorfish® advertising agency. Those who clicked on the banner were transported to a plain website with three links: the first took users to a map with links to virtual art galleries around the world, the second to a listing of AT&T websites, and the third to a survey about the ad itself. The only real "targeting" behind the ad was making a conscious decision to place it in the arts section of HotWired's page. With the explosion of user-generated content in the decades that followed and the development of predictive analytics, advertising would never be this wholesome or straightforward again.

SEE ALSO First Mass-Market Web Browser (1992), E-Commerce (1995)

Banner ads, which generally appear on the top or sides of web pages, direct those who click on them to an advertiser's website or special "landing" pages.

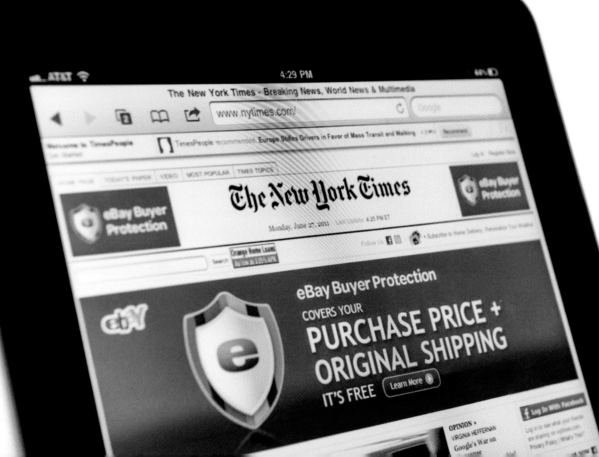

RSA-129 Cracked

Ronald L. Rivest (b. 1947), **Martin Gardner** (1914–2010),
Derek Atkins (b. 1971)

The math behind the RSA public key cryptography system first appeared in Martin Gardner's column Mathematical Games, in the August 1977 issue of *Scientific American*. In addition to the algorithm, Gardner presented a challenge to readers: an encrypted message that could be cracked only by factoring a 129-digit number that had just two factors, a 64-digit prime and a 65-digit prime. Ronald Rivest, one of the inventors of the RSA algorithm, told Gardner that it would take 40 quadrillion years to factor the number, an estimate that was apparently based on 1977-era factoring technology. Rivest offered a $100 prize to the successful codebreaker.

Unlike other encryption algorithms, RSA-encrypted messages could be made arbitrarily difficult to crack, simply by using longer prime numbers to create a key. The number that Gardner published was 129 decimal digits, or 426 bits in length. By the early 1990s, it was clear that wasn't strong enough for commercial communications: experts were recommending keys of 512 bits, and 1024 bits for high-security applications.

In 1992, Derek Atkins, then a 21-year-old computer science student at MIT, decided to attack the 129-digit number. Atkins realized the number was within reach if he could enlist hundreds of people from all over the internet to help, all contributing time to help factor the number. He assembled a group of collaborators, who took existing factoring software and modified it to work on this larger problem. On August 19, 1993, the group announced on a Usenet group that they were looking for help.

Over the next few months, more than 600 people contributed computer time to cracking RSA-129. After eight months, the 129-digit number was factored, revealing the secret message from 1977: *THE MAGIC WORDS ARE SQUEAMISH OSSIFRAGE*

Factoring RSA-129 didn't take 40 quadrillion years after all, but it did require about 100 quadrillion calculations. The $100 prize was donated to the Free Software Foundation, the nonprofit developing the open source GNU operating system.

SEE ALSO Public Key Cryptography (1976), RSA Encryption (1977), Quantum Computer Factors "15" (2001)

After he encrypted these words in 1977, MIT professor Ron Rivest thought that he would never see them again.

THE MAGIC WORDS ARE SQUEAMISH OSSIFRAGE • RSA-129 = 114381625757888867669235779976146612010218296721242362562561842935706935245733897830597123563958705058989075147599290026879543541 = 34905295108476509491478496199038981334177646384933878439908205577 × 32769132993266709549961988190834461413177642967992942539798288533

DVD

Warren Lieberfarb (b. 1943)

The DVD (digital video disc) was invented to solve the technical challenge of fitting an entire feature-length movie onto a hard plate that could be read by a computer. The film industry wanted to evolve its business model from one where people traditionally rented movies on analog tape (VHS or Beta) to one where people were encouraged to purchase, not rent, their favorite films in digital format. The industry would reap higher profit margins with movie purchases while increasing viewing quality with a digitized product.

The trailblazer that the movie industry wanted to emulate was the music industry, which had made the successful transition from cassette and vinyl to compact disc (CD) years earlier. With a storage capacity of 700 megabytes, CDs had been intentionally designed to have enough room to contain all of Beethoven's Ninth Symphony—the lengthiest recording that PolyGram records had in its title library—but the technology of the early 1980s just wasn't up to storing the amount of data required for video, and the electronics of the early 1980s weren't fast enough to handle playback of digital video. That all changed with the DVD.

Like CDs, DVDs have tiny pits or craters etched into the smooth surface of the disc. These indentations and the adjacent smooth space between them is the actual movie encoded in binary format. A laser scans or "reads" the surface of the disc and interprets spots with an indentation as a 1 and spots without an indentation as a 0. The DVD technology breakthrough was the realization that the movie data could be encoded with much smaller indentations, packed more tightly together, if the laser used to read and interpret the bumps was a shorter wavelength than that used for a CD. Denser encoded material means vastly more data could be packed into the disc. Although the DVD had no one inventor—it was an amalgamation of technology developed over three decades—the person who is generally regarded as the driving force behind the DVD is Warren Lieberfarb, who at the time was the president of Warner Home Video® at Warner Bros. Entertainment.

SEE ALSO MPEG (1988), CD-ROM (1988)

The smaller wavelength of the blue-violet laser in a DVD drive allows the drive to read substantially more information than the infrared laser of a CD-ROM drive.

E-Commerce

Buyer behavior is a finicky thing, no matter what service or product the merchant may be offering. Conducting commerce online presents its own set of challenges—practical, technical, and social—that until 1995 were simply not solved well enough to ignite virtual buying and selling on a massive scale. That changed in 1995, however, when a number of seminal events finally made e-commerce "click" for consumers and truly launched the era of online commerce.

The first important ingredient was security. In 1994, Netscape unveiled Secure Sockets Layer (SSL), which let consumers send their credit card numbers over the internet without fear of having them stolen along the way; in April 1995, Verisign® opened for business, selling digital certificates for certifying the authenticity and credibility of online businesses. At the same time, the National Science Foundation (NSF), which had altered the NSFNET acceptable use policy in March 1991 to allow commercial traffic, was now slowly transferring its role as steward of the internet's infrastructure to commercial operators. In 1995, the NSF authorized Network Solutions, Inc. (which had a five-year agreement with NSF) to start charging a domain name registration fee as the volume of commercial requests skyrocketed. These events, along with websites that were quickly becoming more professional, with visual sophistication and technical functions that enabled direct engagement with customers, set the stage for e-commerce.

In 1995, Pierre Omidyar (b. 1967) started eBay, then called *AuctionWeb*, to establish a mechanism for everyday people to sell their stuff to one another. He knew he was onto something when his first sale—$14.83 for his own broken laser pointer—was purchased by someone who collected—you guessed it—broken laser pointers. Amazon launched this year as well, as did DoubleClick®, an early advertising network (now owned by Google). But what actually was the first secure, online transaction? That is up for debate, but among the colorful contenders is Pizza Hut®, which began selling pizza online in August 1994.

SEE ALSO NSFNET (1985), First Mass-Market Web Browser (1992)

Today, many people choose to do their shopping primarily from their laptops or phones.

AltaVista Web Search Engine

Internet search engines existed before the World Wide Web—they searched other services, like File Transfer Protocol (FTP) repositories. And indeed, even after the launch of the World Wide Web, it would be many years before there was enough online content to warrant a search engine.

As the number of web pages exploded following the introduction of NCSA Mosaic, so did the need to improve the automated indexing of the web's expanding virtual geography. Early search engines such as W3 Catalog were limited and results inconsistent. Enter AltaVista®, created by Digital Equipment Corporation, then known as Digital and later as DEC, as a marketing tool to prove the speed and accuracy of its AlphaServer 8400 TurboLaser supercomputer.

AltaVista is credited with a variety of pioneering capabilities, including queries with natural language; indexing the web using data it had found with its web crawler, "Scooter," rather than forcing websites to provide aggregated data of keywords and terms; full text indexing of web pages; expansion of Boolean operators, including "near" and parentheses; search in languages other than English (Malay and Spanish); and searches that could find video, images, and audio.

AltaVista gained popularity rapidly, as it was significantly faster and more sophisticated than other search technologies. The website's user base grew explosively, from 300,000 users a day in 1996 to 80 million a day by 1997. What was created as a byproduct for demonstrating supercomputing power had turned into a capability that pioneered information efficiency on the web, using 20 multiprocessors at one point to do so.

A confluence of business decisions, questionable user-interface design choices, and Google's development of the PageRank® algorithm to address the rising prevalence of website spam caused AltaVista to lose relevance (and customers) within just a few years. AltaVista changed ownership through multiple acquisitions, starting with Compaq's acquisition of Digital in 1998. In 2003, AltaVista was acquired by Overture® for $140 million and then finally sold to Yahoo!®, which shuttered AltaVista in 2013.

SEE ALSO World Wide Web (1989), First Mass-Market Web Browser (1992), Google (1998)

AltaVista's pioneering capabilities included queries with natural language and searches that could find videos, images, and audio recordings.

AltaVista® The most powerful and useful guide to the Net

Ask AltaVista™ a question. Or enter a few words in [any language ◆]

[information architecture] [Search]

Example: **Where can I download sports shareware for Windows?**

Gartner Hype Cycle

The Gartner® Hype Cycle is a pictorial representation of a technology's visible and expected value at one of five points in time: the *technology trigger*, characterized by a technology's emergence into public view; the *peak of inflated expectations*, symbolizing the maximum stage of overenthusiasm and buzz about a technology's potential; the *trough of disillusionment* after the inflated expectations have gone bust; and the *slope of enlightenment* as the once-hyped, then-scorned technology slowly comes into wide adoption, ironically realizing the early expectations. The fifth phase is the *plateau of productivity*, which symbolizes the start of mainstream adoption and expanded innovation around the technology or product. Press becomes more positive, the technology's practical application to real-world needs is clearly defined, and the use of its name may morph into a general description of activity rather than the product itself, such as *Xerox* and *Google* being used as verbs.

Created in 1995 by the Gartner Group, a well-known consultancy and IT advisory company, the Hype Cycle has had remarkable staying power, because it accurately characterizes the boom/bust/slow-boom nature of today's media-driven technology interest and deployment. The model has become a popular source of reference with technology decision-makers—and even the public—who want to understand a technology's potential as it moves through the cycle of innovation. Today the model informs investment decisions, IT strategy planning, and general acumen about emerging technologies, using a mix of quantitative and qualitative data.

Of course, not every technology follows this cycle. Some hyped technologies fail to pan out, and other technologies—such as DVDs and the PC—never really went through the trough of disillusionment. But most technologies do, and their adherence to the cycle is evident in patent filings, press stories, the number of online searches, investment activity, blog posts, and other indicators.

The Gartner Hype Cycle has not been without its critics, and many have challenged the accuracy of its underlying methodologies. It has, however, proved to be both an easy-to-use and an incredibly popular tool that puts a dense, often complicated landscape of technology into an understandable graphic that is useful for both the novice and the expert.

SEE ALSO E-Commerce (1995)

Despite attempts to anticipate a technology's value at a certain point in time, it does not always follow a predictable path of innovation and popularity.

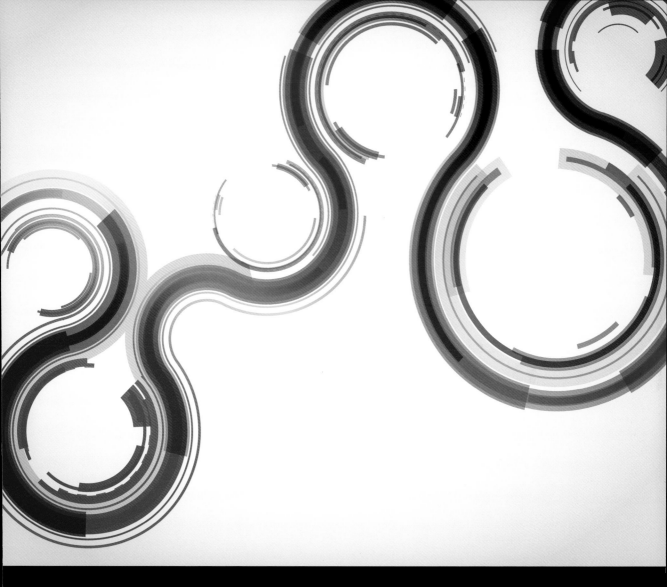

Universal Serial Bus (USB)

By the mid-1990s, the back of most computers looked like a rat's nest of cables and connectors. There was a serial port for connecting to a telephone modem, PS/2 connectors to connect to a keyboard and mouse, a 25-pin parallel cable for the printers, and of course the power and video cables as well.

The Universal Serial Bus (USB) was envisioned to end the madness, providing a single standard for transmitting data and power over cables. It was designed by a group of seven companies: Compaq®, DEC, IBM, Intel, Microsoft, NEC, and Nortel®. The standard was first published beyond the group of creators in January 1996. Its inventors expected the computer industry to make a slow, gradual transition from the era of legacy ports, presumably with several generations of computers that offered both legacy and USB ports.

But USB made its consumer debut with the introduction of the Apple iMac®. Apple, whose Macintosh computers had used the proprietary Apple Desktop Bus (ADB) since 1986, was eager to adopt a new technology that would make it easy for its users to purchase commodity keyboards, mice, and other specialty devices. So Apple jumped in headfirst: when the iMac went on sale in August 1998, it had USB but no legacy ports at all. Legacy devices and cables on the PC side, however, would take a decade to phase out completely.

By 2010, USB had replaced not just all legacy data connectors, but power as well: except for Apple's iPhone, virtually all cell phones and many other low-power devices had adopted USB mini microconnectors for charging. Equally ubiquitous were USB "thumb" drives that packed gigabytes of portable, permanent storage.

A problem with USB is that its cables are not symmetrical: USB cables have an A-side that plugs into computers and a B-side the plugs into the "downstream" device (typically a printer or a phone). The plugs themselves can be plugged in only one way. USB Type-C solves both problems, with connectors that can be flipped and reversible cables. Type-C can also carry up to 100 watts of power. In 2015, Apple introduced a MacBook with a single USB Type-C connector; later Apple laptops featured two or four Type-C connectors.

SEE ALSO Recommended Standard 232 (1960)

The Universal Serial Bus provides a single standard for transmitting data and power over cables.

Computer Is World Chess Champion

Garry Kasparov (b. 1963)

Ever since Alan Turing wrote the first computer chess program in 1950, computer scientists (and the general public) had viewed proficiency at chess as a litmus test for machines' intelligence. Machines, the thinking went, would be truly intelligent if they could beat a human at chess. When that happened, the challenge then subtly changed: would computers ever be able to beat *every human* at chess, even a grand master?

That happened nearly 50 years later in 1996, when IBM's Deep Blue computer beat world chess champion Garry Kasparov.

Kasparov and Deep Blue played two matches—the first took place in February 1996 in Philadelphia. Kasparov lost two games to Deep Blue but still won the match. The rematch occurred a year later in May 1997, when Kasparov lost to Deep Blue with a final score of 3.5 to 2.5 (one game was a draw). In an unusual twist, Deep Blue made an unexpected play during game two of the second match, rattling Kasparov and throwing him off his strategy. Kasparov did not know what to make of the move and considered it a sign of superior intelligence. While counterintuitive, Kasparov's interpretation of Deep Blue's capabilities highlights the power and weakness of relying on human intuition when playing games of skill.

In fact, Deep Blue's advantage was brute force, pure and simple. Deep Blue was really a massively parallel program coded in C, running on a UNIX cluster, and capable of computing 200 million possible board positions each second. Deep Blue's "evaluative function," which decided which board positions were better, was based on assessing four human-programmed variables: *material*, the value of each piece; *position*, the number of squares that buffer a player's piece from attack; *king safety*, a number that represents how safe the king is, given his location on the board and the position of the other pieces; and *tempo*, the success of a player advancing his or her position over time. Given these factors and the relatively constrained size of the board, chess became a "quantifiable" equation for Deep Blue. As such, the computer can win by simply seeking the best board positions—something it can do faster, and better, than any human.

SEE ALSO Computer Beats Master at Go (2016)

Viewers watch world chess champion Garry Kasparov on a television monitor at the start of the sixth and final match against IBM's Deep Blue computer in New York.

4 c6 2. d4 d5 3. Nc3 dxe4

PalmPilot

Jeff Hawkins (b. 1957)

To design the PalmPilot, Jeff Hawkins cut a block of wood that would fit in a man's shirt pocket and carried it around for several months, pretending to use it to look up phone numbers, check his schedule, and put things on his to-do list. It was a pure user-centered design, unencumbered by what technology could produce.

Based on his experience building two previous portable computers, and the practice of pretending to use a little wooden block in his pocket, Hawkins realized that a portable computer didn't need to replace a traditional desktop; it just needed to fill in the gaps. Specifically, the portable computer needed to instantly turn on and let users find the information they were looking for—a person's name or address, for example—or to access a calendar. There was limited need for data input—more important was some way to rapidly synchronize the portable's database with the desktop's.

Because its function was not text entry, there was no need for a keyboard. Instead, there was a small rectangular area at the bottom of the touchscreen where users could enter letters in a stylized alphabet he called *Graffiti*. Similar to traditional Roman characters, Graffiti characters were easier for the device's software to recognize.

It took a team of 27 people just 18 months to develop the product. But with no money to manufacture or market the device, in 1995 Palm Computing was sold to U.S. Robotics Corporation (USR), a modem manufacturer. Two years later, in 1997, USR brought the PalmPilot to market, selling it for a list price of $299. The Palm was a breakthrough. More than 2 million units were sold in just two years; more than 20 million would be sold by 2003.

SEE ALSO Touchscreen (1965), Apple Newton (1993)

The PalmPilot made it easy for users to have instant access to important information, such as calendar items or a person's address.

E Ink

JD Albert (b. 1975), **Barrett Comiskey** (b. 1975),
Joseph Jacobson (dates unavailable)

The electronic paper display (EPD)—a paper-thin, reflective display that's visible in direct sunlight and requires practically no energy at all—traces its origins back to research at Xerox's Palo Alto Research Center (PARC) in the 1970s. But it was JD Albert and Barrett Comiskey, two MIT undergraduates under the direction of their professor, Joseph Jacobson, who made the technological breakthrough that commercialized and launched the industry. Called *E Ink*® (also the name of the company they founded to sell it), the term has subsequently been adopted into the general lexicon as a name for this type of technology.

Despite the spectacular visuals of LCD screens, there are many situations that would be better served by a technology that is as easy to read as paper (no eye strain from glare and backlighting), has virtually no power consumption, and is durable enough to withstand rough environments. A few examples today include e-readers (devices for reading electronic books), wearables such as watches, and dynamically changing price tags on clothing or store shelves. Within a few years, E Ink is expected to appear in tablets, signs, and even walls.

The technology itself is called *microencapsulated electrophoretic display*, and it works like this: sandwiched between two extremely thin pieces of glass (or plastic) are electrically charged microcapsules filled with either titanium oxide (white) or carbon (black) suspended in oil. A change in charge from positive to negative sends the right combination of capsules to the top (visible) part of the screen, reflecting either white or black to make images such as text. The only time power is used is when the capsules are changing state (such as "turning" the page in an e-reader).

In 2013, the United Nations installed the world's largest E Ink display, called the *eWall*, at its headquarters in New York. Covering a 6-meter-wide space with a resolution of 26,400 × 3,360 pixels, it provides news, schedules, and any other information that might be useful to those in its vicinity. In 2016, Albert, Comiskey, and Jacobson were inducted into the National Inventors Hall of Fame, which includes the likes of Thomas Edison and the Wright Brothers.

SEE ALSO RAND Tablet (1964), Touchscreen (1965)

Today's e-readers benefit from the advantages of E Ink, including the low power consumption and lack of eye strain from glare and backlighting.

Diamond Rio MP3 Player

The Rio PMP300, a digital music player from Diamond Multimedia®, was introduced on September 15, 1998. The size of a deck of cards, it sold for $200 and held 32 megabytes of storage. Users could skip, shuffle, repeat, and randomly play tracks. It ran on one AA battery and could play for about 10 hours before needing a new battery or recharge. Music was transferred to it from a personal computer via a proprietary connector to the computer's parallel port.

The real claim to fame of the Diamond Rio®, though, was the historic role it played in clearing the path for the establishment of a new digital music ecosystem, initially dominated by Apple's iTunes and the iPod digital music player.

The Rio played music in a relatively new audio compression format called *MP3*. Created by German engineers, MP3 made possible widespread music sharing and spawned a new industry. Uncompressed, audio files take up a large amount of space— 32 megabytes of storage could hold only a few minutes of music. The new format solved the practical challenge of storing and sharing music files by dramatically shrinking (compressing) the size of the file without a significant loss in sound quality: that same 32 megabytes could hold nearly an hour of music. Within a year, peer-to-peer music file-sharing services such as Napster® emerged, making it possible for people to freely share their digitized music with thousands of others over the internet, rapidly creating a cultural and legal environment that the music industry viewed as dangerous to its bottom line.

In 1999, the Recording Industry Association of America (RIAA) sued Diamond Multimedia for creating a device that allegedly violated the Audio Home Recording Act of 1992 (AHRA) because it did not implement a copyright management system. The RIAA claimed the company had not paid royalties required by law on the sales of the device—and as such, the RIAA alleged that the device was facilitating music piracy. The US Court of Appeals for the Ninth Circuit eventually ruled in favor of the Rio in *Recording Industry Association of America, Inc. v. Diamond Multimedia Systems, Inc.*, based in part on the argument that computer users have the right to "space-shift" their lawfully acquired music files from one location to another, just as television viewers could "time-shift" shows that they recorded on video players for viewing later.

After the lawsuit, the Rio's sales took off and the company launched RioPort, one of the first online music stores where users could legally purchase music.

SEE ALSO iTunes (2001)

The Diamond Rio PMP300 MP3 player, pictured here, offered users innovative new features, such as the ability to skip, shuffle, and randomly play tracks.

DIGITAL
DIAMOND
VOL.
RANDOM
REPEAT
A↔B
DIGITAL STEREO
Rio PMP300

Google

Larry Page (b. 1973), **Sergey Brin** (b. 1973)

The seed for what would become Google started with Stanford graduate student Larry Page's curiosity about the organization of pages on the World Wide Web. Web links famously point forward. Page wanted to be able to go in the other direction.

To go backward, Page built a web crawler to scan the internet and organize all the links, named *BackRub* for the backlinks it sought to map out. He also recognized that being able to qualify the importance of the links would be of great use as well. Sergey Brin, a fellow graduate student, joined Page on the project, and they soon developed an algorithm that would not only identify and count the links to a page but also rank their importance based on quality of the pages from where the links originated. Soon thereafter, they gave their tool a search interface and a ranking algorithm, which they called *PageRank*. The effort eventually evolved into a full-blown business in 1998, with revenue coming primarily from advertisers who bid to show advertisements on search result pages.

In the following years, Google acquired a multitude of companies, including a video-streaming service called *YouTube*®, an online advertising giant called *DoubleClick*, and cell phone maker Motorola, growing into an entire ecosystem of offerings providing email, navigation, social networking, video chat, photo organization, and a hardware division with its own smartphone. Recent research has focused on deep learning and AI (DeepMind®), gearing up for the tech industry's next battle—not over speed, but intelligence.

Merriam-Webster's Collegiate Dictionary and the *Oxford English Dictionary* both added the word *Google* as a verb in 2006, meaning to search for something online using the Google search engine. At Google's request, the definitions refer explicitly to the use of the Google engine, rather than the generic use of the word to describe any internet search.

On October 2, 2015, Google created a parent company to function as an umbrella over all its various subsidiaries. Called *Alphabet Inc.*, the American multinational conglomerate is headquartered in Mountain View, California, and has more than 70,000 employees worldwide.

SEE ALSO First Banner Ad (1994)

Google's self-described mission is to "organize the world's information and make it universally accessible and useful."

Collaborative Software Development

Despite the reputation of software developers as solitary, introverted people, much of their time is spent socializing and collaborating with colleagues and like-skilled experts to solve common problems or work on common projects. By the late 1990s, a combination of factors led to the emergence of collaborative development environments (CDEs), wherein geographically dispersed developers, some connected by corporations, others simply by challenges, would collaborate in virtual space using a variety of features to advance open source projects and develop code.

As software development efforts for web-based platforms grew, so did the need for greater productivity and innovation in meeting the growing demands of these systems and their ever-changing requirements. CDEs evolved in part to meet these demands and to help coders realize the network effects of leveraging expertise and social engagement beyond one's own community or organization. The company that led the charge in this era was SourceForge®, a free service for software developers to manage their code development that came on the scene in 1999. A number of other platforms entered the market soon after.

Collaborative software development has dramatically accelerated the pace of developing open source projects. Without these capabilities, the rate of evolution would have been much slower, and without the benefit of as many perspectives and diverse inputs, the quality would not be nearly so high. One example of this is the Apache Software Foundation's big data software stack, including Hadoop, Apache Spark, and others—which was collaboratively developed by programmers at dozens of different corporations and universities. In large part, the success and vibrancy of these projects is measured not just by their adoption but also by the number of active developers who are improving the code base.

Over time, CDEs incorporated additional features into their platforms beyond simple version control, including threaded discussion forums, calendaring and scheduling, electronic document routing and workflow, projects dashboards, and configuration control of shared artifacts, among others.

SEE ALSO GNU Manifesto (1985), Wikipedia (2001)

Services like SourceForge and GitHub make it possible for many people to work on the same piece of software at the same time, dramatically increasing the rate of software innovation.

Blog Is Coined

Jorn Barger (b. 1953), **Peter Merholz** (dates unavailable),
Evan Clark Williams (b. 1972)

In 1997, cyberspace was an increasingly viable and attractive outlet for humans to do what they have always done—communicate personal opinions and ideas, broadcast their expertise, and share sources of interesting information with others. Active users of the World Wide Web collectively contributed to an ever-increasing volume of online material that covered every topic imaginable.

The origin of the word *blog* started with essayist and active Usenet contributor Jorn Barger, who came up with the word *weblog* to describe the *logging* and curating of links he discovered and routinely attached to his website, *Robot Wisdom*. Then in April or May of 1999, designer Peter Merholz on his website Peterme.com put the following note in the sidebar of the page: "For What It's Worth I've decided to pronounce the word 'weblog' as [']wee'-blog. Or 'blog' for short." Merholz started using his newly coined word in his posts; other people did too.

A few months later, Pyra Labs released Blogger® software for creating weblogs, this time with the word *blog* used by Evan Williams, the company's cofounder. Blogger was immensely successful—Google acquired Blogger and Pyra Labs in 2003. Blogger's success helped institutionalize the word and format, and it provided a new self-publishing tool people could use, in fact, to blog. Other popular platforms for publishing blogs, including WordPress® and Movable Type®, also got their start around this time.

So, who wrote the first "blog"? That depends upon how one defines *blog*—by the personalized content, the chronological format, the software that gives the content its look and feel . . . and so on. There were many prolific commentators and online diarists long before the word was coined. Some of these authors adopted the word *blog* as it was popularized, while others did not.

Regardless of who can legitimately claim credit for being the first blogger, a variety of factors influenced the popularity and usability of what is now defined as a blog. The genre has encouraged and enabled many individual voices around the world to be heard and to contribute to a global knowledge base. Oh, and in 2004, the editors at the Merriam-Webster chose *blog* as their favorite word of the year.

SEE ALSO *Cyberspace* Coined—and Re-Coined (1968), Usenet (1980), Desktop Publishing (1985)

Blogs allow individual voices around the world to be heard and to contribute to a global knowledge base.

Napster

Shawn Fanning (b. 1982), Sean Parker (b. 1979)

Free digital music with a user-friendly interface. That's what Napster promised, and what it delivered to its users. By 1999, software could readily transform digital music from audio CDs into the compact MP3 format, and with high-speed internet connections, users could transmit or receive a compressed song in less than a minute. Digital music piracy had long been a concern for the music industry.

Shawn Fanning's Napster program made the industry's concern real. Copyright law had long had an exemption for "fair use," and courts had never ruled on the legality of teenagers making mix tapes for their friends. So Fanning created a kind of electronic matchmaking service that let people share music over the internet with pretty much anybody.

Fanning's friend Sean Parker was also a software genius and had launched a number of successful companies while in high school. Parker raised money for Napster, which made it possible to run a large, centralized server that could store an index of every Napster user currently online and the music they had available for sharing. New users could download the free Napster client, type the name of a song or artist, and instantly get a list of music available for download. The Napster software would then transfer the music directly from one user to the other—*peer to peer*. It was kind of like making a lot of mix tapes.

The industry saw Napster as a massive copyright-violation machine. And because a single leaked song could end up on Napster and be copied millions of times without cost, music started showing up on Napster before it was available in stores for legitimate purchase. On April 13, 2000, Metallica filed suit against Napster in the Northern District of California for copyright infringement and racketeering, the first such lawsuit against a maker of peer-to-peer file-sharing software. Metallica demanded $100,000 for every song that had been illegally downloaded, with a minimum of $10 million in damages. A year later, on March 5, 2001, the court issued a preliminary injunction requiring Napster to identify and remove all Metallica songs from its system—an all-but-impossible task. After briefly trying to sell the company, Napster's executives declared bankruptcy, and the company was liquidated.

SEE ALSO MIDI Computer Music Interface (1983), Diamond Rio MP3 Player (1998)

Chief lobbyist for the Recording Industry Association of America, Mitch Glazier, and Napster lobbyist Manus Cooney during a debate sponsored by the Senate Republican High-Tech Task Force.

Mitch Glazier
V.P., RIAA

Manus Cooney
V.P., Napster

Senate Republican

HIGH
TECH
TASK
FORCE

USB Flash Drive

Inside the typical USB flash drive you'll find two integrated circuits: a flash memory chip and a Universal Serial Bus (USB) controller. These two different technologies, invented at different times, were paired together by Israeli company M-Systems, which applied for a patent in April 1999. About the size of an adult thumb, the device for US patent 6,148,354 A was described as a "Universal Serial Bus-based PC flash disk." The patent was issued on November 14, 2000. By that time, USB flash drives were being sold by multiple companies.

To understand the significance of this pairing, it's useful to understand the function of the two separate technologies from which it derives. The USB is an industry standard for a common connection interface between devices that connect to computers. It was created in the mid-1990s by a consortium of leading technology companies. Flash memory, invented in 1980, is microelectronics that require little power to operate and can retain data with no power at all.

The marriage of USB and flash memory made data portability and offline sharing much easier while increasing the amount of data that could be stored and moved between devices and personal computers. Previously, using removable flash storage required having a special reader—something that most computers didn't have. But by the year 2000, practically every desktop and laptop computer sold had multiple USB connectors—it had become the standard way for connecting keyboards, mice, printers, and other peripheral devices.

Suddenly any computer could have extra storage that was fast, portable, and didn't require a power supply. This was a leap in convenience for consumers. For many uses, USB drives instantly replaced floppy disks, writable optical discs, Zip drives, and other storage devices.

The inventor of the USB flash drive remains controversial. While M-Systems had the first patent, IBM filed an invention disclosure by one of its employees who came up with the idea. There were also competing patents from Singaporean company Trek Technology and Chinese company Netac Technology. In 2000, Trek Technology became the first to commercially sell the USB flash drive using the trademarked name ThumbDrive®. That same year, IBM was the first to sell the USB flash drive in the US. Called the *DiskOnKey*, its capacity was 8 megabytes. Today, storage capacity for USB flash drives can exceed 512 gigabytes.

SEE ALSO Flash Memory (1980), Universal Serial Bus (USB) (1996)

Most USB flash drives consists of just two chips: a flash memory chip that stores the data, and a microcontroller that transfers data between the USB interface and the flash chip.

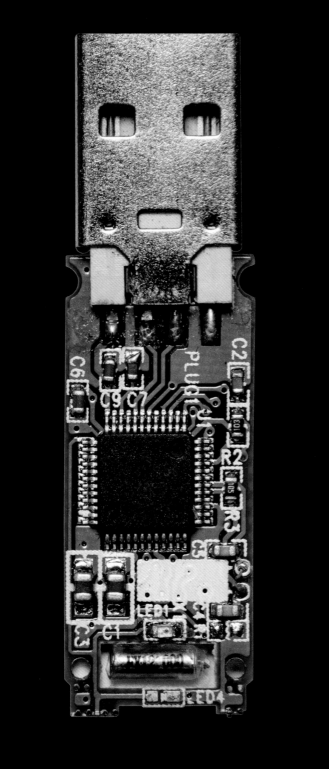

Wikipedia

Jimmy Wales (b. 1966), **Larry Sanger** (b. 1968)

Wikipedia® is an online encyclopedia represented in 287 languages, containing more than 30 million articles. It is owned by the Wikimedia Foundation®, a nonprofit organization. Wikipedia leverages volunteers from anywhere in the world to contribute content or create new entries on largely any topic they wish. Content and individual articles represent the organic contributions of those participating in their chosen language. The guiding principle of Wikipedia is that information is kept current and made more accurate through continuous crowdsourcing, which achieves both authenticity and factuality. The result is a mass-collaboration effort that, while sometimes containing mistakes and bias, provides a general reference on common and uncommon topics for people around the world. It is one of the most popular sites on the web.

Information appearing in Wikipedia articles is supposed to be cited and referenced. Readers can click on the links at the bottom of the page that are supposed to connect back to the originating material and platform. One feature that results from this operating model is "edit wars" between contributors over controversial topics such as politics and religion.

Wikipedia and its cofounders have had their share of controversies and critics as well, but they have also had many supporters and benefactors. Wikipedia's notion of truth and reliability—something is true if it appears in a reliable secondary source— can result in circular reporting, because many journalists have relied on unsourced Wikipedia articles for their references, and then the Wikipedia articles are updated to reference the journalistic article as a source.

Wikipedia requires a significant amount of ongoing and repetitive editing and formatting to correct misspellings and maintain proper arrangement of links and reference material. As such, Wikipedia also uses automated bots to augment human volunteers doing this work. Interestingly, as the number of bots has increased, so has the phenomenon of bot-on-bot edit wars as well, where bots go back and forth undoing and redoing changes to each other's content. Increasingly, entire articles in Wikipedia can be attributed to bots as technical sophistication grows with advancements in artificial intelligence.

SEE ALSO GNU Manifesto (1985), Collaborative Software Development (1999)

The online encyclopedia Wikipedia has more than 30 million crowdsourced articles.

WIKIPEDIA
The Free Encyclopedia

iTunes

Steve Jobs (1955–2011), **Jeff Robbin** (dates unavailable),
Bill Kincaid (b. 1956), **Dave Heller** (dates unavailable)

The music business at the end of the 20th century was in an epic fight to maintain its profitable business model. Music had become 1s and 0s and was being widely shared, without compensation, among users through online services such as Napster. The industry was filing suit against both the services and their users to protect copyrights.

Apple cofounder Steve Jobs saw an opportunity and in 2000 purchased SoundJam MP, a program that functioned as a music content manager and player. It was developed by two former Apple software engineers, Bill Kincaid and Jeff Robbin, along with Dave Heller, who all took up residence at Apple and evolved the product into what would become iTunes.

iTunes debuted on January 9, 2001, at Macworld Expo in San Francisco. For the first two years, iTunes was promoted as a software jukebox that offered a simple interface to organize MP3s and convert CDs into compressed audio formats. In October 2001, Apple released a digital audio player, the iPod, which would neatly sync with a user's iTunes library over a wire. This hardware release set the stage for the next big evolution, which came with iTunes version 4 in 2003—the iTunes Music Store, which launched with 200,000 songs. Now users could buy licensed, high-quality digital music from Apple.

Buying music from a computer company was a radical concept. It flipped the traditional business model and gave the music industry an organized, legitimate mechanism in the digital space to profit from, and protect, their intellectual property.

The music labels agreed to participate in the iTunes model and allowed Jobs to sell their inventory in part because he agreed to copy-protect their songs with Digital Rights Management (DRM). (Apple significantly eased the DRM-based restrictions for music in 2009.) Consumers embraced iTunes in part because they could buy single songs again—no longer did they have to purchase an entire album to get one or two tracks.

In the following years, iTunes would snowball into a media juggernaut adding music videos, movies, television shows, audio books, podcasts, radio, and music streaming—all of which were integrated with new products and services from Apple, including Apple TV, the iPhone, and the iPad.

SEE ALSO MPEG (1988), Diamond Rio MP3 Player (1998)

Apple's Steve Jobs announces the release of new upgrades to iTunes and other Apple products at a press conference in San Francisco, California, on September 1, 2010.

Advanced Encryption Standard

Vincent Rijmen (b. 1970), **Joan Daemen** (b. 1965)

After the US government adopted the Data Encryption Standard (DES) in 1977, it quickly became the most widely used encryption algorithm in the world. But from the start, there were concerns about the algorithm's security. DES had an encryption key of just 56 bits, which meant there were only 72,057,594,037,927,936 possible encryption keys, leaving experts to speculate whether anyone with the means had built special-purpose computers for cracking DES-encrypted messages.

DES had other problems. Designed to be implemented in hardware, software implementations were surprisingly slow. As a result, many academic cryptographers proposed new ciphers in the 1980s and 1990s. These algorithms found increasing use — in web browsers, for instance — but none had the credence that came with having gone through the government's standards-making process.

So, in 1997, the US National Institute of Standards and Technology (NIST) announced a multiyear competition to decide upon the nation's next encryption standard. NIST invited cryptographers all over the world to submit not only their best algorithms, but their recommendations for how the algorithms should be evaluated.

Adding another nail to the DES coffin, in 1998 the Electronic Frontier Foundation (EFF), a tiny civil liberties organization, announced that it had built one of those mythical DES-cracking machines, and for less than $250,000. Called *Deep Crack*, the machine could try 90 billion DES keys a second, allowing it to crack, on average, a DES-encrypted message in just 4.6 days.

In total, there were 15 credible submissions from nine different countries to the NIST contest. After considerable public analysis and three public conferences, the winner was decided in 2001: an algorithm called *Rijndael*, developed by two Belgian cryptographers, Vincent Rijmen and Joan Daemen. Rijndael is now called the *Advanced Encryption Standard (AES)*. It can be run with 128-bit, 192-bit, or 256-bit keys, allowing for unprecedented levels of security. It can run on tiny 8-bit microcontrollers, and nearly all modern microprocessors now have special AES instructions, allowing them to encrypt at blindingly fast speeds.

SEE ALSO Data Encryption Standard (1974)

One of the 29 circuit boards from the Electronic Frontier Foundation's encryption breaking machine, Deep Crack.

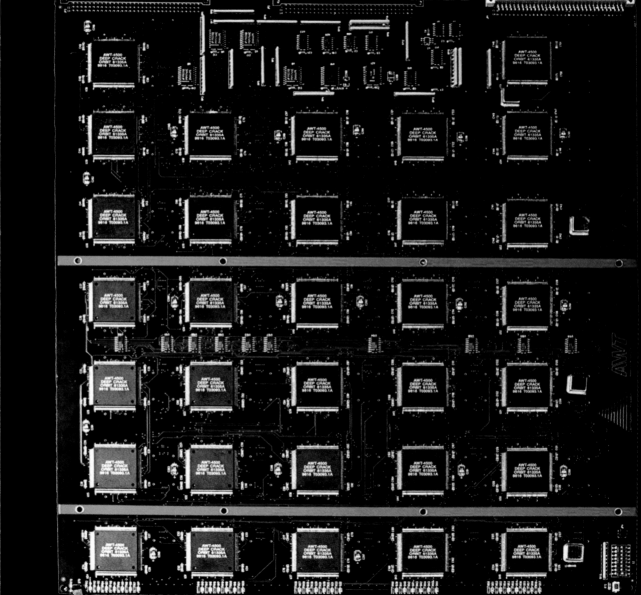

Quantum Computer Factors "15"

Peter Shor (b. 1959), **Isaac Chuang** (b. 1968)

Speed is the promise of quantum computers—not just faster computations, but mind-blowingly, seemingly impossibly fast computations.

That's a problem for anyone who sends encrypted information over the internet, because the public key cryptography algorithms that secure the majority of the internet's data depend on the difficulty of factoring large numbers. There is no known algorithm for efficiently factoring large numbers on a conventional computer. But in 1994, mathematician Peter Shor devised an algorithm for efficiently factoring large numbers on a quantum computer. This means that an organization with a working quantum computer could decrypt practically every message sent over the internet—provided the organization had a large enough quantum computer.

One way to measure a quantum computer is by the number of qubits it can process at a time. In 2001, a team of scientists at IBM's Almaden Research Center led by Isaac Chuang successfully factored the number 15, yielding the factors 3 and 5, with a quantum computer that had 7 qubits.

Although factoring the number 15 might not seem like a big deal, the IBM researchers proved that quantum computers aren't just theoretical—they actually work. Now the race was on to make a quantum computer that was large enough to compute something that could not be computed on a conventional machine.

Since then, quantum computers have gotten steadily bigger, and the factoring algorithms have also improved. In 2012, Shor's algorithm was used on a 10-qubit machine to factor the number 21. That same year, a team in China factored the number 143 with a 4-qubit computer using an improved algorithm. Astonishingly, two years after the Chinese team published its findings, a group of researchers at Kyoto University pointed out that the Chinese system had also factored the numbers 3,599, 11,663, and 56,153, without the authors even being aware of it!

Cryptographers at the US National Institute of Standards and Technology are now racing to develop "post-quantum" encryption algorithms that aren't based on factoring and, as a result, won't be vulnerable to anyone who has a quantum computer.

SEE ALSO The Qubit (1983)

In 2001, a team of scientists used a quantum computer with 7 qubits to factor the number 15, yielding the factors 3 and 5. Since then, quantum computers have factored much larger numbers.

15

Prime factors: 3x5

91

Prime factors: 7x13

103958915
0853266275273
092357067028374082375087178
582405702837592830185764821742857183848584792737483

Prime factors: ????x????

Home-Cleaning Robot

Colin Angle (dates unavailable), **Helen Greiner** (b. 1967),
Rodney Brooks (b. 1954)

In 2002, consumers were introduced to the Roomba®—an autonomous robot vacuum cleaner that not only kept its owners' domiciles clean but directly connected them to a high-tech industry that had previously been associated with science fiction and pop culture characters such as R2D2 and robot house cleaner Rosie Jetson.

The Roomba was created by a company called *iRobot*®, founded in 1990 by MIT roboticists Colin Angle and Helen Greiner and their professor, Rodney Brooks. Until the Roomba, iRobot was largely focused on military and research robots such as Genghis, a robot for space exploration; Ariel, a robot for detecting and removing mines in beach surf zones; and PackBot, a robot that assisted in searches at the World Trade Center following the September 11 attacks and deployed with US troops to Afghanistan the following year.

The Roomba's arrival is one of the earliest and perhaps best-known instances of commercialization of robotics research for the consumer market. Who would have predicted that the domestic minutiae of removing dirt would be a direct application for state-of-the-art robotics research in visual mapping, intelligent navigation, sensors, 3-D manipulation, and artificial intelligence? Not to mention a vehicle for one's cat to joyride through the house.

Early versions of the Roomba cleaned floors by moving in a series of randomized patterns that were designed to cover most rooms, most of the time. Instead of mapping the room, Roomba's sensors were designed to prevent the robot from falling down stairs and detect when it bumped into an object—so it could back up, turn, and keep going. In 2015, iRobot released a Wi-Fi-enabled Roomba incorporating machine vision and a robotic navigation algorithm that visually maps out a room and determines the Roomba's place in it, for more efficient cleaning.

Roomba has taken the connection between high-tech research and the everyday person to another arena as well—allowing consumers to create their own robots. There is a version of Roomba designed intentionally for tinkering. Home hobbyists can add new hardware, software, and sensors for additional functionality and experimentation. Now anyone can have a robot deliver the morning paper and serve breakfast in bed.

SEE ALSO Robby the Robot (1956), Unimate: First Mass-Produced Robot (1961)

Roomba, the robotic vacuum cleaner, keeps consumers' homes clean without the tedium of manual vacuuming.

CAPTCHA

CAPTCHAs are tests administered by a computer to distinguish a human from a bot, or a piece of software that is pretending to be a person. They were created to prevent *programs* (more correctly, *people using programs*) from abusing online services that were created to be used by *people*. For example, companies that provide free email services to consumers sometimes use a CAPTCHA to prevent scammers from registering thousands of email addresses within a few minutes. CAPTCHAs have also been used to limit spam and restrict editing to internet social media pages.

CAPTCHA stands for **C**ompletely **A**utomated **P**ublic **T**uring test to tell **C**omputers and **H**umans **A**part. The term was coined in 2003 by computer scientists at Carnegie Mellon; however, the technique itself dates to patents filed in 1997 and 1998 by two separate teams at Sanctum®, an application security company later acquired by IBM, and AltaVista that describe the technique in detail.

One clever application of CAPTCHAs is to improve and speed up the digitization of old books and other paper-based text material. The ReCAPTCHA program takes words that are illegible to OCR (Optical Character Recognition) technology when scanned and uses them as the puzzles to be retyped. Licensed to Google, this approach helps improve the accuracy of Google's book-digitizing project by having humans provide "correct" recognition of words too fuzzy for current OCR technology. Google can then use the images and human-provided recognition as training data for further improving its automated systems.

As AI has improved, the ability of a machine to solve CAPTCHA puzzles has improved as well, creating a sort of arms race, as each side tries to improve. Different approaches have evolved over the years to create puzzles that are hard for computers but easy for people. For example, one of Google's CAPTCHAs simply asks users to click a box that says "I am not a robot"—meanwhile, Google's servers analyze the user's mouse movements, examine the cookies, and even review the user's browsing history to make sure the user is legitimate. Techniques to break or get around CAPTCHA puzzles also drive the improvement and evolution of CAPTCHA. One manual example of this is the use of "digital sweatshop workers" who type CAPTCHA solutions for human spammers, reducing the effectiveness of CAPTCHAs to limit the abuse of computer resources.

SEE ALSO The Turing Test (1951), First Internet Spam Message (1978)

CAPTCHAs require human users to enter a series of characters or take specific actions to prove they are not robots.

Type the text

?

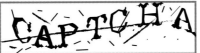

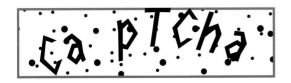

Product Tracking

Sanjay Sarma (b. 1968), **Kevin Ashton** (b. 1968), **David Brock** (b. 1961)

The Universal Product Code (UPC) revolutionized retail sales in the 1970s and '80s. A 12-digit code that could be represented as a string of numbers or a barcode, the UPC gave every kind of item that a consumer could purchase a unique 6-digit manufacturer code and a 6-digit product code. Printed on consumer packages, these codes made it possible for retailers to deploy checkout scanners, which decreased errors, sped checkouts, gave retailers better inventory management, and generally helped to lower the cost of consumer goods.

The next phase in automation was to extend this kind of information awareness all the way back through the supply chain. An Electronic Product Code (EPC), readable by radio, would make it possible to track every object from its point of manufacture through shipping and distribution, and finally to the consumer's house. Tags put on pharmaceuticals would help to stamp out both product duplication and counterfeiting. Tags on groceries could be read by refrigerators to warn of impending spoilage, or by microwave ovens to determine cooking directions, and could even be used for product recalls. To make this work, the tags had to be cheap—ideally less than 5 cents each.

Such was the EPC system developed at the MIT Auto-ID Center and ratified as a standard in 2004. It uses technology similar to a wireless building-access card but at a different part of the radio frequency (RF) spectrum. The EPC system allows every tag to have a company prefix, an item product code, and an electronic serial number. Each EPC can also be turned into an internet Uniform Resource Identifier (URI), potentially giving every product its own web address. More advanced tags are equipped with read-write memory, onboard sensors that can record temperature and pressure, and even a "kill" command for consumers wanting to protect their privacy.

Companies are now starting to equip products with EPC tags, just as they started printing barcodes on product packages in the 1970s. Today, passive 96-bit EPC tags with an integrated antenna generally cost from 7 to 15 cents, according to RFID Journal. Readers typically cost $500 to $2,000.

SEE ALSO Smart Homes (2011)

The wireless Electronic Product Code allows an entire box of inventory or a whole shelf of products to be scanned in seconds.

Facebook

Mark Zuckerberg (b. 1984)

Facebook®, the 800-pound gorilla of the social networking world, is one of the most significant communications platforms of the modern era. While the site was not the first online service that enabled people to exchange information about themselves or publicly promote their interests, it was *the* service that took the phenomenon global. Facebook raised public awareness about what constituted "social networking" and brought into focus how a simple piece of software could enable the everyday person to have a voice disproportionate to his or her economic position, geographic location, or access to sources of community organization and influence—a voice bounded only by the strength of what he or she had to say.

Facebook also served as a widespread wake-up call to traditional media outlets that their business models were ripe for disruption, as the mass media's audience flipped and went from being consumers of media to creators. Now suddenly people were their own storytellers, editors, publishers, neighborhood leaders, or global trailblazers with a platform for instantaneous projection of information around the world.

Facebook was founded in 2004 by Mark Zuckerberg and fellow Harvard students who developed a centralized website to connect students across the university. The early roots of the Facebook site are generally believed to have originated with Zuckerberg's short-lived "FaceMash" site, which gamified the choice of who was more attractive among pairs of people. The launch of "TheFacebook," as it was first called, demonstrated that there was an unfulfilled desire to connect with and learn about other people—at least among the Harvard student body. Along the way there were various legal challenges and allegations of idea theft, including a lawsuit by two brothers named Cameron and Tyler Winklevoss, who ended up with a settlement worth $65 million—chump change, compared to Facebook's 2017 market value of more than $500 billion.

Facebook quickly expanded beyond Harvard and opened to other universities, eventually turning into a business that was inclusive of anyone who wanted to join. In March 2017, the site had 1.94 billion monthly active user accounts.

SEE ALSO *Blog* Is Coined (1999), Social Media Enables the Arab Spring (2011)

Facebook CEO Mark Zuckerberg appears before a joint hearing of the Commerce and Judiciary Committees on Capitol Hill, on April 10, 2018, about the use of Facebook data in the 2016 US presidential election.

First International Meeting on Synthetic Biology

Adam Arkin (dates unavailable), **Drew Endy** (b. 1970), **Tom Knight** (b. 1948)

Over the past half century, scientists have been using advanced computer technology to investigate and seek knowledge about the biological world at its most fundamental level. Many of the discoveries in biology have been made possible by information processing for analyzing the large amounts of data that can be generated by a biological study, computational science for modeling biological systems, and, increasingly, advanced laboratory robotics for carrying out the experiments themselves.

Perhaps because of their familiarity with computers, many biologists have come to think of cells not just as metabolic systems, but also as information-processing systems. This has led to a basic question at the foundation of the new field of synthetic biology: is it possible to apply what has been learned in the creation of computers to the design and programming of cells?

In the 1990s, MIT computer scientist Tom Knight set up a biology laboratory in the MIT Laboratory for Computer Science with the goal of doing just that. One of his first milestones: creating bioluminescent bacteria that could be programmed to blink. In 1999, synthetic biologists Drew Endy and Adam Arkin proposed in a white paper that there should be "a standard parts list for biological circuitry." By 2003, these parts had become a reality, with the creation of Tom Knight's "BioBrick" standard for combining synthetic biology parts.

MIT hosted the First International Meeting on Synthetic Biology in 2004. Over the next few years, the number of researchers in the field multiplied. A key milestone followed: oscillator circuits that could be built into cells, coupled with other circuits that could be used to count the number of oscillations and signal the results at distances up to several centimeters.

SEE ALSO DNA Data Storage (2012)

The vision for synthetic biology is to enable engineers to program cells the way they can program computers.

Video Game Enables Research into Real-World Pandemics

In September 2005, programmers of the popular *World of Warcraft*® (WoW) game introduced a blood pathogen to an area accessible only to high-level players. Due to a programming bug, the disease jumped beyond the intended area, infecting and killing many lower-level players. With a dramatic spurt of digitized blood, those who were in range of the carriers died. It did not take long before entire cities were decimated, littered with the computerized corpses of players' characters, with survivors either running about in pandemonium or disappearing into the countryside.

The event was triggered by a programming glitch that allowed players' "pets" and "minions" to catch the so-called Corrupted Blood disease and pass it back to other humans, combined with the ability of characters to teleport out of the restricted area— an interaction that had not been anticipated by the game's programmers.

In the chaos and shock that ensued, several unexpected behaviors emerged as players had to make decisions about the safety and survival of their characters, and opportunities arose to pursue murderous and destructive activity. Blizzard Entertainment®, the game's developer, set up voluntary quarantine areas when it realized what was happening, but players ignored them. Some characters with healing powers tried to help those who were sick and dying, while others who were infected but immune intentionally spread the disease by teleporting to densely populated places along with their infected pets. Thousands of virtual characters died.

These observable behaviors were incredibly useful to epidemiologists who rely on computer simulations to model pandemics. Such simulations are difficult to construct and cannot reliably predict human behavior. Infected chickens propagate diseases differently than vengeful people intent on using their infection as a weapon. Realizing that teleportation was the equivalent of modern-day air travel, epidemiologists studied the Corrupted Blood incident, exploring how the digital disease could be used to model other diseases that can jump from animals to humans, such as SARS and avian flu.

Back at Blizzard, the plague itself was contained only when programmers created a "spell" to cure the patients. If only fighting disease were so easy in the real world.

SEE ALSO Morris Worm (1988)

A programming bug in World of Warcraft *unleashed the "Corrupted Blood" disease, which spread through the online community much like a real virus in an actual pandemic.*

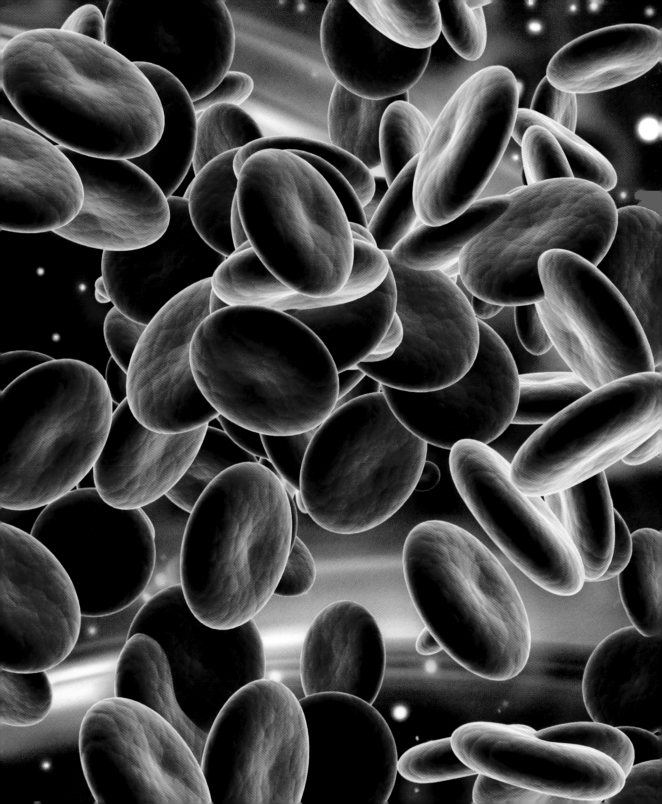

Hadoop Makes Big Data Possible

Doug Cutting (dates unavailable)

Parallelism is the key to computing with massive data: break a problem into many small pieces and attack them all at the same time, each with a different computer. But until the early 2000s, most large-scale parallel systems were based on the scientific computing model: they were one-of-a-kind, high-performance clusters built with expensive, high-reliability components. Hard to program, these systems mostly ran custom software to solve problems such as simulating nuclear-weapon explosions.

Hadoop takes a different approach. Instead of specialty hardware, Hadoop lets corporations, schools, and even individual users build parallel processing systems from ordinary computers. Multiple copies of the data are distributed across multiple hard drives in different computers; if one drive or system fails, Hadoop replicates one of the other copies. Instead of moving large amounts of data over a network to super-fast CPUs, Hadoop moves a copy of the program to the data.

Hadoop got its start at the Internet Archive, where Doug Cutting was developing an internet search engine. A few years into the project, Cutting came across a pair of academic papers from Google, one describing the distributed file system that Google had created for storing data in its massive clusters, and the other describing Google's MapReduce system for sending distributed programs to the data. Realizing that Google's approach was better than his, he rewrote his code to match Google's design.

In 2006, Cutting recognized that his implementation of the distribution systems could be used for more than running a search engine, so he took 11,000 lines of code out of his system and made them a standalone system. He named it "Hadoop" after one of his son's toys, a stuffed elephant.

Because the Hadoop code was open source, other companies and individuals could work on it as well. And with the "big data" boom, many needed what Hadoop offered. The code improved, and the systems' capabilities expanded. By 2015, the open source Hadoop market was valued at $6 billion and estimated to grow to $20 billion by 2020.

SEE ALSO Connection Machine (1985), GNU Manifesto (1985),

Although the big-data program Hadoop is typically run on high-performance clusters, hobbyists have also run it, as a hack, on tiny underpowered machines like these Cubieboards.

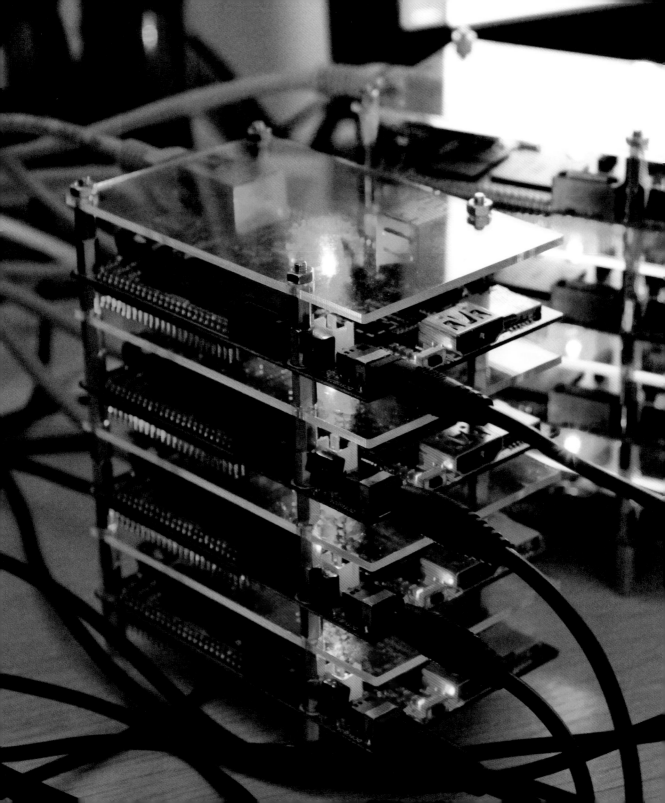

Differential Privacy

Cynthia Dwork (b. 1958), **Frank McSherry** (b. 1976), **Kobbi Nissim** (b. 1965), **Adam Smith** (b. 1977)

Differential privacy was conceived in 2006 by Cynthia Dwork and Frank McSherry, both at Microsoft Research; Kobbi Nissim at Ben-Gurion University in Israel; and Adam Smith at Israel's Weizmann Institute of Science to solve a common problem in the information age: how to use and publish statistics based on information about individuals, without infringing on those individuals' privacy.

Differential privacy provides a mathematical framework for understanding the privacy loss that results from data publications. Starting with a mathematical definition of privacy—the first ever—it provides information custodians with a formula for determining the amount of privacy loss that might result to an individual as a consequence of a proposed data release. Building on that definition, the inventors created mechanisms that allow statistics about a dataset to be published while retaining some amount of privacy for those in the dataset. How much privacy is retained depends on the accuracy of the intended data release: differential privacy gives data holders a mathematical knob they can use to decide the balance between accuracy and privacy.

For example, using differential privacy, a hypothetical town could publish "privatized" statistics that were mathematically guaranteed to protect individual privacy, while still producing aggregate statistics that could be used for traffic planning.

In the years following the discovery, there were a number of high-profile incidents in which data and statistics were published that were supposedly aggregated or deidentified, but for which the data contributed by specific individuals could be disaggregated and reidentified. These cases, combined with undeniable mathematical proofs about the ease of recovering individual data from aggregate releases, sparked interest in differential privacy among businesses and governments. In 2017, the US Census Bureau announced that it would use differential privacy to publish the statistical results of the 2020 census of population and households.

SEE ALSO Public Key Cryptography (1976), Zero-Knowledge Proofs (1985)

Differential privacy addresses how to maintain the privacy of individuals while using and publishing statistics based on their data.

iPhone

Steve Jobs (1955–2011)

Rarely do consumers line up two days before the release of a product—armed with sleeping bags and changes of clothes—to make sure they can buy it. But that is exactly what preceded the launch of the Apple iPhone on June 29, 2007.

The iPhone's design and functionality changed the entire smartphone concept by bundling together capabilities that had never been married before: telephony, messaging, internet access, music, a vibrant color screen, and an intuitive, touch-based interface. Without the physical buttons that were common on other smartphones at the time, the entire surface was available for presenting information. The keyboard appeared only when needed—and it was much easier to type accurately, thanks to behind-the-scenes AI that invisibly adjusted the sensitive area around each key in response to what letters the user was forecast to press next.

The following year, Apple introduced its next big thing: specialized programs called *apps*, downloadable over the air. The original iPhone shipped with a few built-in apps and a web browser. Apple CEO Steve Jobs had envisioned that only third-party developers would be able to write web apps. Early adopters, however, started overcoming Apple's security mechanisms by "jailbreaking" their phones and installing their own native apps. Jobs realized that if users were *that* determined to run native apps, Apple might as well supply the content and make a profit.

The Apple iTunes App Store® opened in 2008 with 500 apps. Suddenly that piece of electronics in your pocket was more than a phone to make calls or check email—it became a super gadget, able to play games, manipulate photographs, track your workout, and much more. In October 2013, Apple announced that there were a million apps available for the iPhone, many of them realizing new location-based services, such as ride-sharing, dating, and localized restaurant reviews, to name a few.

While the iPhone has largely been celebrated, it has also been accused of ushering in the era of "smartphone addiction," with the average person, according to a 2016 study, now checking his or her smartphone 2,617 times a day. Since the original release in 2007, more than 1 billion iPhones have been sold worldwide, and it still holds the record for taking only three months to get to 1 million units sold.

SEE ALSO Touchscreen (1965), Augmented Reality Goes Mainstream (2016)

More than 1 billion iPhones have been sold worldwide since the product's release in 2007.

Bitcoin

Satoshi Nakamoto (pseudonym)

Bitcoin was the first digital currency to gain mainstream use and demonstrate a practical application for blockchain, the powerful concept on which Bitcoin is based. Invented in 2008 by "Satoshi Nakamoto," a pseudonym, Bitcoin immediately caught the interest of cypherpunks and cryptographers but was slow to gain broader adoption.

In the world economic system, most transactions don't involve the exchange of cash but rather the movement of bits in banks' computers. Bitcoin works much the same way, except that cooperating computers, rather than countries, mint the money. Every customer's balance is public; the Bitcoin system is based upon an open, common ledger that records every single Bitcoin transaction that has ever occurred. Collections of transactions—called *blocks*—make up the links in this ledger, which is called the *blockchain*.

If Jean wants to send Pat five bitcoins, Jean sends a message to the Bitcoin network, which is made up of computers called *miners*. The miners verify the proposed transaction is legitimate, using the parties' digital signatures and reading the entire blockchain to make sure that Jean has at least five bitcoins in the ledger. Next, the miners race to be the first to solve a complex math puzzle that includes Jean's transaction and every other transaction in the network's pool. The first miner that solves the puzzle sends the solution to the other miners, in the process confirming the pending transaction, minting 50 bitcoins for the miner, adding the completed puzzle to the Bitcoin blockchain, and starting all of the miners on the next puzzle.

On May 22, 2010, Laszlo Hanyecz paid 10,000 bitcoins to have someone deliver him two pizzas. It was the first Bitcoin transaction for a physical object. At the time, those bitcoins were worth about $40; by 2017, they were worth more than $20 million. May 22 is now known as *Bitcoin Pizza Day*.

Bitcoin is an open source project, and numerous digital currencies over the years have mimicked or improved upon the original concept. Recently there have been efforts to separate the blockchain concept from the financial system and use it as a public record to memorialize contracts, healthcare records, and other kinds of information.

SEE ALSO Digital Money (1990)

Bitcoin, in which cooperating computers "mint" money, is an increasingly popular payment method.

Maestro

VISA

VISA
ELECTRON

V
PAY

ADUNO
payment services

tripadvisor®
Die weltweit größte Reise-Website

AMERICAN
EXPRESS

UnionPay
银联

Klimamenu
Mehr Genuss – weniger CO₂
www.klimamenu.ch
eaternity   myblueplanet
                   today together for tomorrow

ZÜR
SÜDI

LUNCH-CHECKS

schäffel

bitcoin
ACCEPTED HERE

Air Force Builds Supercomputer with Gaming Consoles

Mark Barnell (dates unavailable), **Gaurav Khanna** (dates unavailable)

It is said that necessity is the mother of invention. In 2010, the Air Force Research Laboratory (AFRL) in Rome, New York, built a "budget" supercomputer called the *Condor Cluster* using commercial, off-the-shelf hardware consisting primarily of 1,716 PlayStation® 3 (PS3®) game consoles. Motivated to save money while advancing his research programs, Mark Barnell, director of AFRL's high-power computing division, took an unorthodox approach to achieve the number-crunching capacity required for AFRL's use of radar data to create images of cities.

The PS3's computer power came from its "cell" processor, which relied upon numerous specialized cores. Barnell's team connected the PS3s along with 168 graphical processing units and 84 coordinating servers in a parallel array to realize a capability that could perform 500 trillion floating-point calculations per second. Put another way, the Condor was 50,000 times faster than an average laptop. At the time, the Condor Cluster was considered the 35th- or 36th-fastest computer in the world and cost just $2 million—almost 30 times cheaper than a proper supercomputer, which would typically set an organization back $50 million to $80 million.

The idea to build a do-it-yourself supercomputer using linked PS3s did not originate inside the Air Force, however. In 2007, an engineer at North Carolina State University created a scientific research cluster with eight connected PS3s that cost $5,000. That same year, University of Massachusetts physicist Dr. Gaurav Khanna networked 16 PS3s to model black-hole collisions. Named the *Gravity Grid*, this was the cluster that caught the attention of the Air Force team. Dr. Khanna would go on to publish a paper in the journal *Parallel and Distributed Computing and Systems* showing how the PS3 processor sped up scientific calculations over traditional processors by a factor of 10.

The high price tag of "proper" supercomputers includes lots of other hardware, of course, such as power conditioners and cooling. Dr. Khanna's team found a cheap, off-the-shelf way of keeping their PS3s cool: they put them in a refrigerated shipping container designed to transport milk.

SEE ALSO Connection Machine (1985), Hadoop Makes Big Data Possible (2006)

Photograph of the supercomputer built by the Air Force Research Lab from PlayStation 3 game consoles.

Cyber Weapons

In June 2010, an antivirus company called *VirusBlokAda* publicly reported a highly sophisticated computer worm expanding its presence on computers running Microsoft Windows, with most of the early infected systems located in Iran.

Interest grew as more security experts looked at the code. Whereas viruses and worms typically spread when a person runs a program, this one spread when a person opened Windows folders containing the infected files. This program could also spread using previously unknown vulnerabilities in the Windows printer subsystem. Once activated, it installed a sophisticated "rootkit," making the program invisible to antivirus programs. It then sought software designed to control industrial systems such as motors, pumps, and compressors, and if it found the correct software, it installed a carefully crafted flaw.

That flaw, analysis later revealed, attacked computer-controlled motors manufactured by two specific vendors located in Finland and Iran, placing significant mechanical strain on the motors and anything connected to them. All this activity would be invisible to software controlling the drive.

Symantec®, a US-based antivirus company, named the malware *Stuxnet*, based in part on files that the malware carried. In the weeks that followed, both Symantec and the Russian antivirus firm Kaspersky published detailed analyses of Stuxnet. The program exploited four previously unreported vulnerabilities in the Windows operating system. The flaws, the attacks on the computer-controlled motors, and the fact that Stuxnet was mostly spreading in Iran, led many observers to conclude that the software had been written by a state sponsor.

Stuxnet is now generally regarded as the world's first cyber weapon that had physical effects—at least, the first that was caught and publicly analyzed. Why was it caught? It seems that a programming error resulted in Stuxnet spreading much further than intended, and important aspects of the program were not encrypted, making it dramatically easier to analyze.

Writing in the journal *International Security*, Cornell professor Rebecca Slayton calculates that in the end, Stuxnet cost its sponsors between $11 million and $67 million to develop, while the total impact on Iran was just $4 million in additional cybersecurity costs, $5 million in lost productivity, and $1.8 million in centrifuge replacement, for a total of $11 million.

SEE ALSO Morris Worm (1988)

Logic bombs, physical bombs . . . In the world of the computer, it's all the same.

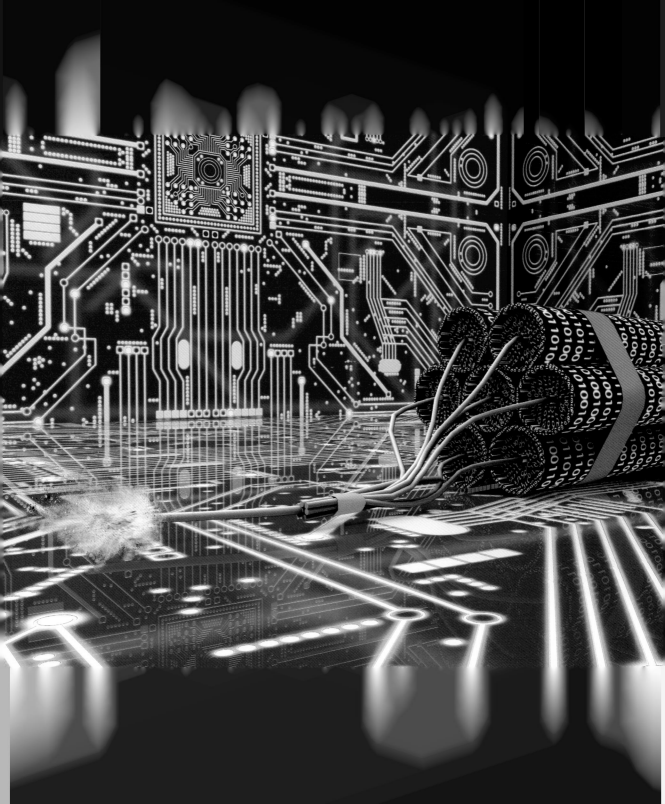

Smart Homes

In 1990, a Boston internet company called *FTP Software* demonstrated its "internet toaster" at Interop, an information technology trade show. But while the toaster could be started and controlled over the network, there was no way to load it with toast. No problem! The following year, FTP showed off its improved internet toaster, complete with a robot crane built from LEGO® bricks for picking up a slice of bread and inserting it into the network-enabled device.

Back in the 1990s, internet-connected devices were little more than a geek's prank. Nobody seriously imagined there would be demand one day for internet-connected appliances.

The debut of the Nest Learning Thermostat® in 2011 showed that conventional wisdom was wrong, and that there was real consumer demand for connected devices—provided that they could do something useful. It also demonstrated how a smart abode fit into the larger ecosystem called the *Internet of Things* that was gaining steam in the media, if not in the living room. Nest was a practical example of how people could "improve" management of their lives in an environment saturated with smart devices.

Unlike FTP's toaster, the Nest was first and foremost a learning device, which means it used advanced machine learning algorithms to learn and eventually anticipate the user's needs. Algorithms that learn schedules are complex, and Nest shipped before the units were all perfect. Because the thermostat was connected to the internet, it could receive software updates and become smarter.

Like FTP's toaster, the Nest could also be controlled over the internet. But in 2011, unlike 1990, this meant it could be controlled from a smartphone. Forget to turn down the thermostat before going on that vacation? No problem: just pull out the smartphone and do it from the airport—or from Aruba.

Early adopters bought the Nest because it was neat and high tech, but the thermostats started flying off the shelves because people who installed the Nest saved money. And that, in the end, was the most useful thing a smart thermostat could possibly do.

SEE ALSO Unimate: First Mass-Produced Robot (1961), *Computers at Risk* (1991)

The Nest Learning Thermostat is an example of how connected devices allow consumers to electronically manage their lives and homes.

Watson Wins *Jeopardy!*

David Ferrucci (b. 1962)

For all of the mathematical accomplishments that computers are capable of, a machine that engages people in conversation is still the work of fiction and computer scientists' dreams. When IBM's Watson® beat the two best-ever *Jeopardy!* players—Ken Jennings and Brad Rutter—the dream seemed a little more real. Indeed, when Jennings realized he had lost, he tweaked a line from an episode of *The Simpsons* to display on his screen: "I, for one, welcome our new computer overlords."

Unlike chess, for which IBM's Deep Blue demonstrated domination when it beat the world's best chess player, Garry Kasparov, in 1996, *Jeopardy!* is not a game governed by clear and objective rules that translate into mathematical calculations and statistical models. It's a game governed by finding answers in language—a messy, unstructured, ambiguous jumble of symbols that humans understand as a result of context, culture, inference, and a vast corpus of knowledge acquired by virtue of being a human and having a lifetime of sensory experiences. Designing a computer that could beat a person at this game was a really big deal.

Watson was designed over several years using a 25-person team of multidisciplinary experts in fields that included natural language processing, game theory, machine learning, informational retrieval, and computational linguistics. The team accomplished much of its work in a common war room where the exchange of diverse ideas and perspectives enabled faster and more incremental progress than may have occurred using a more traditional research approach. The goal was not to model the human brain but to "build a computer that can be more effective in understanding and interacting in natural language, but not necessarily the same way humans do it," according to David Ferrucci, Watson's lead designer.

Watson's success was not due to any one breakthrough, but rather incremental improvements in cognitive computing along with other factors, including the massive supercomputing capabilities of speed and memory that IBM could direct to the project, more than 100 algorithms the team had working in parallel to analyze questions and answers, and the corpus of millions of electronic documents Watson ingested, including dictionaries, literature, news reports, and Wikipedia.

SEE ALSO Computer Is World Chess Champion (1997), Wikipedia (2001), Computer Beats Master at Go (2016)

Contestants Ken Jennings and Brad Rutter compete against Watson at a press conference before the "Man v. Machine" *Jeopardy!* competition at the IBM Thomas J. Watson Research Center in Yorktown Heights, New York.

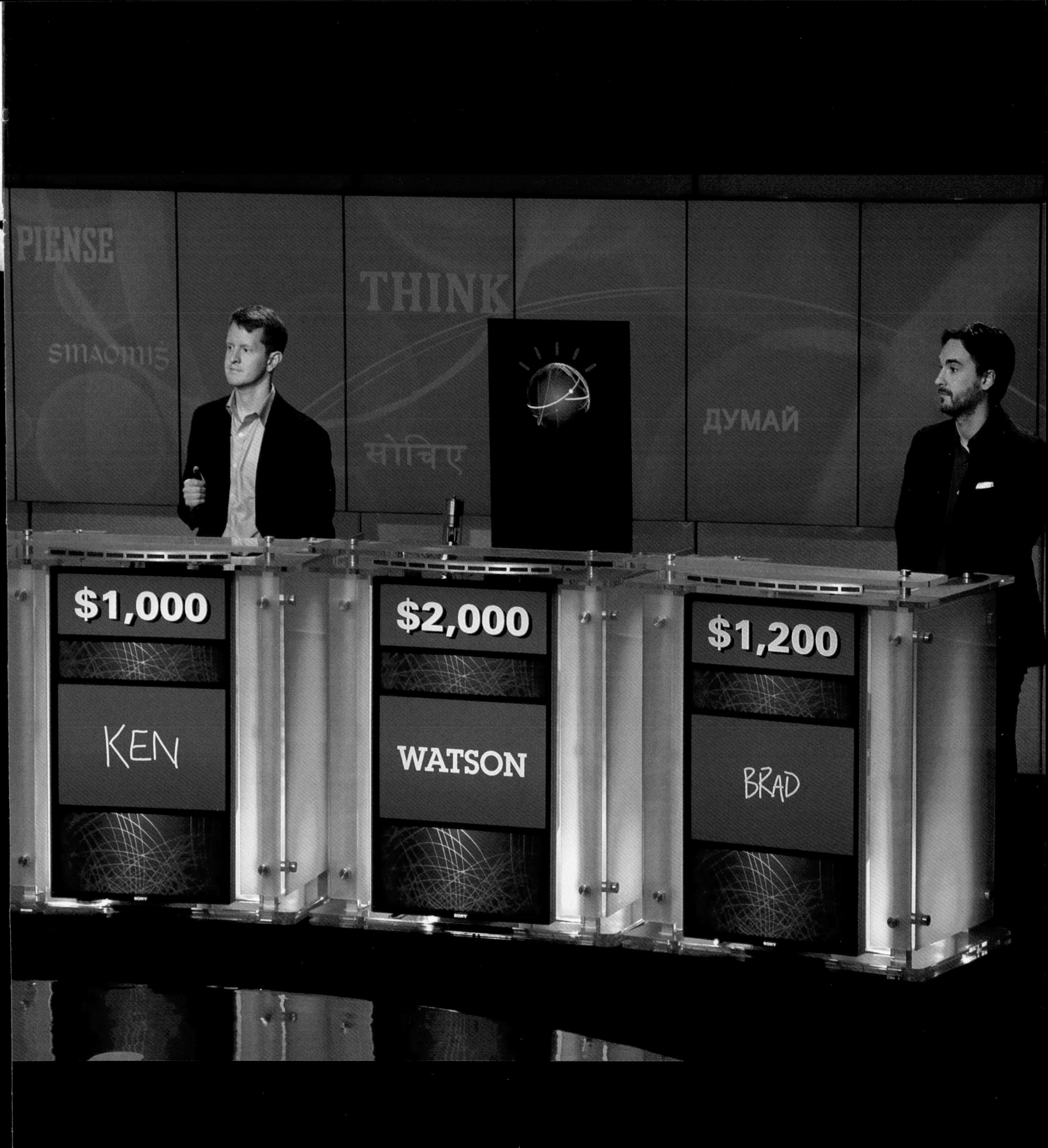
PIENSE
THINK
सोचिए
ДУМАЙ
$1,000
$2,000
$1,200
KEN
WATSON
BRAD

World IPv6 Day

Every computer on the internet has an Internet Protocol (IP) address, a number that the internet uses to route network packets to the computer. When internet engineers adopted Internet Protocol Version 4 (IPv4) in 1984, they thought that 32 bits would be sufficient, because it allowed for $2^{32} = 4,294,967,296$ possible computers. Back then, that seemed like enough.

As things turned out, 4 billion addresses were nowhere near enough. Many early internet adopters got unreasonably large blocks of addresses—MIT got $2^{24} = 16,777,216$ of them! But to realize the dreams of a fully networked society, every cell phone—indeed, every light bulb—would potentially need its own address. Even properly allocated, 32 bits just wouldn't be enough.

Throughout the 1990s, internet infrastructure engineers periodically warned that the internet was running out of address space. In 1998, the Internet Engineering Task Force (IETF) officially published version 6 of the IP specification (IPv6). The new protocol used 128-bit addresses, allowing a maximum of 2^{128} addresses. To get an idea of how fantastically large this number is, it is considerably larger than the number of grains of sand on the earth (estimated at 2^{63}) or stars in the sky (2^{76}).

IPv6 is similar to IPv4, but it is fundamentally incompatible. Thousands of programs had to be rewritten, and millions of computers needed to be upgraded.

Early efforts to turn on IPv6 failed: so many systems were misconfigured or simply missing IPv6 support that flipping the switch resulted in users losing service.

Then, in January 2011, there were no new IPv4 addresses to hand out.

On January 12, 2011, more than 400 companies, including the internet's largest providers, enabled IPv6 for the first time on their primary servers. It was the final test, and this time it (mostly) worked. Called *World IPv6 Day*, the event lasted 24 hours. After analyzing the data, the leading participants declared that no serious service interruptions had been experienced, but more work needed to be done. The following year, they turned it on for good.

Today IPv4 and IPv6 coexist on the internet, and when you connect to a host such as Google or Facebook, there's a good chance your connection is traveling over IPv6.

SEE ALSO IPv4 Flag Day (1983)

With IPv6, there are enough internet addresses for all of the stars in the sky and all of the grains of sand on the earth.

2011

Social Media Enables the Arab Spring

Mohamed Bouazizi (1984–2011)

On December 17, 2010, a 26-year-old Tunisian street vendor named Mohamed Bouazizi set himself on fire in protest over the harassment and public humiliation he received at the hands of the local police for refusing to pay a bribe. Protests following the incident were recorded by participants' cell phones. In the era of globally networked information systems, Bouazizi's story spread much further than the people at the rally or Bouazizi's family could have mustered on their own. After the video was uploaded to Facebook, it was shared and reshared across the social media sphere. The Facebook pages and the subsequent social media pathways the video appeared on were critical enablers that galvanized people to action on a scale they otherwise could not have achieved.

This incident is often cited as the catalyst for the Tunisian Revolution. Leveraging the digital diffusion capabilities of computer technology, including Facebook and Twitter, individuals and groups organized and coordinated real-world demonstrations, then posted the results for the world to see. The digital platform amplified and extended the emotional impact of these experiences to many others, who in turn shared the content even further.

Videos of protests might have been the spark that lit the uprising, but the pressure had been building since November 28, 2010, when allegedly leaked US government cables that discussed the corruption of Tunisia's ruling elites started circulating on the internet. The government of president Zine El Abidine Ben Ali, who had been in office since 1987, fell on January 14, 2011, triggering demonstrations, coups, and civil wars elsewhere in northern Africa and the Middle East, in what is now known as the *Arab Spring*.

Immediate causalities of the Arab Spring included Egypt's president, Hosni Mubarak, who was pushed out of office by a popular revolt on February 11, 2011, after nearly 30 years of authoritarian rule, and Libya's ruler, Muammar Mohammed Abu Minyar Gaddafi, who had seized power in 1969 and was executed by rebels on October 20, 2011.

SEE ALSO Facebook (2004)

Protesters charging their all-important mobile phones in Tahrir Square, in Cairo, Egypt.

494

DNA Data Storage

George Church (b. 1954), **Yuan Gao** (dates unavailable),
Sriram Kosuri (dates unavailable), **Mikhail Neiman** (1905–1975)

In 2012, George Church, Yuan Gao, and Sriram Kosuri, all with the Harvard Medical School's Department of Genetics, announced that they had successfully stored 5.27 megabits of digitized information in strands of deoxyribonucleic acid (DNA), the biological molecule that is the carrier of genetic information. The stored information included a 53,400-word book, 11 JPEG images, and a JavaScript program. The following year, scientists at the European Bioinformatics Institute (EMBL-EBI) successfully stored and retrieved an even larger amount of data in DNA, including a 26-second audio clip of Martin Luther King's "I Have a Dream" speech, 154 Shakespeare sonnets, the famous Watson and Crick paper on DNA structure, a picture of EMBL-EBI headquarters, and a document that described the methods the team used to accomplish the experiment.

Although first demonstrated in 2012, the *concept* of using DNA as a recording, storage, and retrieval mechanism goes back to 1964, when a physicist named Mikhail Neiman published the idea in the Soviet journal *Radiotekhnika*.

To accomplish this storage and retrieval, first a digital file represented as 1s and 0s is converted to the letters A, C, G, and T. These letters are the four chemical bases that make up DNA. The resulting long string of letters is then used to manufacture synthetic DNA molecules, with the sequence of the original bits corresponding to the sequence of nucleic acids. To decode the DNA and reconstitute the digital file, the DNA is put through a sequencing machine that translates the letters back into the original 1s and 0s of the original digital files. Those files can then be displayed on a screen, played through a speaker, or even run on a computer's CPU.

In the future, DNA could allow digital archives to reliably store vast amounts of digitized data: a single gram of DNA has the potential to store 215 million gigabytes of data, allowing all the world's information to be stored in a space the size of a couple of shipping containers.

SEE ALSO Magnetic Tape Used for Computers (1951), DVD (1995)

To store information in DNA, a digital file represented as 1s and 0s is converted to the letters A, C, G, and T, the four chemical bases that make up DNA.

```
CTGAGTCCCCTGGAACGGGGCGCCCATAGAGGGTGAGAGCCCCGTATAGTCGG
TCTAAGTTCCTTGGAACAGGACGTCATAGAGGGTGAGAATCCCGTATGTGAC
TCTAAGTTCCTTGGAACAGGACGTCACAGAGGGGTGAGAATCCCGTATGTGAT
GTCTAAGTTCCTTGGAACAGGACGTCATAGAGGGTGAGAATCCCGTATGTGA
CCGAGTTCCCTGGAACGGGACGCCACAGAGGGTGAGAGCCCCGTATGGTTGG
CCGCCCGAGTTCCCTGGAACGGGACGCCCACAGAGGGTGAGAGCCCCGTCTGG
CCCGGGTTAAATTTCTTGGAACAGAATGTCATAGAGGGTGAGAATCCCGTCT
CCCCATCTAAGTGCCCTGGAACGGGACGTCATAGAGGGTGAGAATCCCGTAT
CCGAGTTCCCTGGAACGGGACGCCCACAGAGGGTGAGAGCCCCGTATGGTTGG
AGCAGTCCAAGTTCTTTGGAACAGGACGTCAGAGAGGGTGAGAATCCCGTAT
AGCAGTCCAAGTTCCTTGGAACAGGACGTCAGAGAGGGTGAGAATCCCGTAC
GCCGGCCTAAGTCCCTTGGAACAGGGCGTCATAGAGGGTGAGAATCCCGTAT
CCTTGGAACAGGACGTCACAGAGGGTGAGAATCCCGTACGTGGTCGCTAGCC
TCTAAGTTCCTTGGAACAGGACGTCATAGAGGGTGAGAATCCCGTATGCGAC
ACCGGTCTAAGTCCCTGGAACGGGGTGTCACAGAGGGTGAGAATCCCGTATC
CCGGTCTAAATTTCTTGGAACAGAATGTCAGAGAGGGTGAGAATCCCGTCTT
TGCAGCCTAAGTTCCTTGGAACAGGTCATCATAGAGGGTGAGAATCCCGTAT
GTCTAAGTTCCTTGGAACAGGACGTCATAGAGGGTGAGAATCCCGTATGTGA
TGTGGTCTAAGTTCCTTGGAACAGGACGTCACAGAGGGTGAGAATCCCGTACC
TTCTGAGTTCCCTGGAACGGGACGCCAGAGAGGGTGAGAGCCCCGTACGGTTC
CCGAGTTCCCTGGAACGGGACGCCACAGAGGGTGAGAGCCCCGTATGGTCGG
CTGAGTCCCTTGGAACAGGGCGCCATAGAGGGTGAGAGCCCCGTATAGTCGG
CCAGTCTATGTTCCTTGGAACAGGACGTCATAGAGGGTGAGAATCCCGTTCAT
GCTGTCCTAAGTTCCTTGGAACAGGATGACATAGAGGGTGAGATCCCCGTGCC
TTCCGAGTTCCCTGGAACGGGACGCCTTACAGGGTGAGAGCCCCGTACGGTT
AGTCTAAGTTCCTTGGAACAGGACGTCATAGAGGGTGAGAATCCCGTATGTG
GCGGCGGTCTAAGTTCCCTGGAACAGGACATCGCAGAGGGTGAGAATCCCGT
GTGGCGTCTAAGTTCCTTGGAACAGGACATCGCAGAGGGTGAGAATCCCGT
GTCTAAGTTCCTTGGAACAGGACGTCATAGAGGGTGAGAATCCCGTATGTGA
TGTCTAAGTTCCTTGGAACAGGACGTCATAGAGGGTGAGAATCCCGTATGTG
GCGGCGGTCTAAGTTCCTTGGAACAGGACATCGCAGAGGGTGAGAATCCCGT
GTCTAAGTCTCTTGGAACAGGGCGTCATAGAGGGTGAGAATCCCGTATGCGA
ACCGACCTAAGTTCCTTGGAACAGGACGTCATAGAGGGTGAGAATCCCGTAT
AGCGGTCCAAGTTCCTTGGAACAGGACGTCACAGAGGGTGAGAATCCCGTAC
CCACGGTGTAAGTCCCTTGGAACAGGGCGTCATAGAGGGTGAGAATCCCGTC
CCACGGCATAGTTCCTTGGAACAGGACGTCATAGAGGGTGAGAATCCCGTC
CCGCGGCGTAAGTCCCTTGGAACAGGGCGTCAGAGAGGGTGAGAATCCCGTC
```

Algorithm Influences Prison Sentence

Eric Loomis was sentenced to six years in prison and five years' extended supervision for charges associated with a drive-by shooting in La Crosse, Wisconsin. The judge rejected Loomis's plea deal, citing (among other factors), the high score that Loomis had received from the computerized COMPAS (Correctional Offender Management Profiling for Alternative Sanctions) risk-assessment system.

Loomis appealed his sentence on the grounds that his due process was violated, as he did not have any information into how the algorithm derived his score. As it turns out, neither did the judge. And the creators of COMPAS—Northpointe Inc.—refused to provide that information, claiming that it was proprietary. The Wisconsin Supreme court upheld the lower court's ruling against Loomis, reasoning that the COMPAS score was just one of many factors the judge used to determine the sentence. In June 2017, the US Supreme Court decided not to give an opinion on the case, after previously inviting the acting solicitor general of the United States to file an amicus brief.

Data-driven decision-making focused on predicting the likelihood of some future behavior is not new—just ask parents who pay for their teenager's auto insurance or a person with poor credit who applies for a loan. What *is* relatively new, however, is the increasingly opaque reasoning that these models perform as a consequence of the increasing use of sophisticated statistical machine learning. Research has shown that hidden bias can be inadvertently (or intentionally) coded into an algorithm. Illegal bias can also result from the selection of data fed to the data model. An additional question in the Loomis case is whether gender was considered in the algorithm's score, a factor that is unconstitutional at sentencing. A final complicating fact is that profit-driven companies are neither required nor motivated to reveal any of this information.

State v. Loomis helped raise public awareness about the use of "black box" algorithms in the criminal justice system. This, in turn, has helped to stimulate new research into development of "white box" algorithms that increase the transparency and understandability of criminal prediction models by a nontechnical person.

SEE ALSO DENDRAL (1965), *The Shockwave Rider* (1975)

Computer algorithms such as the COMPAS risk-assessment system can influence the sentencing of convicted defendants in criminal cases.

Subscription Software

In 2013, Adobe stopped selling copies of its tremendously popular Photoshop and Illustrator programs and instead started to rent them. Microsoft and others would soon follow. The era of "subscription software" had arrived.

Despite providing many economically sound reasons why this move was in the interest of its customers (and of course equally good for the company's bottom line), Adobe's announcement was met with a wave of negativity and petitions to reinstate the traditional purchase model. Why? Because many customers didn't upgrade their software every year, and they resented being put in the position of having to pay up annually or have their software stop working.

Purchasing subscriptions for digital services was not new—cable TV, streaming video, and telephone service are all sold by subscription. Software as a product, however, had been different since the birth of the microcomputer. Even though Adobe's Photoshop is as much a series of 1s and 0s as a streaming movie, consumers did not experience it that way, because they traditionally did not receive it that way. Since it first went on sale in 1988, Photoshop had been sold as a physical object, packaged on floppy disk, CD, or DVD. It was a physical, tactile, or otherwise visible exchange of money for goods. But once the CD or DVD "packaging" of those 1s and 0s was replaced with the delivery of bits over a network connection, it was only a matter of time until the publisher decided to attach a time limit to that purchase. People were not just confused; they were downright furious.

Over time the advantages for most customers became clear: subscription software can be updated more often, and publishers can easily sell many different versions at different price points. The subscription model also gives consumers the flexibility to make small, incremental purchases without a big up-front investment. Now people who were interested in, but not committed to using, a professional photo-editing suite can spend $40 to try it out for a month, rather than spending thousands of dollars up front for a product suite that might not precisely align with their needs or interests. The advent of subscription software brought with it a new model for enabling fast evolution and innovation in software products that in turn drove competition across a landscape of ecommerce services.

SEE ALSO Over-the-Air Vehicle Software Updates (2014)

Purchasing subscriptions for popular software programs has become increasingly popular, replacing the previous model of buying and owning the software.

Microsoft
1 YEAR
NEW OR
RENEWAL
Office 365
Home
Subscription
Includes
Word
Excel
PowerPoint
OneNote
Outlook
Ongoing upgrades
1 TB OneDrive storage
For Windows & Mac
5 PCs/Macs + 5 tablets
PRODUCT KEY INSIDE • NO DISC
Office 365
Home
Office 365

Data Breaches

In 2014, data breaches touched individuals on a scale not seen before, in terms of both the amount and the sensitivity of the data that was stolen. These hacks served as a wake-up call to the world about the reality of living a digitally dependent way of life—both for individuals and for corporate data masters.

Most news coverage of data breaches focused on losses suffered by corporations and government agencies in North America—not because these systems were especially vulnerable, but because laws required public disclosure. High-profile attacks affected millions of accounts with companies including Target® (in late 2013), JPMorgan Chase®, and eBay. Midway through the year, the United States Office of Personnel Management revealed that highly personal (and sensitive) information belonging to 18 million former, current, and prospective federal employees allegedly had been stolen. Meanwhile, information associated with at least half a billion user accounts at Yahoo! was being hacked, although this information wouldn't come out until 2016.

Data from organizations outside the US was no less immune. The European Central Bank, HSBC Turkey, and others were hit. These hacks represented millions of victims across a spectrum of industries, such as banking, government, entertainment, retail, and health. While some of the industry and government datasets ended up online, available to the highest bidder in the criminal underground, many other datasets did not, fueling speculation and public discourse about why and what could be done with such data.

The 2014 breaches also expanded the public's understanding about the value of certain types of hacked data beyond the traditional categories of credit card numbers, names, and addresses. The November 24, 2014, hack of Sony Pictures, for example, didn't just temporarily shut down the film studio: the hackers also exposed personal email exchanges, harmed creative intellectual property, and rekindled threats against the studio's freedom of expression, allegedly in retaliation for the studio's decision to participate in the release of a Hollywood movie critical of a foreign government.

Perhaps most importantly, the 2014 breaches exposed the generally poor state of software security, best practices, and experts' digital acumen across the world. The seams between the old world and that of a world with modern, networked technology were not as neatly stitched as many had assumed.

SEE ALSO Morris Worm (1988), Cyber Weapons (2010)

Since 2014, high-profile data breaches have affected billions of people worldwide.

Over-the-Air Vehicle Software Updates

Elon Musk (b. 1971)

In January 2014, the National Highway Traffic Safety Administration (NHTSA) published two safety recall notices for components in cars that could overheat and potentially cause fires. The first recall notice was for General Motors (GM) and required owners to physically take their cars to a dealership to correct the problem. The second was for Tesla Motors, and the recall was performed wirelessly, using the vehicle's built-in cellular modem.

The remedy described by the NHTSA required Tesla to contact the owners of its 2013 Model S vehicles for an over-the-air (OTA) software update. The update modified the vehicle's onboard charging system to detect any unexpected fluctuations in power and then automatically reduce the charging current. This is a perfectly reasonable course of action for what is essentially a 3,000-pound computer on wheels, but an OTA fix for a car? It was a seismic event for the automotive industry, as well as for the general public.

Tesla's realization of OTA updates as the new normal for car maintenance was a big deal in and of itself. But the "recall" also provided an explicit example of how a world of smart, interconnected things will change the way people go about their lives and take care of the domestic minutiae that are part and parcel to the upkeep of physical stuff. It was also a glimpse into the future for many, including those whose jobs are to roll up their sleeves and physically repair cars. The event also called into question the relevance of NHTSA using the word *recall*, because no such thing actually took place, according to Tesla CEO Elon Musk. "The word 'recall' needs to be recalled," Musk tweeted.

This was not the first time Tesla had pushed an update to one of its vehicles, but it was the most public, because it was ordered by a government regulatory authority. It also served as a reminder of the importance of computer security in this brave new connected world—although Tesla has assured its customers that cars will respond only to authorized updates.

Indeed, OTA updates will likely become routine for all cars for that very reason—timely security updates will be needed when hackers go after those 3,000-pound computers on wheels.

SEE ALSO *Computers at Risk* (1991), Smart Homes (2011), Subscription Software (2013)

Don't think of the Tesla as a car with a computer; think of it as a computer that has wheels.

Google Releases TensorFlow

Makoto Koike (dates unavailable)

Cucumbers are a big culinary deal in Japan. The amount of work that goes into growing them can be repetitive and laborious, such as the task of hand-sorting them for quality based on size, shape, color, and prickles. An embedded-systems designer who happens to be the son of a cucumber farmer (and future inheritor of the cucumber farm) had the novel idea of automating his mother's nine-category sorting process with a sorting robot (that he designed) and some fancy machine learning (ML) algorithms. With Google's release of its open source machine learning library, TensorFlow®, Makoto Koike was able to do just that.

TensorFlow, a deep learning neural network, evolved from Google's DistBelief, a proprietary machine learning system that the company used for a variety of its applications. (Machine learning allows computers to find relationships and perform classifications without being explicitly programmed regarding the details.) While TensorFlow was not the first open source library for machine learning, its release was important for a few reasons. First, the code was easier to read and implement than most of the other platforms out there. Second, it used Python, an easy-to-use computer language widely taught in schools, yet powerful enough for many scientific computing and machine learning tasks. TensorFlow also had great support, documentation, and a dynamic visualization tool, and it was as practical to use for research as it was for production. It ran on a variety of hardware, from high-powered supercomputers to mobile phones. And it certainly didn't hurt that it was a product of one of the world's behemoth tech companies whose most valuable asset is the gasoline that fuels ML and AI—data.

These factors helped to drive TensorFlow's popularity. The greater the number of people using it, the faster it improved, and the more areas in which it was applied. This was a good thing for the entire AI industry. Allowing code to be open source and sharing knowledge and data from disparate domains and industries is what the field needed (and still needs) to move forward. TensorFlow's reach and usability helped democratize experimentation and deployment of AI and ML applications. Rather than being exclusive to companies and research institutions, AI and ML capabilities were now in reach of individual consumers—such as cucumber farmers.

SEE ALSO GNU Manifesto (1985), Computer Beats Master at Go (2016), Artificial General Intelligence (AGI) (~2050)

TensorFlow's hallucinogenic images show the kinds of mathematical structures that neural networks construct in order to recognize and classify images.

Augmented Reality Goes Mainstream

John Hanke (b. 1967), **Satoru Iwata** (1959–2015), **Tsunekazu Ishihara** (b. 1957)

Pokémon GO® was a watershed event for the computer game industry and for augmented reality (AR). In the first two months, the game had 500 million downloads—roughly 7 percent of the world's population, and one out of every seven people use a cell phone. It popularized the use of a smartphone's camera to integrate real views of the physical world with superimposed computer-generated images associated with the phone's actual physical location.

The game works by overlaying Pokémon characters onto live video images captured from the phone's camera. Relying on smartphones' GPS and motion sensors, characters appear in different physical locations as the player moves about the real world trying to find them. The basic objective is to find and catch as many of these characters as possible for points, privileges within the game, and opportunities for collaboration. Think of it as a scavenger hunt that connects the virtual and physical worlds.

Before *Pokémon GO*, it was often hard to explain AR to a person unfamiliar with the concept. The game's design, popularity, and incessant news coverage alleviated that ambiguity, giving the general public an understanding of the technology—both its potential benefits, and its dangers. Imagine trying to refill the fluids in a car—standing there with your phone showing images of the engine compartment overlaid with arrows and indicators, perhaps fed directly from the car's own computer. On the other hand, in Missouri, armed robbers used a move in the game to lure Pokémon players to a secluded location, where they were ambushed.

Satoru Iwata and Tsunekazu Ishihara at the Pokémon Company in Japan came up with the idea of the game; the program uses crowdsourced data from *Ingress*, a geo-based AR game from Niantic®, a company founded by John Hanke. The original *Ingress* game was an alternate-reality world created by game designer and screenwriter Flint Dille along with E. Daniel Arey.

SEE ALSO Head-Mounted Display (1967), VPL Research, Inc. (1984), GPS Is Operational (1990)

A Pokémon player goes "fishing" for the Pokémon species Magikarp in Lake Windermere in the United Kingdom.

Computer Beats Master at Go

The path for machine victory over the humans who play the ancient Chinese game of Go was not achieved through mathematical superiority, because Go is a very different game from chess.

Rather than the 8 × 8 grid for chess, Go is played on a 19 × 19 board, with each player having dozens of black or white stones. Each stone has the same value—unlike chess, in which the pieces are not all equal. The rules of Go are fairly straightforward—the two players try to surround each other's stones and take territory from each other. However, because of the size of the grid, the number of potential positions in Go is staggering—considerably larger than the number of atoms in the Universe.

This sheer complexity is why intuition is so often cited as a key factor in winning the game, and why a computer program beating one of the best Go players that ever lived was considered so significant. As players add more stones to the board, the number of possible countermoves and counter-countermoves grows exponentially. As a result, brute-force "look-ahead" computing approaches to solving Go just can't look far enough ahead: computers aren't big enough. The Universe isn't big enough.

AlphaGo® is the AI program that beat South Korean Go master Lee Sedol (b. 1983) in March 2016, in four out of five games, by adopting the same sort of strategic search strategies a human would. The program was created by the Google DeepMind team that evolved from Google's acquisition of British company DeepMind Technologies, a British AI company that built a neural network to play video games like a human.

Lee Sedol did win once, however, so the computer did not dominate the match. In game four, white move 78, Lee Sedol found AlphaGo's Achilles' heel and made a move that so thoroughly confused the system that it started to make rookie mistakes, not recovering in time to save the game. The irony is that Sedol placed the stone where he did because AlphaGo had put him in a position where he saw no alternative move to make.

SEE ALSO Computer Is World Chess Champion (1997)

Go is played on a 19 × 19 board, with one player using black stones and the other using white stones, all possessing the same value.

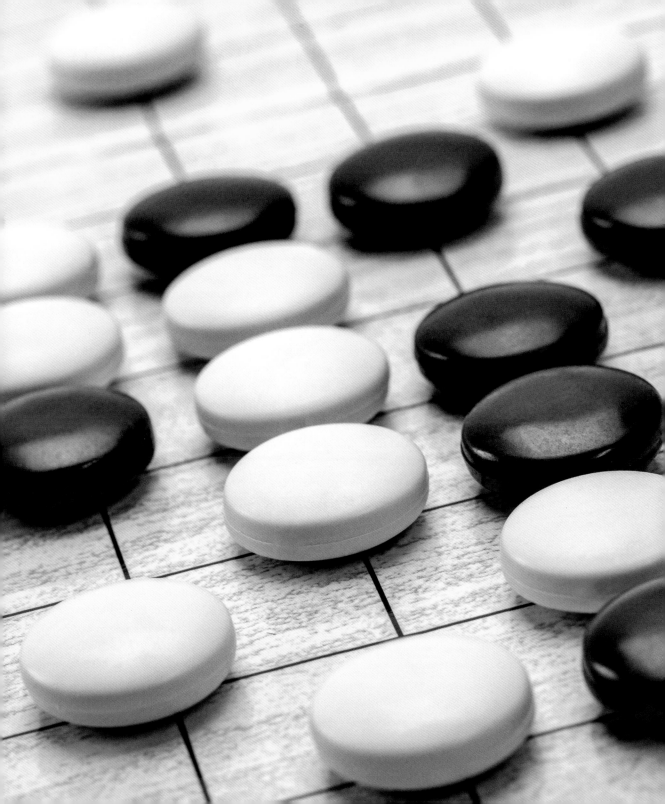

Artificial General Intelligence (AGI)

The definition and metric that determines whether computers have achieved human intelligence is controversial among the AI community. Gone is the reliance on the Turing test—programs can pass the test today, and they are clearly not intelligent.

So how can we determine the presence of true intelligence? Some measure it against the ability to perform complex intellectual tasks, such as carrying out surgery or writing a best-selling novel. These tasks require an extraordinary command of natural language and, in some cases, manual dexterity. But none of these tasks require that computers be sentient or have sapience—the capacity to experience wisdom. Put another way, would human intelligence be met only if a computer could perform a task such as carrying out a conversation with a distraught individual and communicating warmth, empathy, and loving behavior—*and then in turn receive feedback from the individual that stimulates those feelings within the computer as well?* Is it necessary to experience emotions, rather than *simulate* the experience of emotions? There is no correct answer to this, nor is there a fixed definition of what constitutes "intelligence."

The year chosen for this entry is based upon broad consensus among experts that, by 2050, many complex human tasks that do not require cognition and self-awareness in the traditional biochemical sense will have been achieved by AI. Artificial general intelligence (AGI) comes next. AGI is the term often ascribed to the state in which computers can reason and solve problems like humans do, adapting and reflecting upon decisions and potential decisions in navigating the world—kind of like how humans rely on common sense and intuition. "Narrow AI," or "weak AI," which we have today, is understood as computers meeting or exceeding human performance in speed, scale, and optimization in specific tasks, such as high-volume investing, traffic coordination, diagnosing disease, and playing chess, but without the cognition and emotional intelligence.

The year 2050 is based upon the expected realization of certain advances in hardware and software capacity necessary to perform computationally intense tasks as the measure of AGI. Limitations in progress thus far are also a result of limited knowledge about how the human brain functions, where thought comes from, and the role that the physical body and chemical feedback loops play in the output of what the human brain can do.

SEE ALSO The "Mechanical Turk" (1770), The Turing Test (1951)

Artificial general intelligence refers to the ability of computers to reason and solve problems like humans do, in a way that's similar to how humans rely on common sense and intuition.

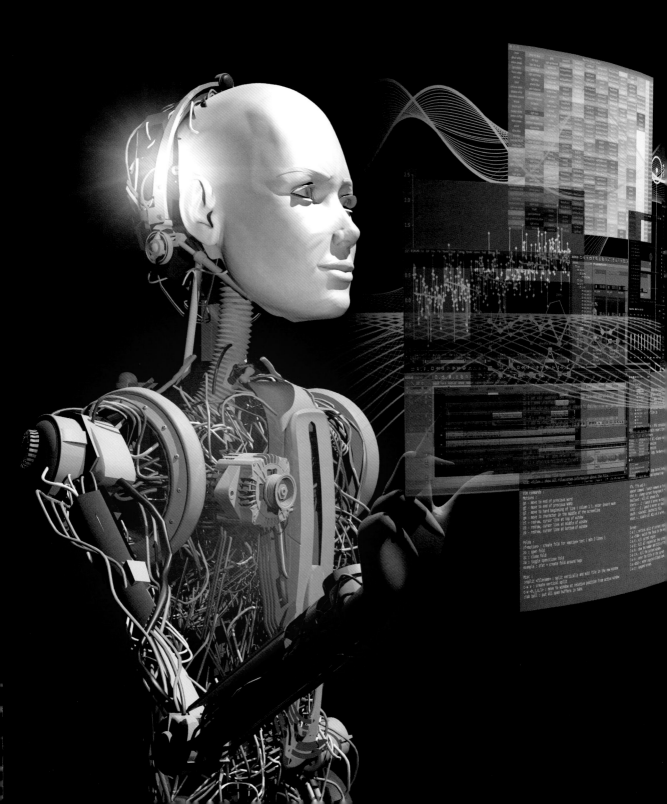

The Limits of Computation?

Seth Lloyd (b. 1960)

Each generation of technology has seen faster computations, larger storage systems, and improved communications bandwidth. Nevertheless, physics may impose fundamental limits on computing systems that cannot be overcome. The most obvious limit is the speed of light: a computer in New York City will never be able to request a web page from a server in London and download the results with a latency of less than 0.01 seconds, because light takes 0.0186 seconds to travel the 5,585 kilometers each direction, consistent with Einstein's Theory of Special Relativity. On the other hand, recently some scientists have claimed that they can send information without sending light particles by using quantum entanglement, something Einstein dismissively called *spooky action at a distance*. Indeed, in 2013, scientists in China measured the speed of information propagation due to quantum entanglement and found that it was at least 10,000 times faster than the speed of light.

Computation itself may also have a fundamental limit, according to Seth Lloyd, a professor of mechanical engineering and physics at MIT. In 2000, Lloyd showed that the ultimate speed of a computer was limited by the energy that it had available for calculations. Assuming that the computations would be performed at the scale of individual atoms, a central processor of 1 kilogram occupying the volume of 1 liter has a maximum speed of 5.4258×10^{50} operations per second—roughly 10^{41}, or a billion billion billion billion times faster than today's laptops.

Such speeds may seem unfathomable today, but Lloyd notes that if computers double in speed every two years, then this is only 250 years of technological progress. Lloyd thinks that such technological progress is unlikely. On the other hand, in 1767, the fastest computers were humans.

Because AI is increasingly able to teach and train itself across all technological and scientific domains—doing so at an exponential rate while sucking in staggering amounts of data from an increasingly networked and instrumented world—perhaps it is appropriate that a question mark be the closing punctuation for the title of this entry.

SEE ALSO Sumerian Abacus (c. 2500 BCE), Slide Rule (1621), The Difference Engine (1822), ENIAC (1943), Quantum Cryptography (1983)

Based on our current understanding of theoretical physics, a computer operating at the maximum speed possible would not be physically recognizable by today's standards. It would probably appear as a sphere of highly organized mass and energy.

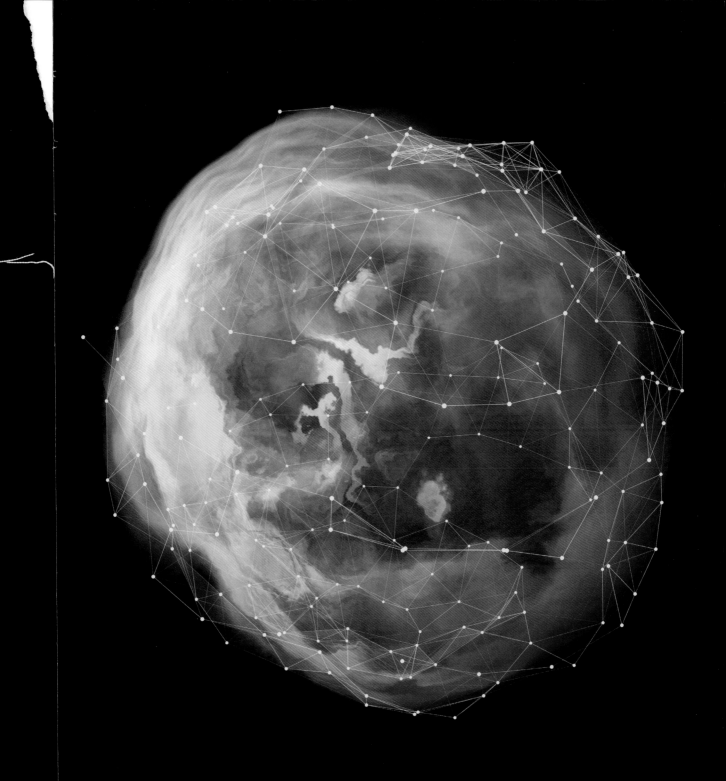

Notes and Further Reading

c. 700 BCE, Scytale

Ellison, Carl. "Cryptography Timeline." Last modified December 11, 2004. http://world.std.com/~cme/html/timeline.html.

Kelly, T. "The Myth of the Skytale." *Cryptologia* 22, no. 3 (1998): 244–60.

Mure, William. *A Critical History of the Language and Literature of Antient [sic] Greece.* 2nd ed., vol. III. London: Longman, Brown, Green, and Longmans, 1854. https://goo.gl/Jd1ByZ.

c. 60, Programmable Robot

Sharkey, N., and A. Sharkey. "Electro-Mechanical Robots Before the Computer." *Proceedings of the Institution of Mechanical Engineers, Part C: J. Mechanical Engineering Science* 223, no. 1 (2009): 235–41.

c. 850, *On Deciphering Cryptographic Messages*

Al-Kadit, Ibrahim A. "Origins of Cryptology: The Arab Contributions." *Cryptologia* 16, no. 2 (1992): 97–126.

Al-Tayeb, Tariq. "Al-Kindi, Cryptography, Code Breaking and Ciphers." Muslim Heritage (website), http://www.muslimheritage.com/article/al-kindi-cryptography-code-breaking-and-ciphers.

c. 1470, Cipher Disk

Kahn, David. *The Codebreakers: The Comprehensive History of Secret Communication from Ancient Times to the Internet.* New York: Scribner, 1996.

1613, First Recorded Use of the Word *Computer*

Gray, Jonathan. "'Let Us Calculate!': Leibniz, Llull, and the Computational Imagination." *The Public Domain Review,* 2016. https://publicdomainreview.org/2016/11/10/let-us-calculate-leibniz-llull-and-computational-imagination.

Kulstad, Mark, and Laurence Carlin. "Leibniz's Philosophy of Mind." *The Stanford Encyclopedia of Philosophy.* Last modified November 11, 2013. https://plato.stanford.edu/archives/win2013/entries/leibniz-mind.

1703, Binary Arithmetic

Leibnitz, Gottfried Wilhelm. "Explication de l'arithmétique binaire, qui se sert des seuls caractères 0 & 1; avec des remarques sur son utilité, & sur ce qu'elle donne le sens des anciennes figues Chinoises de Fohy." *Des Sciences* (1703): 85–89.

Norman, Jeremy. "Leibniz on Binary Arithmetic." Jeremy Norman's HistoryofInformation.com. Last modified July 26, 2014. http://www.historyofinformation.com/expanded.php?id=454.

1758, Human Computers Predict Halley's Comet

Grier, David Alan. "Human Computers: The First Pioneers of the information Age." *Endeavour* 25, no. 1 (2001): 28–32.

– – –. *When Computers Were Human.* Princeton, NJ: Princeton University Press, 2005.

1770, The "Mechanical Turk"

Charney, Noah. "Are Robots Moving Sculptures? On Art, Illusion and Artificial Intelligence." *Salon* online, August 13, 2017. http://www.salon.com/2017/08/13/only-human-after-all.

England, Jason. "The Turk Who Beat Napoleon." *Cosmos* online, October 6, 2014. https://cosmosmagazine.com/technology/turk-who-beat-napoleon.

1792, Optical Telegraph

Hearfield, John. "The Chappe Semaphore Telegraph." John & Marion Hearfield: Research into Questions We Find Interesting (website). http://www.johnhearfield.com/Radar/Chappe.htm.

Holzmann, Gerald, and Björn Pehrson. *The Early History of Data Networks.* Hoboken, NJ: Wiley-IEEE Computer Society, 1994.

Schofield, Hugh. "How Napoleon's Semaphore Telegraph Changed the World." *BBC News Magazine,* June 17, 2013. http://www.bbc.com/news/magazine-22909590.

Sterling, Christopher H. *Military Communications: From Ancient Times to the 21st Century.* Santa Barbara, CA: ABC-CLIO, 2007.

1822, The Difference Engine

Park, Edwards. "What a Difference the Difference Engine Made: From Charles Babbage's Calculator Emerged Today's Computer." *Smithsonian Magazine* online, February 1996. https://www.smithsonianmag.com/history/what-a-difference-the-difference-engine-made-from-charles-babbages-calculator-emerged-todays-computer-109389254.

1843, Ada Lovelace Writes a Computer Program

Menabrea, L. F. "Sketch of the Analytical Engine Invented by Charles Babbage." Translation and notes by Ada Augusta. *Bibliothèque Universelle de Genève,* no. 82 (1842). http://www.fourmilab.ch/babbage/sketch.html.

Sherman, Nat, dir. *Calculating Ada: The Countess of Computing.* BBC Four. September 17, 2015.

1843, Fax Machine Patented

Rensen, Marius. "The Bartlane System (Coded System)." HF-Fax Image Communication (website). http://www.hffax.de/history/html/bartlane.html.

1851, Thomas Arithmometer

Johnston, Stephen. "Making the Arithmometer Count." *Bulletin of the Scientific Instrument Society* 52 (1997): 12–21. http://www.mhs.ox.ac.uk/staff/saj/arithmometer.

1854, Boolean Algebra

University College Cork. "George Boole 200: Celebrating George Boole's Bicentenary." https://georgeboole200.ucc.ie.

1890, Tabulating the US Census

da Cruz, Frank. "Hollerith 1890 Census Tabulator." Columbia University History of Computing (website), last modified March 28, 2011. http://www.columbia.edu/cu/computinghistory/census-tabulator.html.

History of Computers. "The Tabulating Machine of Herman Hollerith." https://history-computer.com/ModernComputer/Basis/TabulatingMachine_Hollerith.html.

United States Census Bureau. "1890 Overview." https://www.census.gov/history/www/through_the_decades/overview/1890.html.

United States Census Bureau. "The Hollerith Machine." https://www.census.gov/history /www/innovations/technology/the_hollerith _tabulator.html.

1917, Vernam Cipher
Bellovin, Steve. "Frank Miller: Inventor of the One-Time Pad." *Cryptologia* 35, no. 3 (July 12, 2011): 203–22.

1927, *Metropolis*
Carnegie Mellon Robot Hall of Fame. http:// www.robothalloffame.org.

Dayal, Geeta. "Recovered 1927 *Metropolis* Film Program Goes Behind the Scenes of a Sci-Fi Masterpiece." *Wired.* July 12, 2012.

1931, Differential Analyzer
Boast, Robin. *The Machine in the Ghost: Digitality and Its Consequences.* London: Reaktion Books, 2017.

1941, Z3 Computer
Rojas, R., F. Darius, C. Goktekin, and G. Heyne. "The Reconstruction of Konrad Zuse's Z3." *IEEE Annals of the History of Computing* 23, no. 3 (July–Sept. 2005): 23–32.

Trautman, Peggy Salz. "A Computer Pioneer Rediscovered, 50 Years On." *New York Times*, April 20, 1994.

Zuse, Konrad. *Der Computer: Mein Lebenswerk*, 3rd ed. Berlin: Springer-Verlag, 1993.

1942, Atanasoff-Berry Computer
Computer History Museum. "The Atanasoff-Berry Computer in Operation." Ames Lab Video Production. Recorded 1999. https:// www.youtube.com/watch?v=YyxGIbtMS9E.

Garfinkel, Simson. "Jane Smiley's Nonfiction Tale of the Invention of the Computer." *Washington Post*, January 6, 2011.

North Carolina State University "The Atanasoff-Berry Computer." http://www4.ncsu.edu /~belail/The_Introduction_of_Electronic _Computing/Atanasoff-Berry_Computer.html.

Smiley, Jane. *The Man Who Invented the Computer: The Biography of John Atanasoff, Digital Pioneer.* New York: Doubleday, 2010.

1943, Colossus
Erskine, Ralph, and Michael Smith (eds.). *The Bletchley Park Codebreakers.* London: Biteback Publishing, Ltd. 2011.

Flowers, Thomas H. "The Design of Colossus." *Annals of the History of Computing* 5, no. 3 (July 1983): 239–52.

1944, Delay Line Memory
University of Cambridge, Computer Laboratory. "EDSAC 99: 15–16 April 1999." Last modified September 30, 1999. http://www.cl.cam.ac.uk /events/EDSAC99.

Manchester, Phil. "How Alan Turing Wanted to Base EDSAC's Memory on BOOZE." *The Register*, June 28, 2013. https://www.theregister. co.uk/2013/06/28/wilkes_centenary_mercury _memory.

1944, Binary-Coded Decimal
Torrey, Volta. "Robot Mathematician Knows All the Answers." *Popular Science*, 145 (October 1944): 85–89, 222, 226, 230. https://goo.gl /P6ovUK.

1945, "As We May Think"
Bush, Vannevar. "As We May Think." *Atlantic Monthly*, July 1945. https://www.theatlantic. com/magazine/archive/1945/07/as-we-may -think/303881.

1945, EDVAC *First Draft* Report
Institute for Advanced Study. "Electronic Computer Project." https://www.ias.edu /electronic-computer-project.

Mauchly, Bill, Jeremy Bernstein, Mark Dowson, and David K. Adams. "Who Gets Credit for the Computer?: An Exchange." *New York Review of Books*, September 27, 2012. http://www .nybooks.com/articles/2012/09/27/who -gets-credit-computer-exchange.

1946, Williams Tube
I Programmer. "Jay Forrester and Whirlwind." November 18, 2016. http://www.i-programmer .info/history/people/439-jay-forrester.html.

IBM. "The IBM 700 Series: Computing Comes to Business." http://www-03.ibm.com/ibm /history/ibm100/us/en/icons/ibm700series /impacts.

1947, Actual Bug Found
The National Museum of American History. "Log Book with Computer Bug." http:// americanhistory.si.edu/collections/search /object/nmah_334663.

1947, Silicon Transistor
APS News. "This Month in Physics History, November 17–December 23, 1947: Invention of the First Transistor." Vol. 9, no. 10 (November 2000). https://www.aps.org /publications/apsnews/200011/history.cfm.

1948, The Bit
Gleick, James. "The Lives They Lived: Claude Shannon." *New York Times Magazine*, December 30, 2001. http://www.nytimes .com/2001/12/30/magazine/the-lives-they-lived -claude-shannon-b-1916-bit-player.html.

O'Neil, William. "An Application of Shannon's Coding Theorem to Information Transmission in Economic Markets." *Information Sciences* 41, no. 2 (March 1987): 171–85.

Overbye, Dennis. "Hawking's Breakthrough Is Still an Enigma." *New York Times*, January 22, 2002.

Shannon, C. E. "A Mathematical Theory of Communication." *The Bell System Technical Journal* 28, no. 3 (July 1948): 379–423.

1948, Curta Calculator
Herzstark, Clark. "Oral History Interview with Clark Herzstark." By Erwin Tomash, Charles Babbage Institute, 1987. https://conservancy .umn.edu/handle/11299/107358.

Szondy, David. "Curta Calculator: The Mechanical Marvel Born in a Nazi Death Camp." *New Atlas*. October 11, 2016. https:// newatlas.com/curta-death-camp -calculator/45506.

Thadeusz, Frank. "The Sad Story of an Inventor at Buchenwald." *Spiegel* online, July 3, 2013. http://www.spiegel.de/international/germany /concentration-camp-inmate-invented-world-s -first-pocket-calculator-a-909062.html.

1948, Manchester SSEM
Napper, Brian. "Computer 50: University of Manchester Celebrates the Birth of the Modern Computer." The University of Manchester. Last modified July 26, 2010. http://curation.cs.manchester.ac.uk /computer50.

Williams, F. C., and T. Kilburn. "Electronic Digital Computers." *Nature* 162, no. 487 (1948).

1950, Error-Correcting Codes
Robertson, Edmond F. "Richard W. Hamming." A.M. Turing Award Citation. https://amturing. acm.org/award_winners/hamming_1000652. cfm.

1951, The Turing Test
Turing, Alan M. "Computing Machinery and Intelligence." *Mind* 49 (1950): 433–60. https:// home.manhattan.edu/~tina.tian/CMPT420 /Turing.pdf.

1951, Magnetic Tape Used for Computers

Ciletti, Eddie. "If I Knew You Were Coming I'd Have Baked a Tape! A Recipe for Tape Restoration." Last modified 2011. http://www.tangible-technology.com/tape/baking1.html.

da Cruz, Frank. "The IBM 1401." Columbia University Computing History (website), last modified September 7, 2015. http://www.columbia.edu/cu/computinghistory/1401.html.

Maxfield, Clive. "How It Was: Paper Tapes and Punched Cards." EE|Times, October 13, 2011. https://www.eetimes.com/author.asp?section_id=14&doc_id=1285484.

Schoenherr, Steven. "The History of Magnetic Recording." Paper presented at the IEEE Magnetics Society Seminar, San Diego, CA, November 5, 2002. http://www.aes-media.org/historical/html/recording.technology.history/magnetic4.html.

1951, Core Memory

"Early Digital Computing." MIT Lincoln Laboratory. Adapted from E. C. Freeman, ed., Technology in the National Interest, Lexington, MA: MIT Lincoln Laboratory, 1995. https://www.ll.mit.edu/about/History/earlydigitalcomputing.html.

1952, Computer Speech Recognition

Pieraccini, Roberto. The Voice in the Machine: Building Computers that Understand Speech. Cambridge. MA: MIT Press, 2012.

1956, First Disk Storage Unit

Cohen, Peter. "A History of Hard Drives." Backblaze (blog), November 17, 2016. https://www.backblaze.com/blog/history-hard-drives.

IBM. "IBM RAMAC." Presented by the Data Processing Division. IBM Promotional Video, 1956. https://archive.org/details/Ibm305ramac.

IBM. "RAMAC: The First Magnetic Hard Disk." http://www-03.ibm.com/ibm/history/ibm100/us/en/icons/ramac.

1956, The Byte

Remington Rand. Manual of Operations: The Central Computer of the UNIVAC System. Prepared by the Training Section, Electronic Computer Department, 1954. http://bitsavers.trailing-edge.com/pdf/univac/univac1/UNIVAC1_Operating_Manual_1954.pdf.

1956, Robby the Robot

Booker, M. Keith. Alternate Americas: Science Fiction Film and American Culture. Westport, CT: Praeger, 2006.

1957, FORTRAN

IBM. Preliminary Report: Specifications for the IBM Mathematical FORmula TRANslating System, FORTRAN. November 10, 1954. http://archive.computerhistory.org/resources/text/Fortran/102679231.05.01.acc.pdf.

1957, First Digital Image

Kirsch, Joan L., and Russell A. Kirsch. "Storing Art Images in Intelligent Computers." Leonardo 23, no. 1 (1990): 99–106.

Roberts, Steven. "Distant Writing: A History of the Telegraph Companies in Britain between 1838 and 1868." Distance Writing (website), 2006–2012. http://distantwriting.co.uk/Documents/Distant%20Writing%202012.pdf.

Schooley, Jim. "NBS Builds a Computer." SAA History Committee, National Institute of Standards and Technology, October 24, 2010. https://www.nist.gov/director/nist-culture-excellence-article-4.

1958, SAGE Computer Operational

Anthony, Sebastian. "Inside IBM's $67 Billion SAGE, the Largest Computer Ever Built." Extremetech, March 28, 2013. https://www.extremetech.com/computing/151980-inside-ibms-67-billion-sage-the-largest-computer-ever-built.

Carter, James. "IBM AN/FSQ-7." Starring the Computer. http://starringthecomputer.com/computer.html?c=73.

IBM. "SAGE: The First National Air Defense Network." http://www-03.ibm.com/ibm/history/ibm100/us/en/icons/sage.

MIT Lincoln Library. "The SAGE Air Defense System." https://www.ll.mit.edu/about/History/SAGEairdefensesystem.html.

1959, IBM 1401

IBM. "1401 Data Processing System." https://www-03.ibm.com/ibm/history/exhibits/mainframe/mainframe_PP1401.html.

1959, PDP-1

Digital Equipment Corporation. Nineteen Fifty-Seven to the Present. Maynard, MA: ComputerHistory.org, 1978. http://s3data.computerhistory.org/pdp-1/dec.digital_1957_to_the_present_(1978).1957-1978.102630349.pdf.

Maloney, Dan. "The PDP-1: The Machine That Started the Hacker Culture." Hackaday (website), June 27, 2017. https://hackaday.com/2017/06/27/the-pdp-1-the-machine-that-started-hacker-culture.

Olsen, Ken. "National Museum of American History, Smithsonian Institute, Oral History Interview." By David Allison, National Museum of American History, September 28, 29, 1988. http://americanhistory.si.edu/comphist/olsen.html.

Schein, Edgar H. DEC Is Dead, Long Live DEC: The Lasting Legacy of Digital Equipment Corporation. San Francisco: Berrett-Koehler Publishers, 2004.

1959, Quicksort

Hoare, C. A. R. "An Interview with C. A. R. Hoare." By Len Shustek, Communications of the ACM 52, no. 3 (March 2009): 38–41.

1959, Airline Reservation System

Smith, R. Blair. "Oral History Interview with R. Blair Smith." By Robina Mapstone, Charles Babbage Institute, 1980. https://conservancy.umn.edu/handle/11299/107637.

1960, COBOL Computer Language

Gürer, Denise. "Pioneering Women in Computer Science." ACM SIGCSE Bulletin 34, no. 2 (June 2002): 175–80.

Sammet, Jean E. "The Early History of COBOL." ACM SIGPLAN Notices—Special Issue: History of Programming Languages Conference 13, no. 8 (August 1978): 121–61.

1961, Time-Sharing

Corbató, Fernando J., Marjorie Merwin Daggett, and Robert C. Daley. "An Experimental Time-Sharing System." Computation Center, MIT. Cambridge, MA, 1962. http://larch-www.lcs.mit.edu:8001/~corbato/sjcc62.

1963, Sketchpad

Sutherland, Ivan Edward. Sketchpad: A Man-Machine Graphical Communication System. Cambridge, MA: MIT, 1963.

1964, RAND Tablet

Davis, Malcolm, and T. O. Ellis. "The RAND Tablet: A Man-Machine Graphical Communication Device." Santa Monica, CA: RAND Corporation, 1964. https://www.rand.org/pubs/research_memoranda/RM4122.html.

1964, IBM System/360

Clark, Gavin. "Why Won't You DIE? IBM's S/360 and Its Legacy at 50." The Register, April 7, 2014. https://www.theregister.co.uk/2014/04/07/ibm_s_360_50_anniversary.

IBM. "System/360 Announcement." Data Processing Division. April 7, 1964. http://www-03.ibm.com/ibm/history/exhibits/mainframe/mainframe_PR360.html.

IBM. "System 360: From Computers to Computer Systems." Icons of Progress, http://www-03.ibm.com/ibm/history/ibm100/us/en/icons/system360/.

Sparkes, Matthew. "IBM's $5Bn Gamble: Revolutionary Computer Turns 50." *Telegraph* online, April 7, 2014. http://www.telegraph.co.uk/technology/news/10719418/IBMs-5bn-gamble-revolutionary-computer-turns-50.html.

1964, BASIC Computer Language

Dartmouth College Computation Center. *BASIC.* October 1, 1964. https://archive.org/details/bitsavers_dartmouthB_2146200.

"Fifty Years of BASIC, the Programming Language That Made Computers Personal." *TIME*, April 29, 2014.

1965, First Liquid-Crystal Display

Gross, Benjamin. "How RCA Lost the LCD." *IEEE Spectrum*, November 1, 2012. https://spectrum.ieee.org/tech-history/heroic-failures/how-rca-lost-the-lcd.

Yardley, William. "George H. Heilmeier, an Inventor of LCDs, Dies at 77." *New York Times* online, May 6, 2014. https://www.nytimes.com/2014/05/06/technology/george-h-heilmeier-an-inventor-of-lcds-dies-at-77.html.

1965, DENDRAL

Feigenbaum, Edward A., and Bruce G. Buchanan. "DENDRAL and Meta-DENDRAL: Roots of Knowledge Systems and Expert Systems Applications." *Artificial Intelligence* 59 (1993): 233–40.

Lindsay, Robert K., Bruce G. Buchanan, Edward A. Feigenbaum, and Joshua Lederberg. "DENDRAL: A Case Study of the First Expert System for Scientific Hypothesis Formation." *Artificial Intelligence* 61 (1993): 209–61.

1966, Dynamic RAM

IBM. "DRAM: The Invention of On-Demand Data." Icons of Progress, http://www-03.ibm.com/ibm/history/ibm100/us/en/icons/dram.

1967, First Cash Machine

Harper, Tom R., and Bernardo Batiz-Lazo. *Cash Box: The Invention and Globalization of the ATM.* Louisville, KY: Networld Media Group, 2013.

McRobbie, Linda Rodriguez. "The ATM Is Dead. Long Live the ATM!" *Smithsonian Magazine* online, January 8, 2015. https://www.smithsonianmag.com/history/atm-dead-long-live-atm-180953838.

1967, Head-Mounted Display

Sutherland, Ivan E. "A Head-Mounted Three Dimensional Display." *Proceedings of the AFIPS Fall Joint Computer Conference* (December 9–11, 1968): 757–64. https://doi.org/10.1145/1476589.1476686.

1968, Software Engineering

Bauer, F. L. L. Bolliet, and H. J. Helms. *Software Engineering.* "Report on a Conference Sponsored by the NATO Science Committee, Garmisch, Germany, 7th to 11th October 1968." January 1969. http://homepages.cs.nc.ac.uk/brian.randell/NATO/nato1968.pdf.

1968, First Spacecraft Guided by Computer

Coldewey, Devin. "Grace Hopper and Margaret Hamilton Awarded Presidential Medal of Freedom for Computing Advances." *TechCrunch*, November 17, 2016. https://techcrunch.com/2016/11/17/grace-hopper-and-margaret-hamilton-awarded-presidential-medal-of-freedom-for-computing-advances.

McMillan, Robert. "Her Code Got Humans on the Moon—and Invented Software Itself." *Wired* online, October 13, 2015. https://www.wired.com/2015/10/margaret-hamilton-nasa-apollo.

"Original Apollo 11 Guidance Computer (AGC) Source Code for the Command and Lunar Modules." Assembled by yaYUL, https://github.com/chrislgarry/Apollo-11.

1968, *Cyberspace* Coined—and Re-Coined

Gibson, William. "Burning Chrome." *Omni*, July 1982.

———. *Neuromancer.* New York: Ace Books, 1984.

Weiner, Norbert. *Cybernetics: Or, Control and Communication in the Animal and the Machine.* 2nd ed. Cambridge, MA: MIT Press, 1961.

Lillemose, Jacob, and Mathias Kryger. "The (Re)invention of Cyberspace." *Kunstkritikk*, August 24, 2015. http://www.kunstkritikk.com/kommentar/the-reinvention-of-cyberspace.

1969, ARPANET/Internet

Hafner, Katie. *Where Wizards Stay Up Late: The Origins of the Internet.* New York: Simon & Schuster, 1996.

Leiner, Barry M., Vinton G. Cerf, David D. Clark, Robert E. Kahn, Leonard Kleinrock, Daniel C. Lynch, Jon Postel, Larry G. Roberts, and Stephen Wolff. "Brief History of the Internet." The Internet Society, 1997. http://www.internetsociety.org/internet/what-internet/history-internet/brief-history-internet.

Pelkey, James. *Entrepreneurial Capitalism and Innovation: A History of Computer Communications: 1968–1988* (self-pub e-book). 2007. http://www.historyofcomputercommunications.info/Book/BookIndex.html.

Robat, Cornelis, eds. "The History of the Internet, 1957–1976." The History of Computing Foundation (website). https://www.thocp.net/reference/internet/internet1.htm.

1969, Digital Imaging

Canon. "CCD and CMOS Sensors." Infobank. http://cpn.canon-europe.com/content/education/infobank/capturing_the_image/ccd_and_cmos_sensors.do.

Sorrell, Charlie. "Inside the Nobel Prize: How a CCD Works." *Wired*, October 7, 2009. https://www.wired.com/2009/10/ccd-inventors-awarded-nobel-prize-40-years-on.

1969, Utility Computing

F. J. Corbató, J. H. Saltzer, and C. T. Clingen. 1971. "Multics: The First Seven Years." *Proceedings of the AFIPS Spring Joint Computer Conference* (May 16–18, 1972): 571–83. http://dx.do.org/10.1145/1478873.1478950.

Multics (Multiplexed Information and Computing Services). http://multicians.org.

1969, UNIX

Gabriel, Richard P. "Worse Is Better." *Dreamsongs* (blog), 2000. https://www.dreamsongs.com/WorseIsBetter.html.

1970, Fair Credit Reporting Act

Garfinkel, Simson. *Database Nation: The Death of Privacy in the 21st Century.* Sebastopol, CA: O'Reilly, 2000.

1971, Laser Printer

Dalkakov, Georgi. "Laser Printer of Gary Starkweather." History of Computers (website), last modified April 9, 2018. http://history-computer.com/ModernComputer/Basis/laser_printer.html.

The Edward A. Feigenbaum Papers. Stanford University Libraries. "Info and Sample Sheets for Xerox Graphics Printer (XPG); Notes on How to Use with PUB." https://exhibits .stanford.edu/feigenbaum/catalog/tf421vy6196.

Gladwell, Malcolm. "Creation Myth: Xerox PARC, Apple, and the Truth about Innovation." *New Yorker*, May 16, 2011.

Philip Greenspun's Homepage. "The Dover." NE43 Memory Project, 1998. http://philip .greenspun.com/bboard/q-and-a-fetch -msg?msg_id=0006XJ&topic_id=27&topic=N E43+Memory+Project.

1971, NP-Completeness

Dennis, Shasha, and Cathy Lazere. *Out of Their Minds: The Lives and Discoveries of 15 Great Computer Scientists*. New York: Springer, 1995.

1972, Cray Research

Breckenridge, Charles W. "A Tribute to Seymour Cray." SRC Computers, Inc. Presented at the Supercomputing conference, November 19, 1996. https://www.cgl.ucsf.edu /home/tef/cray/tribute.html.

Russell, Richard M. "The CRAY-1 Computer System." *Communications of the ACM* 21, no. 1 (January 1978): 63–72.

Semiconductors Central. "Seymour Cray: The Father of Super Computers." Four Peaks Technologies, http://www. semiconductorscentral.com/seymour_cray _page.html.

1972, HP-35 Calculator

Hewlett-Packard. "HP-35 Handheld Scientific Calculator, 1972." Virtual Museum, http:// www.hp.com/hpinfo/abouthp/histnfacts /museum/personalsystems/0023/.

1973, First Cell Phone Call

Alfred, Randy. "April 3, 1973: Motorola Calls AT&T . . . By Cell." *Wired* online, April 3, 2008. https://www.wired.com/2008/04 /dayintech-0403/.

Bloomberg. "The First Cell Phone Call Was an Epic Troll." Digital Originals (video), April 24, 2015. https://www.bloomberg.com/news /videos/2015-04-24/the-first-cell-phone-call-was -an-epic-troll.

Greene, Bob. "38 Years Ago He Made the First Cell Phone Call." CNN online, April 3, 2011. http://www.cnn.com/2011/OPINION/04/01 /greene.first.cellphone.call/index.html.

1973, Xerox Alto

Perry, Tekla S. "The Xerox Alto Struts Its Stuff on Its 40th Birthday." *IEEE Spectrum* online, November 15, 2017. https://spectrum.ieee.org /view-from-the-valley/tech-history/silicon -revolution/the-xerox-alto-struts-its-stuff-on-its -40th-birthday.

Shirriff, Ken. "Y Combinator's Xerox Alto: Restoring the Legendary 1970s GUI Computer." *Ken Shirriff's Blog*, June 2016. http://www.righto.com/2016/06/y-combinators -xerox-alto-restoring.html.

Smith, Douglas K., and Robert C. Alexander. *Fumbling the Future: How Xerox Invented, Then Ignored, the First Personal Computer*. New York: William Morrow, 1988.

1975, *The Shockwave Rider*

Brunner, John. *The Shockwave Rider*. New York: Harper & Row, 1975.

Toffler, Alvin. *Future Shock*. New York: Bantam Books, 1970.

1975, AI Medical Diagnosis

Buchanan, Bruce G. *Rule Based Expert Systems: The MYCIN Experiments of the Stanford Heuristic Programming Project*. Boston: Addison-Wesley, 1984

1976, Public Key Cryptography

Corrigan-Gibbs, Henry. "Keeping Secrets." *Stanford Magazine*, November/December 2014.

Diffie, Whitfield. "The First Ten Years of Public-Key Cryptography." *Proceedings of the IEEE* 76, no. 5 (May 1988): 560–77.

Merkle, Ralph. "Secure Communications over Insecure Channels (1974), by Ralph Merkle, With an Interview from the Year 1995." Arnd Weber, ed. Institut für Technikfolgenabschätzung und Systemanalyse (ITAS), January 16, 2002. http://www.itas.kit .edu/pub/m/2002/mewe02a.htm.

Savage, Neil. "The Key to Privacy." *Communications of the ACM* 59, no. 6 (June 2016): 12–14.

1977, RSA Encryption

Garfinkel, Simson. *PGP: Pretty Good Privacy*. Sebastopol, CA: O'Reilly, 1994.

1979, Secret Sharing

Shamir, Adi. "How to Share a Secret." *Communications of the ACM* 22, no. 11 (November 1979): 612–13.

1980, Flash Memory

Fulford, Benjamin. "Unsung Hero." *Forbes* online, June 24, 2002. https://www.forbes.com /global/2002/0624/030.html#314fe2023da3.

Yinug, Falan. "The Rise of the Flash Memory Market: Its Impact on Firm Behavior and Global Semiconductor Trade Patterns." *Journal of International Commerce and Economics* (July 2007). https://www.usitc.gov/publications/332 /journals/rise_flash_memory_market.pdf.

1981, IBM PC

IBM. "The Birth of the IBM PC." IBM Archives, https://www-03.ibm.com/ibm/history/exhibits /pc25/pc25_birth.html.

Pollac, Andrew. "Next, a Computer on Every Desk." *New York Times*, August 23, 1981. https://www.nytimes.com/1981/08/23 /business/next-a-computer-on-every-desk.html.

Yardley, William. "William C. Lowe, Who Oversaw the Birth of IBM's PC, Dies at 72." *New York Times*, October 28, 2013. https:// www.nytimes.com/2013/10/29/business /william-c-lowe-who-oversaw-birth-of-the-ibm -pc-dies-at-72.html.

1981, Japan's Fifth Generation Computer Systems

Feigenbaum, Edward A., and Pamela McCorduck. *The Fifth Generation: Japan's Computer Challenge to the World*. Reading, MA: Addison-Wesley, 1983.

Hewitt, Carl. "The Repeated Demise of Logic Programming and Why It Will Be Reincarnated." In *Technical Report SS-06-08*. AAAI Press. March 2006.

Nielsen, Jakob. "Fifth Generation 1988 Trip Report." Nielsen Norman Group, December 10, 1988. https://www.nngroup.com/articles /trip-report-fifth-generation.

Pollack, Andrew. "'Fifth Generation' Became Japan's Lost Generation." *New York Times*, June 5, 1992. https://www.nytimes .com/1992/06/05/business/fifth-generation -became-japan-s-lost-generation.html.

1982, PostScript

Reid, Brian. "PostScript and Interpress: A Comparison." USENET Archives, March 1, 1985.

1982, First CGI Sequence in Feature Film

Lucasfilm Ltd. *Making of the Genesis Sequence from Star Trek II*. Graphics Project Computer Division, 1982. https://www.youtube.com /watch?v=Qe9qSLYK5q4.

Meyer, Nicholas, dir. *Star Trek II: The Wrath of Khan*. 1982. Hollywood, CA: Paramount Pictures.

1982, *National Geographic* Moves the Pyramids
The National Press Photographers Association, "Code of Ethics." https://nppa.org/code-ethics.
New York Times. "New Picture Technologies Push Seeing Still Further from Believing." July 2, 1989. https://www.nytimes.com /1989/07/03/arts/new-picture-technologies -push-seeing-still-further-from-believing.html.

1982, Secure Multi-Party Computation
Maurer, Ueli. "Secure Multi-Party Computation Made Simple." *Discrete Applied Mathematics* 154, no. 2 (February 1, 2006): 370–81.

1982, Home Computer Named Machine of the Year
McCracken, Harry. "*TIME*'s Machine of the Year, 30 Years Later." *TIME* online, January 4, 2013. http://techland.time.com/2013/01/04 /times-machine-of-the-year-30-years-later.
"The Computer, Machine of the Year." *TIME* cover, January 3, 1983. http://content.time .com/time/covers/0,16641,19830103,00.html.

1983, The Qubit
MIT Technology Review. "Einstein's 'Spooky Action at a Distance' Paradox Older Than Thought." Emerging Technology from the arXiv, *Physics arXiv Blog*, March 8, 2012. https://www.technologyreview.com/s/427174 /einsteins-spooky-action-at-a-distance-paradox -older-than-thought/.
Preskill, John. "Who Named the Qubit." *Quantum Frontiers* (blog), Institute for Quantum Information and Matter @ Caltech, June 9, 2015. https://quantumfrontiers .com/2015/06/09/who-named-the-qubit.
Schumacher, Benjamin. "Quantum Coding." *Physical Review A* 51, no. 4 (April 1995).

1983, 3-D Printing
Hull, Charles W. Apparatus for production of three-dimensional objects by stereolithography. US Patent 4,575,330, filed August 8, 1984, issued March 11, 1986.
NASA. "3D Printing in Zero-G Technology Demonstration." December 6, 2017. https:// www.nasa.gov/mission_pages/station/research /experiments/1115.html.

Prisco, Jacopo. "'Foodini' Machine Lets You Print Edible Burgers, Pizza, Chocolate." CNN online, December 31, 2014. https://www.cnn. com/2014/11/06/tech/innovation/foodini -machine-print-food/index.html.
US Department of Energy. "Transforming Wind Turbine Blade Mold Manufacturing with 3D Printing." 2016. https://www.energy.gov/eere /wind/videos/transforming-wind-turbine-blade -mold-manufacturing-3d-printing.

1983, MIDI Computer Music Interface
MIDI Association. https://www.midi.org.

1984, Macintosh
Stanford University. *Making the Macintosh: Technology and Culture in Silicon Valley.* https://web.stanford.edu/dept/SUL/sites/mac.

1984, VPL Research, Inc.
Lanier, Jaron. *Dawn of the New Everything: Encounters with Reality and Virtual Reality.* New York: Henry Holt, 2017.

1984, Quantum Cryptography
Mann, Adam. "The Laws of Physics Say Quantum Cryptography is Unhackable. It's Not." *Wired*, June 7, 2013.
Powell, Devin. "What is Quantum Cryptography?" *Popular Science* online, March 3, 2016. https://www.popsci.com/what -is-quantum-cryptography.

1985, Connection Machine
Hillis, W. Danny. "The Connection Machine." Dissertation, Cambridge University, 1985.
Kahle, Brewster, and W. Daniel Hillis. "The Connection Machine Model CM-1 Architecture." *IEEE Transactions on Systems, Man, and Cybernetics* 19, no. 4 (July/August 1989): 707–13.

1985, Zero-Knowledge Proofs
Goldwasser, S. , S. Micali, and C. Rackoff. "The Knowledge Complexity of Interactive Proof Systems." *SIAM Journal on Computing* 18, no. 1 (1989): 186–208.
Green, Matthew. "Zero Knowledge Proofs: An Illustrated Primer." *A Few Thoughts on Cryptographic Engineering* (blog), November 27, 2014. https://blog.cryptographyengineering .com/2014/11/27/zero-knowledge-proofs -illustrated-primer/.
MIT. "Interactive Zero Knowledge 3-Colorability Demonstration." http://web.mit .edu/~ezyang/Public/graph/svg.html.

1985, NSFNET
Photo: "T1 Backbone and Regional Networks Traffic, 1991," representing data collected by Merit Network, Inc. Visualization by Donna Cox and Robert Patterson, National Center for Supercomputing Applications, University of Illinois at Urbana-Champaign.

1985, Field-Programmable Gate Array
Lanzagorta, Marco, Stephen Bique, and Robert Rosenberg. "Introduction to Reconfigurable Supercomputing." *Synthesis Lectures on Computer Architecture* 4, no. 1 (2009): 1–103.

1985, GNU Manifesto
Williams, Sam. *Free as in Freedom: Richard Stallman's Crusade for Free Software.* Beijing: O'Reilly, 2002.

1986, Software Bug Fatalities
Leveson, Nancy G., and Clark S. Turner. "An Investigation of the Therac-25 Accidents." *IEEE Computer Society* 26, no. 7 (July 1993): 18–41.

1986, Pixar
Museum of Science. *The Science Behind Pixar.* http://sciencebehindpixar.org.

1987, GIF
Battilana, Mike. "The GIF Controversy: A Software Developer's Perspective." *mike.pub* (blog), last modified June 20, 2004. https:// mike.pub/19950127-gif-lzw.

1988, CD-ROM
Coldewey, Devin. "30 Years Ago, the CD Started the Digital Music Revolution." *NBC News* online, last modified September 28, 2012. https://www.nbcnews.com/tech/gadgets/30 -years-ago-cd-started-digital-music-revolution -fna6167906.
Dalakov, Georgi. "Compact Disk of James Russel." (online) August 23, 2017. http:// history-computer.com/ModernComputer /Basis/compact_disc.html.

1990, Digital Money
Chaum, David. "Blind Signatures for Untraceable Payments." In *Advances in Cryptology*, 199–203. Boston: Springer, 1983.

1991, Pretty Good Privacy (PGP)

Garside, Juliette."Philip Zimmermann: King of Encryption Reveals His Fears for Privacy." *The Guardian* online, May 25, 2015. https://www .theguardian.com/technology/2015/may/25 /philip-zimmermann-king-encryption-reveals -fears-privacy.

McCullagh, Declan. "PGP: Happy Birthday to You." *Wired*, June 5, 2001. https://www.wired .com/2001/06/pgp-happy-birthday-to-you/.

Ranger, Steve. "Defending the Last Missing Pixels: Phil Zimmerman Speaks Out on Encryption, Privacy, and Avoiding a Surveillance State." *TechRepublic*, https://www .techrepublic.com/article/defending-the-last -missing-pixels-phil-zimmermann.

Zimmermann, Philip R. "Why I Wrote PGP." Philip Zimmermann (website), 1991. https:// www.philzimmermann.com/EN/essays /WhyIWrotePGP.html.

1991, *Computers at Risk*

National Research Council. *Computers at Risk: Safe Computing in the Information Age.* System Security Study Committee. Washington, DC: National Academic Press, 1991.

1991, Linux Kernel

Torvalds, Linus, and David Diamond. *Just for Fun: The Story of an Accidental Revolutionary.* New York: HarperBusiness, 2002.

"LINUX's History." A collection of postings to the group comp.os.minix, July 1991–May 1992. https://goo.gl/LrSjih.

1992, JPEG

Pessina, Laure-Anne. "JPEG Changed Our World." Phys.org, December 12, 2014. https:// phys.org/news/2014-12-jpeg-world.html.

1992, First Mass-Market Web Browser

Isaacson, Walter. *The Innovators: How a Group of Hackers, Geniuses, and Geeks Created the Digital Revolution.* New York: Simon & Schuster, 2014.

1992, Unicode

Unicode Inc. "Unicode History Corner." History of Unicode, last modified November 18, 2015. https://www.unicode.org/history.

1993, Apple Newton

Honan, Mat. "Remembering the Apple Newton's Prophetic Failure and Lasting Impact." *Wired* online, August 5, 2013. https:// www.wired.com/2013/08/remembering-the -apple-newtons-prophetic-failure-and-lasting -ideals.

Hormby, Tom. "The Story Behind Apple's Newton." *Gizmodo*, January 19, 2010. https:// gizmodo.com/5452193/the-story-behind -apples-newton.

1994 RSA-129 Cracked

Atkins, Derek, Michael Graff, Arjen K. Lenstra, and Paul C. Leyland. "The Magic Words Are Squeamish Ossifrage." *Advances in Cryptology* 917 (ASIACRYPT 1994): 261–77.

Kolata, Gina. "100 Quadrillion Calculations Later, Eureka!" *New York Times*, April 27, 1994. https://www.nytimes.com/1994/04/27 /us/100-quadrillion-calculations-later-eureka .html.

Levy, Steven. "Wisecrackers." *Wired*, March 1, 1996. https://www.wired.com/1996/03 /crackers.

Mulcahy, Colm. "The Top 10 Martin Gardner Scientific American Articles." *Scientific American* (blog), October 21, 2014. https:// blogs.scientificamerican.com/guest-blog /the-top-10-martin-gardner-scientific-american -articles.

1997, Computer Is World Chess Champion

Levy, Steven. "What Deep Blue Tells Us About AI in 2017" *Wired* online, May 23, 2017. https://www.wired.com/2017/05/what-deep -blue-tells-us-about-ai-in-2017.

Parnell, Brid-Aine. "Chess Algorithm Written by Alan Turing Goes Up against Kasparov." *The Register*, June 26, 2012. https://www.theregister .co.uk/2012/06/26/kasparov_v_turing.

1997, PalmPilot

Hormby, Tom. "A History of Palm." Parts 1 through 5, Welcome to Low End Mac (website), July 19–25, 2014.

Krakow, Gary. "Happy Birthday, Palm Pilot." *NBC News* online, last modified March 22, 2006. http://www.nbcnews.com/id/11945300 /ns/technology_and_science-tech_and_ gadgets/t/happy-birthday-palm-pilot/#. WsPlItPwZTY.

1998, Google

Batelle, John. "The Birth of Google." *Wired*, August 1, 2005. https://www.wired.com /2005/08/battelle.

Brin, Sergey, and Lawrence Page. "The Anatomy of a Large-Scale Hypertextual Web Search Engine." In *Proceedings of the Seventh International Conference on World Wide Web* 7. Brisbane, Australia: Elsevier, 1998, 107–17.

1999, Collaborative Software Development

Brown, A. W., and Grady Booch. "Collaborative Development Environments." *Advances in Computers* 53 (June 2003): 1–29.

1999, *Blog* Is coined

Merholz, Peter. "Play with Your Words." petermescellany (website), May 17, 2002. http://www.peterme.com/archives/00000205 .html.

Wortham, Jenna. "After 10 Years of Blog, the Future's Brighter than Ever." *Wired*, December 17, 2007. https://www.wired .com/2007/12/after-10-years-of-blogs-the -futures-brighter-than-ever/.

2000, USB Flash Drive

Ban, Amir, Dov Moran, and Oron Ogdan. Architecture for a universal serial bus-based PC flash disk. US Patent 6,148,354, filed April 5, 1999, and issued November 14, 2000.

Buchanan, Matt. "Object of Interest: The Flash Drive." *New Yorker* online, June 14, 2013. http://www.newyorker.com/tech/elements /object-of-interest-the-flash-drive.

2001, Quantum Computer Factors "15"

Chirgwin, Richard. "Quantum Computing Is So Powerful It Takes Two Years to Understand What Happened." *The Register*, December 4, 2014. https://www.theregister.co.uk/2014/12 /04/boffins_we_factored_143_no_you_ factored_56153.

Chu, Jennifer. "The Beginning of the End for Encryption Schemes?" *MIT News* release, March 3, 2016. http://news.mit.edu/2016 /quantum-computer-end-encryption -schemes-0303.

IBM. "IBM's Test-Tube Quantum Computer Makes History." News release, December 19, 2001. https://www-03.ibm.com/press/us/en /pressrelease/965.wss.

2004, First International Meeting on Synthetic Biology

Cameron, D. Ewen, Caleb J. Bashor, and James Collins. "A Brief History of Synthetic Biology." *Nature Reviews, Microbiology* 12, no. 5 (May 2014): 381–90.

Knight, Helen. "Researchers Develop Basic Computing Elements for Bacteria." *MIT News* release, July 9, 2015. http://news.mit.edu/2015/basic-computing-for-bacteria-0709.

Sleator, Roy D. "The Synthetic Biology Future." *Bioengineered* 5, no. 2 (March 2014): 69–72.

2005, Video Game Enables Research into Real World Pandemics

Balicer, Ran D. "Modeling Infectious Diseases Dissemination through Online Role-Playing Games." *Journal of Epidemiology* 18, no. 2 (March 2007): 260–1.

Lofgren, Eric T., and Nina H. Hefferman. "The Untapped Potential of Virtual Game Worlds to Shed Light on Real World Epidemics." *The Lancet, Infectious Diseases* 7, no. 9 (September 2007): 625–29.

2006, Hadoop Makes Big Data Possible

Dean, Jeffrey, and Sanjay Ghemawat. "MapReduce: Simplified Data Processing on Large Clusters." In *Proceedings of the Sixth Symposium on Operating System Design and Implementation (OSDI '04): December 6–8, 2004, San Francisco, CA.* Berkeley, CA: USENIX Association, 2004.

Ghemawat, Sanjay, Howard Gobioff, and Shun-Tak Leung. "The Google File System." In *SOSP '03: Proceedings of the Nineteenth ACM Symposium on Operating Systems Principles,* 29–43. Vol. 37, no. 5 of *Operating Systems Review.* New York: Association for Computing Machinery, October, 2003.

2006, Differential Privacy

Dwork, Cynthia, and Aaron Roth. *The Algorithmic Foundations of Differential Privacy.* Breda, Netherlands: Now Publishers, 2014.

2007, iPhone

9to5 Staff. "Jobs' Original Vision for the iPhone: No Third-Party Native Apps." *9To5 Mac* (website), October 21, 2011. https://9to5mac.com/2011/10/21/jobs-original-vision-for-the-iphone-no-third-party-native-apps/.

2010, Cyber Weapons

Falliere, Nicolas, Liam O Murchu, and Eric Chien. "W32.Stuxnet Dossier." Symantec Corp., Security Center white paper, February 2011.

Kaspersky, Eugene. "The Man Who Found Stuxnet—Sergey Ulasen in the Spotlight." *Nota Bene* (blog), November 2, 2011. https://eugene.kaspersky.com/2011/11/02/the-man-who-found-stuxnet-sergey-ulasen-in-the-spotlight/.

Kushner, David. "The Real Story of Stuxnet." *IEEE Spectrum* online, February 26, 2013. https://spectrum.ieee.org/telecom/security/the-real-story-of-stuxnet.

O'Murchu, Liam. "Last-Minute Paper: An In-Depth Look into Stuxnet." *Virus Bulletin,* September 2010.

Stewart, Holly, Peter Ferrrie, and Alexander Gostev. "Last-Minute Paper: Unravelling Stuxnet." *Virus Bulletin,* September 2010.

Zetter, Kim. *Countdown to Zero Day: Stuxnet and the Launch of the World's First Digital Weapon.* New York: Crown, 2014.

2011, Watson Wins *Jeopardy!*

Thompson, Clive. "What is I.B.M.'s Watson?" *New York Times* online, June 16, 2010. http://www.nytimes.com/2010/06/20/magazine/20Computer-t.html.

2013, Algorithm Influences Prison Sentence

Angwin, Julia, Jeff Larson, Surya Mattu, and Lauren Kirchner. "Machine Bias" *ProPublica,* May 23, 2016. https://www.propublica.org/article/machine-bias-risk-assessments-in-criminal-sentencing.

Eric L. Loomis v. State of Wisconsin, 2015AP157-CR (Supreme Court of Wisconsin, October 12, 2016).

Harvard Law Review. "State v. Loomis: Wisconsin Supreme Court Requires Warning Before Use of Algorithmic Risk Assessments in Sentencing." Vol. 130 (March 10, 2017): 1530–37.

Liptak, Adam. "Sent to Prison by a Software Program's Secret Algorithms." *New York Times* online, May 1, 2017. https://www.nytimes.com/2017/05/01/us/politics/sent-to-prison-by-a-software-programs-secret-algorithms.html.

Pasquale, Frank. "Secret Algorithms Threaten the Rule of Law." *MIT Technology Review,* June 1, 2017. https://www.technologyreview.com/s/608011/secret-algorithms-threaten-the-rule-of-law/.

State of Wisconsin v. Eric L. Loomis 2015AP157-CR (Wisconsin Court of Appeals District IV, September 17, 2015). https://www.wicourts.gov/ca/cert/DisplayDocument.pdf?content=pdf&seqNo=149036.

2013, Subscription Software

Pogue, David. "Adobe's Software Subscription Model Means You Can't Own Your Software." *Scientific American* online, October 13, 2013. https://www.scientificamerican.com/article/adobe-software-subscription-model-means-you-cant-own-your-software.

Whitler, Kimberly A. "How the Subscription Economy Is Disrupting the Traditional Business Model." *Forbes* online, January 17, 2016.

2015, Google Releases TensorFlow

Knight, Will. "Here's What Developers Are Doing with Google's AI Brain." *MIT Technology Review,* December 8, 2015. https://www.technologyreview.com/s/544356/heres-what-developers-are-doing-with-googles-ai-brain.

Metz, Cade. "Google Just Open Sources TensorFlow, Its Artificial Intelligence Engine." *Wired* online, November 9, 2015. https://www.wired.com/2015/11/google-open-sources-its-artificial-intelligence-engine.

2016, Computer Beats Master at Go

Byford, Sam. "Why Is Google's Go Win Such a Big Deal?" *The Verge,* March 9, 2016. https://www.theverge.com/2016/3/9/11185030/google-deepmind-alphago-go-artificial-intelligence-impact.

House, Patrick. "The Electronic Holy War." *New Yorker* online, May 25, 2014. https://www.newyorker.com/tech/elements/the-electronic-holy-war.

Koch, Christof. "How the Computer Beat the Go Master." *Scientific American* online, March 19, 2016. https://www.scientificamerican.com/article/how-the-computer-beat-the-go-master.

Moyer, Christopher. "How Google's AlphaGo Beat a World Chess Champion." *Atlantic* online, March 28, 2016. https://www.theatlantic.com/technology/archive/2016/03/the-invisible-opponent/475611.

~9999, The Limits of Computation?

Lloyd, Seth. "Ultimate Physical Limits to Computation." *Nature* 406, no. 8 (August 2000): 1047–54.

Yin, Juan, et al. "Bounding the Speed of 'Spooky Action at a Distance.'" *Physical Review Letters* 110, no. 26 (2013).

Index